Introduction to Mathematical Methods for Environmental Engineers and Scientists

Scrivener Publishing
100 Cummings Center, Suite 541J
Beverly, MA 01915-6106

Publishers at Scrivener
Martin Scrivener (martin@scrivenerpublishing.com)
Phillip Carmical (pcarmical@scrivenerpublishing.com)

Introduction to Mathematical Methods for Environmental Engineers and Scientists

Charles Prochaska
Louis Theodore

Scrivener Publishing

WILEY

This edition first published 2018 by John Wiley & Sons, Inc., 111 River Street, Hoboken, NJ 07030, USA and Scrivener Publishing LLC, 100 Cummings Center, Suite 541J, Beverly, MA 01915, USA

For more information about Scrivener publications please visit www.scrivenerpublishing.com.

Wiley Global Headquarters
111 River Street, Hoboken, NJ 07030, USA

For details of our global editorial offices, customer services, and more information about Wiley products visit us at www.wiley.com.

Library of Congress Cataloging-in-Publication Data

ISBN 978-1-119-36349-1

Cover image: Pixabay.Com
Cover design by Russell Richardson

Set in size of 10pt and Minion Pro by Exeter Premedia Services Private Ltd., Chennai, India

10 9 8 7 6 5 4 3 2 1

To

my parents who have provided immeasurable support in every aspect of my life (CP)

and

Arthur Lovely, a truly great guy and the world's greatest sports historian... ever (LT)

Contents

Preface

It is no secret that in recent years the number of people entering the environmental field has increased at a near exponential rate. Some are beginning college students and others had earlier chosen a non-technical major/career path. A large number of these individuals are today seeking technical degrees in environmental engineering or in the environmental sciences. These prospective students will require an understanding and appreciation of the numerous mathematical methods that are routinely employed in practice. This technical steppingstone to a successful career is rarely provided at institutions that award technical degrees. This introductory text on mathematical methods attempts to supplement existing environmental curricula with a sorely needed tool to eliminate this void.

The question often arises as to the educational background required for meaningful analysis capabilities since technology has changed the emphasis that is placed on certain mathematical subjects. Before computer usage became popular, instruction in environmental analysis was (and still is in many places) restricted to simple systems and most of the effort was devoted to solving a few derived elementary equations. These cases were mostly of academic interest, and because of their simplicity, were of little practical value. To this end, a considerable amount of time is now required to acquire skills in mathematics, especially in numerical methods, statistics, and optimization. In fact, most environmental engineers and scientists are given courses in classical mathematics, but experience shows that very little of this knowledge is retained after graduation for the simple reason that these mathematical methods are not adequate for solving most systems of equations encountered in industry. In addition, advanced mathematical skills are either not provided in courses or are forgotten through sheer disuse.

As noted in the above paragraph, the material in this book was prepared primarily for beginning environmental engineering and science students *and*, to a lesser extent, for environmental professionals who wish to obtain a better understanding of the various mathematical methods that can be employed in solving technical problems. The content is such that it is suitable both for classroom use and for individual study. In presenting the text material, the authors have stressed the pragmatic approach in the application of mathematical tools to assist the reader in grasping the role of mathematical skills in environmental problem solving situations.

In effect, this book serves two purposes. It may be used as a textbook for beginning environmental students or as a "reference" book for practicing engineers, scientists, and technicians involved with the environment. The authors have assumed that the reader has already taken basic courses in physics and chemistry, and should have a minimum background in mathematics through elementary calculus. The authors' aim is to offer the reader the fundamentals of numerous mathematical methods with accompanying practical environmental applications. The reader is encouraged through references to continue his or her own development beyond the scope of the presented material.

As is usually the case in preparing any text, the question of what to include and what to omit has been particularly difficult. The material in this book attempts to address mathematical calculations common to both the environmental engineering and science professionals. The book provides the reader with nearly 100 solved illustrative examples. The interrelationship between both theory and applications is emphasized in nearly all of the chapters. One key feature of this book is that the solutions to the problems are presented in a stand-alone manner. Throughout the book, the illustrative examples are laid out in such a way as to develop the reader's technical understanding of the subject in question, with more difficult examples located at or near the end of each set.

The book is divided up into five (V) parts (see also the Table of Contents):

I. Introduction
II. Analytical Analysis
III. Numerical Analysis
IV. Statistical Analysis
V. Optimization

Most chapters contain a short introduction to the mathematical method in question, which is followed by developmental material, which in turn, is followed by one or more illustrative examples. Thus, this book offers material not only to individuals with limited technical background but also to those with extensive environmental industrial experience. As noted above, this book may be used as a text in either a general introductory environmental engineering/ science course and (perhaps) as a training tool in industry for challenged environmental professionals.

Hopefully, the text is simple, clear, to the point, and imparts a basic understanding of the theory and application of many of the mathematical methods employed in environmental practice. It should also assist the reader in helping master the difficult task of explaining what was once a very complicated subject matter in a way that is easily understood. The authors feel that this delineates this text from the numerous others in this field.

It should also be noted that the authors have long advocated that basic science courses - particularly those concerned with mathematics - should be taught to engineers and applied scientists by an engineer or applied scientist. Also, the books adopted for use in these courses should be written by an engineer or an applied scientist. For example, a mathematician will lecture on differentiation - say dx/dy - not realizing that in a real-world application involving an estuary y could refer to concentration while x could refer to time. The reader of this book will not encounter this problem.

The reader should also note that parts of the material in the book were drawn from one of the author's notes of yesteryear. In a few instances, the original source was not available for referencing purposes. Any oversight will be corrected in a later printing/edition.

The authors wish to express appreciation to those who have contributed suggestions for material covered in this book. Their comments have been very helpful in the selection and presentation of the subject matter. Special appreciation is extended to Megan Menzel for her technical contributions and review, Dan McCloskey for preparing some of the first draft material in Parts II and III, and Christopher Testa for his contributions to Chapters 13 and 14. Thanks are also due to Rita D'Aquino, Mary K. Theodore, and Ronnie Zaglin.

Finally, the authors are especially interested in learning the opinions of those who read this book concerning its utility and serviceability in meeting the needs for which it was written. Corrections, improvements and suggestions will be considered for inclusion in later editions.

Chuck Prochaska
Lou Theodore
April 2018

Part I
INTRODUCTORY PRINCIPLES

Webster defines introduction as ... "the preliminary section of a book, usually explaining or defining the subject matter..." And indeed, that is exactly what this Part I of the book is all about. The chapters contain material that one might view as a pre-requisite for the specific mathematical methods that are addressed in Parts II–V.

There are seven chapters in Part I. The chapter numbers and accompanying titles are listed below.

Chapter 1: Fundamentals and Principles of Numbers
Chapter 2: Series Analysis
Chapter 3: Graphical Analysis
Chapter 4: Flow Diagrams
Chapter 5: Dimensional Analysis
Chapter 6: Economics
Chapter 7: Problem Solving

1

Fundamentals and Principles of Numbers

The natural numbers, or so-called counting numbers, are the positive integers: 1, 2, 3, … and the negative integers: −1, −2, −3,… The following applies to real numbers:

$$a > b \text{ means that } a - b \text{ is a positive real number} \tag{1.1}$$

$$\text{If } a < b \text{ and } b < c, \text{ then } a < c \tag{1.2}$$

$$\text{If } a < b \text{ and } c > 0, \text{ then } ac < bc \tag{1.3}$$

$$\text{If } a < b \text{ and } c < d, \text{ then } a + c < b + d \tag{1.4}$$

$$\text{If } a < b \text{ and } ab < 0, \text{ then } \frac{1}{a} > \frac{1}{b} \tag{1.5}$$

$$a + b = b + a;\ a \cdot b = b \cdot a = (a)(b) \tag{1.6}$$

$$a + (b + c) = (a + b) + c;\ a \cdot (bc) = (ab) \cdot c \tag{1.7}$$

$$a(b+c) = a \cdot b + a \cdot c \tag{1.8}$$

$$\text{If } |x| = a, \text{ where } a > 0, \text{ then } x = a, \text{ or } x = -a \tag{1.9}$$

$$\text{If } |x| < c, \text{ then} -c < x < c \text{ where } c > 0 \tag{1.10}$$

$$a^{-n} = \frac{1}{a^n}, \quad a \neq 0 \tag{1.11}$$

$$(ab)^n = a^n b^n \tag{1.12}$$

$$(a^n)^m = a^{nm}, \quad a^n a^m = a^{n+m} \tag{1.13}$$

$$a^0 = 1, \quad a \neq 0 \tag{1.14}$$

$$\log ab = \log a + \log b, \quad a > 0, b > 0 \tag{1.15}$$

$$\log a^n = n \log a \tag{1.16}$$

$$\log\left(\frac{a}{b}\right) = \log a - \log b \tag{1.17}$$

$$\log \sqrt[n]{a} = \left(\frac{1}{n}\right) \log a \tag{1.18}$$

$$ln^a = 2.3026 \log a \tag{1.19}$$

Based on the above, one may write

$$(4)(9) = (3.60)(10)^1$$

$$(6)^3 = (2.16)(10)^2$$

$$\sqrt[3]{3375} = (1.50)(10)^1$$

$$(0.916)^{\frac{3}{4.15}} = (9.39)(10)^{-1}$$

$$\log_{10}(2.45) = 0.3892$$

$$\ln (245) = (2.3892)(2.3026) = 5.5013$$

$$\log_{10}(0.245) = 9.3892 - 10 = -0.6108$$

Given any quadratic equation of the general form

$$ax^2 + bx + c = 0 \tag{1.20}$$

a number of methods of solution are possible depending on the specific nature of the equation in question. If the equation can be factored, then the solution is straightforward. For instance, consider

$$x^2 - 3x = 10 \tag{1.21}$$

Put into the standard form,

$$x^2 - 3x - 10 = 0 \tag{1.22}$$

this equation can be factored as follows:

$$(x - 5)(x + 2) = 0 \tag{1.23}$$

This condition can be met, however, only when the individual factors are zero, i.e., when $x = 5$ and $x = -2$. That these are indeed the solutions to the equation may be verified by substitution.

If, upon inspection, no obvious means of factoring an equation can be found, an alternative approach may exist. For example, in the equation

$$4x^2 + 12x = 7 \tag{1.24}$$

the expression

$$4x^2 + 12x \tag{1.25}$$

could be factored as a perfect square if it were

$$4x^2 + 12x + 9 \tag{1.26}$$

which equals

$$(2x + 3)^2 \tag{1.27}$$

This can easily be achieved by adding 9 to the left side of the equation. The same amount must then, of course, be added to the right side as well, resulting in:

$$4x^2 + 12x + 9 = 7 + 9 \tag{1.28}$$

so that,

$$(2x + 3)^2 = 16 \tag{1.29}$$

This can be reduced to

$$(2x + 3) = \sqrt{16} \tag{1.30}$$

or

$$2x + 3 = +4 \tag{1.31}$$

and

$$2x + 3 = -4 \tag{1.32}$$

Since $\sqrt{16}$ above has two solutions, i.e., +4 and −4, the first equation leads to the solution $x = 0.5$ while the second equation leads to the solution $x = -7/2$, or $x = -3.5$.

If the methods of factoring or completing the square are not possible, any quadratic equation can always be solved by the *quadratic formula*. This provides a method for determining the solution of the equation if it is in the form

$$ax^2 + bx + c = 0 \tag{1.33}$$

In all cases, the two solutions of x are given by the formula

$$x = \frac{-b \pm \sqrt{b^2 - 4ac}}{2a} \tag{1.34}$$

For example, to find the roots of

$$x^2 - 4x = -3 \tag{1.35}$$

the equation is first put into the standard form of Equation (1.33)

$$x^2 - 4x + 3 = 0 \tag{1.36}$$

As a result, $a = 1$, $b = -4$, and $c = 3$. These terms are then substituted into the quadratic formula presented in Equation (1.34).

$$x = \frac{-(-4) \pm \sqrt{(-4)^2 - 4(1)(3)}}{2(1)} = \frac{4 \pm \sqrt{16-12}}{2} \tag{1.37}$$

$$= \frac{4 \pm \sqrt{4}}{2} = \frac{4 \pm 2}{2} = 3 \text{ and } 1 \tag{1.38}$$

The practicing environmental engineer and scientist occasionally has to solve not just a single equation but several at the same time. The problem is to find the set of all solutions that satisfies both equations. These are called simultaneous equations, and specific algebraic techniques may be used to solve them. For example, a simple solution exists given two linear equations and two unknowns:

$$3x + 4y = 10 \tag{1.39}$$

$$2x + y = 5 \tag{1.40}$$

The variable y in Equation (1.40) is isolated ($y = 5 - 2x$), and then this value of y is substituted into Equation (1.39).

$$3x + 4(5 - 2x) = 10 \tag{1.41}$$

This reduces the problem to one involving the single unknown x and it follows that

$$3x + 20 - 8x = 10$$

or

$$-5x = -10 \tag{1.42}$$

so that

$$x = 2 \tag{1.43}$$

When this value is substituted into either equation above, it follows that

$$y = 1 \tag{1.44}$$

A faster method of solving simultaneous equations, however, is obtained by observing that if both sides of Equation (1.40) are multiplied by 4, then

$$8x + 4y = 20 \tag{1.45}$$

If Equation (1.39) is subtracted from Equation (1.45), then $5x = 10$, or $x = 2$. This procedure leads to another development in mathematics, i.e., matrices, which can help to produce solutions for any set of linear equations with a corresponding number of unknowns (refer also to Chapter 13).

Four sections compliment the presentation of this chapter. Section numbers and subject titles follow:

1.1: Interpolation and Extrapolation
1.2: Significant Figures and Approximate Numbers
1.3: Errors
1.4: Propagation of Errors

1.1 Interpolation and Extrapolation

Experimental data (and data in general) in environmental engineering and science may be presented using a table, a graph, or an equation. Tabular presentation permits retention of all significant figures of the original numerical data. Therefore, it is the most numerically accurate way of reporting data. However, it is often difficult to interpolate between data points within tables.

Tabular or *graphical* presentation of data is usually used if no theoretical or empirical equations can be developed to fit the data. This type of presentation of data is one method of reporting experimental results. For example, heat capacities of benzene might be tabulated at various temperatures. This data may also be presented graphically. One should note that graphs are inherently less accurate than numerical tabulations. However, they are useful for visualizing variations in data and for *interpolation* and *extrapolation.*

Interpolation is of practical importance to the environmentalist because of the occasional necessity of referring to sources of information expressed in the form of a table. Logarithms, trigonometric functions, water properties of steam, liquid water and ice vapor pressures, and other physical and chemical data are commonly given in the form of tables in the standard reference works. Although these tables are sometimes given in sufficient detail so that interpolation may not be necessary, it is important to be able to interpolate properly when the need arises.

Assume that a series of values of the dependent variable y are provided for corresponding tabulated values of the independent variable x. The goal of interpolation is to obtain the correct value of y at any value of x. (Extrapolation refers to a value of x lying outside the range of tabulated values of x.) Clearly, interpolation or extrapolation may be accomplished by using data for x and y to develop a linear relationship between the two variables. The general method would be to fit two points (y_1, x_1), and (y_2, x_2) by means of

$$y = a + bx \tag{1.46}$$

and then employ this equation to calculate y for some value of x lying between x_1 and x_2. Most practitioners do this mentally when reading values from a table, e.g., steam tables. If a number of points are used, a polynomial of a correspondingly higher degree may be employed. Thus, interpolation may be viewed as the process of finding the value of a function at some arbitrary point when the function is not known but is represented over a given range as a table of discrete points. (See also Table 1.1 where y represents a reservoir's height as a function of time in days during a rainy season.) Interpolation is thus necessary to find y when x is some value not given in the table. For instance, one may be interested in finding y when $x = 11$. (The process of finding x when y is known is referred to as inverse interpolation). Given a table such as Table 1.1, one can draw a picture and write the equation of the straight line through the points (x_1, y_1) and (x_2, y_2) for y.

$$y - y_1 = \left(\frac{y_2 - y_1}{x_2 - x_1}\right)(x - x_1) \tag{1.47}$$

Equation (1.47) can be solved for y in terms of x

$$y = y_1 + \frac{(y_2 - y_1)(x - x_1)}{x_2 - x_1} \tag{1.48}$$

or

$$y = \frac{y_0(x - x_0) + (y_1 - y_0)(x - x_0)}{x_1 - x_0} \tag{1.49}$$

Illustrative Example 1.1

Refer to Table 1.1. Find y at $x = 11$.

Table 1.1 Reservoir height vs. time in days.

x, height	*y, days*
0	30
3	31
6	33
9	35
12	39
15	46
18	52

Solution

Proceed as follows. Set up calculations as shown in Table 1.2

Table 1.2 Information for Illustrative Example 1.1.

Data Point *i*	x_i	y_i	$x - x_i$
1	9	35	2
2	12	39	1

Apply Equation (1.48). Therefore,

$$y = 39 + \left(\frac{1}{3}\right)[(35)(1) - (39)(2)] = 35 + \frac{(4)(2)}{3} = 35 + \frac{8}{3}$$

As noted above, *inverse interpolation* involves estimating x which corresponds to a given value of y and extrapolation involves estimating values of y outside the interval in which the data x_0, …, x_n fall. It is generally unwise to extrapolate any empirical relation significantly beyond the first and last data points. If, however, a certain form of equation is predicted by theory and substantiated by (other) available data, reasonable extrapolation is ordinarily justified.

1.2 Significant Figures and Approximate Numbers [1]

Significant figures provide an indication of the precision with which a quantity is measured or known. The last digit represents in a qualitative sense, some degree of doubt. For example, a measurement of 8.32 nm (nanometers) implies that the actual quantity is somewhere between 8.315 and 8.325 nm. This applies to calculated and measured quantities; quantities that are known exactly (e.g., pure integers) have an infinite number of significant figures. Note, however, that there is an upper limit to the accuracy with which physical measurements can be made.

The method for counting the significant digits of a number follows one of two rules depending on whether there is or is not a decimal point present. The significant digits of a number always start from the first nonzero digit on the left to either:

1. the last digit (whether it is nonzero or zero) on the right if there is a decimal point present, or
2. the last nonzero digit on the right of the number if there is no decimal point present.

For example:

370	has 2 significant figures
370	has 3 significant figures
370.0	has 4 significant figures
28,070	has 4 significant figures
0.037	has 2 significant figures
0.0370	has 3 significant figures
0.02807	has 4 significant figures

Whenever quantities are combined by multiplication and/or division, the number of significant figures in the result should equal the lowest number of significant figures of any of the quantities. In *long* calculations, the final result should be rounded off to the correct number of significant figures. When quantities are combined by addition and/or subtraction, the final result cannot be more precise than any of the quantities added or subtracted. Therefore, the position (relative to the decimal point) of the last significant digit in the number that has the lowest degree of precision is the position of the last permissible significant digit in the result. For example, the sum of 3702, 370, 0.037, 4, and 37 should be reported as 4110 (without a decimal). The least precise of the five number is 370, which has its last significant digit in the *tens* position. Therefore, the answer should also have its last significant digit in its *tens* position.

Unfortunately, environmental engineers and scientists rarely concern themselves with significant figures in their calculations. However, it is recommended that the reader attempt to follow the calculational procedure set forth in this section.

In the process of preforming engineering/scientific calculations, very large and very small number are often encountered. A convenient way to represent these numbers is to use *scientific notation*. Generally, a number represented in scientific notation is the product of a number *and* 10 raised to an integer power. For example,

$$28{,}070{,}000{,}000 = 2.807 \times 10^{10} = (2.807)(10)^{10}$$

$$0.000002807 = 2.807 \times 10^{-6}$$

A positive feature of using scientific notation is that only the significant figures need appear in the number.

Thus, when approximate numbers are added, or subtracted, the results are presented in terms of the least precise number. Since this is a relatively simple rule to master, note that the answer in Equation (1.50) follows the aforementioned rule of precision.

$$6.04L + 2.8L - 4.173L = 4.7L \tag{1.50}$$

(The result is 4.667L.) The expressions in Equation (1.50) have two, one, and three decimal places respectively. The least precise number (least decimal places) in the problem is 2.8, a value carried only to the tenths position. Therefore, the answer must be calculated to the tenths position only. Thus, the correct answer is 4.7L. (The last 6 and the 7 are dropped from the 4.667L, and the first 6 is rounded up to provide 4.7L.)

In multiplication and division of approximate numbers, finding the number of significant digits is used to determine how many digits to keep (i.e., where to truncate). One must first understand *significant digits* in order to determine the correct number of digits to keep or remove in multiplication and division problems. As noted earlier in this section, *the digits 1 through 9 are considered to be significant.* Thus, the numbers 123, 53, 7492, and 5 contain three, two, four and one significant digits respectively. The digit zero must be considered separately.

Zeroes are significant when they occur between significant digits. In the following example, all zeroes are significant: 10001, 402, 1.1001, 500.09 with five, three, five, and four significant figures, respectively. *Zeroes are not significant when they are used as place holders.* When used as a place holder, a zero simply identifies where a decimal is located. For example, each of the following numbers has only one significant digit: 1000, 500, 60, 0.09, 0.0002. In the numbers 1200, 540, and 0.0032 there are two significant digits, and the zeroes are not significant. *When zeroes follow a decimal and are preceded by a significant digit, the zeroes are significant.* In the following examples, all zeroes are significant: 1.00, 15.0, 4.100, 1.90, 10.002, 10.0400. For 10.002, the zeroes are significant because they fall between two significant digits. For 10.0400, the first two zeroes are significant because they fall between two significant digits; the last two zeroes are significant because they follow a decimal and are preceded by a significant digit. As noted above, when approximate numbers are multiplied or divided, the result is expressed as a number having the same number of significant digits as the number in the problem with the least number of significant digits.

When truncating (removing final, unwanted digits), rounding is normally applied to the last digit to be kept. Thus, *if the value of the first digit to be discarded is less than 5, one should retain the last retained digit with no change. If the value of the first digit to be discarded is 5 or greater, one should increase the last kept digit's value by one.* Assume, for example, only the first two decimal places are to be kept for 25.0847 (the 4 and 7 are to be dropped). The number is then 25.08. Since the first digit to be discarded (4) is less than 5, i.e., the 8 is not rounded up. If only the first two decimal places are to be kept for 25.0867 (the 6 and the 7 are to be dropped), it should be rounded to 25.09. Since the first digit to be discarded (6) is 5 or more, the 8 is rounded up to 9.

When adding or subtracting approximate numbers, a rule based upon precision determines how many digits are kept. In general, precision relates to the decimal significance of a number. When a measurement is given as 1.005 cm, one can say that the number is precise to the *thousandth* of a centimeter. If the decimal is removed (1005 cm), the number is precise to *thousands* of centimeters.

In some water pollution studies, a measurement in gallons or liters may be required. Although a gallon or liter may represent an exact quantity, the measuring instruments that are used are only capable of producing approximations. Using a standard graduated flask in liters as an example, can one determine whether there is exactly one liter? Not likely. In fact, one would be pressed to verify that there was a liter to within ±1/10 of a liter. Therefore, depending upon the instruments used, the precision of a given measurement may vary.

If a measurement is given as 16.0L, the zero after the decimal indicates that the measurement is precise to within 1/10L i.e., 0.1L. A given measurement of 16.00L, indicates precision to the 1/100L. As noted, the digits following the decimal indicate how precise the measurement is. Thus, precision is used to determine where to truncate when approximate numbers are added or subtracted.

1.3 Errors

This is the first of two sections devoted to errors. This section introduces the various classes of errors while the next section demonstrates the propagation of some of these errors. As one might suppose, numerous books have been written on the general subject of "errors." Different definitions for errors appear in the literature but what follows is the authors' attempt to clarify the problem [1].

Any discussion of errors would be incomplete without providing a clear and concise definition of two terms: the aforementioned precision and accuracy. The term precision is used to describe a state or system or measurement for which the word precise implies little to no variation; some refer to this as reliability. Alternatively, accuracy is used to describe something free from the matter of errors. The accuracy of a value, which may be represented in either absolute or relative terms, is the degree of agreement between the measured value and the *true value*.

All measurements and calculations are subject to two broad classes of errors: determinate and indeterminate. The error is known as a "determinate error" if an error's magnitude and sign are discovered and accounted for in the form of a correction. All errors that either cannot be or are not properly allowed for in magnitude and sign are known as "indeterminate errors."

A particularly important class of indeterminate errors is that of accidental errors. To illustrate the nature of these, consider the very simple and direct measurement of temperature. Suppose that several independent readings are made and that temperatures are read to 0.1 °F. When the results of the different readings are compared, it may be found that even though they have been performed very carefully, they may differ from each other by several tenths of a degree. Experience has shown that such deviations are inevitable in all measurements and that these result from small unavoidable errors of observation due to the sensitivity of measuring instruments and the keenness of the sense of perception. Such errors are due to the combined effect of a large number of undetermined causes and they can be defined as "accidental errors."

Regarding the words precision and accuracy, it is also important to note that a result may be extremely precise and at the same time inaccurate. For instance, the temperature readings just mentioned might all agree within 1 °F. From this it would not be permissible to conclude that the temperature is accurate to 1 °F until it can be definitively shown that the combined effects of uncorrected constant errors and known errors are negligible compared with 1 °F. It is quite conceivable that the calibration of the thermometer might be grossly incorrect. Errors such as these are almost always present and can never be detected individually. Such errors can be detected only by obtaining the readings with several different thermometers and, if possible, several independent methods and observers.

It should also be understood at the onset that most *numerical* calculations are by their very nature inexact. The errors are primarily due to one of three sources: inaccuracies in the original data, lack of precision in carrying out calculations, or inaccuracies introduced by approximate or incorrect methods of solution. Of particular significance are the aforementioned errors due to "round-off" and the inability to carry more than a certain number of significant figures. The errors associated with the method of solution are usually the area of greatest concern [1]. These usually arise as a result of approximations and assumptions made in the development of an equation used to calculate a desired result and should not be neglected in any error analysis.

Finally, many list the following three errors associated with a computer (calculator).

1. *Truncation error.* With the truncation of a series after only a few terms, one is committing a generally known error. This error is not machine-caused but is due to the method.
2. *Round-off error.* The result of using a finite number of digits to represent a number. In reality, numbers have an infinite number of digits extending past the decimal point. For example, the integer 1 is really 1.000…0 and π is 3.14159… but numbers are rounded to allow for calculation and representation. This rounding is a form of error known as *round-off error.*
3. *Propagation or inherited error.* This is caused by sequential calculations that include points previously calculated by the computer which already are erroneous owing to the two errors above. Since the result is already off the solution curve, one cannot expect any new values computed to be on the correct solution curve. Adding the round-off errors and truncation errors into the calculation causes further errors to *propagate,* adding more error at each step.

1.4 Propagation of Errors

When a desired quantity W is related to several directly measured *independent* quantities W_1, W_2, W_3, …, W_n by the equation

$$W = W(W_1, W_2, W_3, \ldots, W_n) \tag{1.51}$$

W becomes an indirectly measured *dependent* quantity. In general, the true value of W cannot be known because the true values of W_1, W_2, W_3, …, W_n are unknown, but the most probable value of W may be calculated by inserting the most probable values of W_1, W_2, W_3, …, W_n, into Equation (1.51). The errors in the directly measured quantities of W_i will result in an error in the calculated quantity W, the value of which is important to ascertain. If the original measurements are available, a method referred to as the "propagation-of-error" could be employed to estimate in the resultant error. The general "propagation-of-error" for a function $W = f(x_1, x_2)$ is described by

$$s^2 = \left(\frac{\partial W}{\partial x_1}\right)^2 s_{x_1}^2 + \left(\frac{\partial W}{\partial x_2}\right)^2 s_{x_2}^2 \tag{1.52}$$

where s = error in the function W
s_{x_1} = error in variable x_1
s_{x_2} = error in variable x_2

Thus, if a linear function $W = ax_1 + bx_2$ is involved, it is found by direct application of Equation (1.52) that

$$s^2 = a^2 s_{x_1}^2 + b^2 s_{x_2}^2 \tag{1.53}$$

where

$$a = \frac{\partial W}{\partial x_1} \tag{1.54}$$

$$b = \frac{\partial W}{\partial x_2} \tag{1.55}$$

For the case where $W = x_1 x_2$ application of Equation (1.52) can be shown to give

$$\frac{s^2}{W^2} = \frac{s_{x_1}^2}{x_1^2} + \frac{s_{x_2}^2}{x_2^2} \tag{1.56}$$

Thus, the square of the fractional error is equal to the sum of the squares of fractional errors of the independent variables. Alternatively, taking the logarithm of a product $c = xy$ reduces it to the form of Equation (1.57):

$$(s_{\log W})^2 = (s_{logx_1})^2 + (s_{logx_2})^2 \tag{1.57}$$

As would be expected, the greatest variation in the approach of different investigators lies in the details of how they propose to obtain values of the error measurement. Basically, these may be obtained by comparing either a series of pairs of estimates and true values from similar previous readings (i.e., "looking at the record") or obtaining a number of independent estimates of the particular value needed.

Illustrative Example 1.2 [2]

Table 1.3 gives basic data on investment and production costs for a unit for coking a heavy crude to produce gasoline and distillate fuel. It is desired to determine the standard deviation of the profit and its significance in predicting the uncertainty involved.

Solution

The standard deviation equation is applied to determine the standard deviation of the profit (income – cost). This is represented as (refer to Table 1.3)

$$\begin{aligned} s^2 &= (480{,}000)^2 + (180{,}900)^2 + (66{,}300)^2 + (3860)^2 + (1060)^2 + (9480)^2 \\ &\quad + (191{,}000)^2 + (24{,}000)^2 \\ &= 30.47 \times 10^{11} \\ s &= \$552{,}000 \end{aligned}$$

Table 1.3 Investment data.

Item	Average value $\overline{X}$, $/yr	Standard deviation
Investment	$5,195,000	$513,000
Direct production cost		
Raw materials	$2,137,000	$180,900
Labor	215,000	66,300
Utilities	26,000	3,800
Operating Supplies	6,100	1,060
Maintenance	29,300	9,480
Royalties	16,500	0
	$2,466,500	–
Indirect production cost	2,383,000	$191,000
Fixed production cost	494,000	24,000
Sales	6,700,000	480,000
Profits	1,356,500	–

Happel also goes on to calculate the percent return of an investment – a calculation beyond the scope of this introductory book [2].

Illustrative Example 1.3

Consider the data taken from an actual experiment performed by McHugh [3]. One step involves determining the number of moles of benzene, N_B, loaded into a round-bottom flask. Moles of benzene are calculated from two pieces of weight data that are measured experimentally. Errors in these weight data accumulate to result in an estimated accumulated error, *s*.

A Mettler electronic balance was used in an actual gravimetric transfer of benzene, C_6H_6, into the round-bottom flask. The following data was obtained:

Weight of volumetric flask + benzene before transfer to round-bottom flask,

$$W_1 = 105.321 \pm 0.001 \text{ g}$$

Weight of volumetric flask + benzene after transfer to round-bottom flask,

$$W_2 = 85.466 \pm 0.001 \text{ g}$$

Solution

The molecular weight of C_6H_6 (MW_g) may be taken as equal to 78.12 g/g-mol. The purity of the C_6H_6 is 99.99%; this percentage is included as a variable (wt%) in developing the equation for N_B. The equation below is the defining equation for moles of benzene, N_B.

$$N_B = \left(\frac{W_1 - W_2}{MW_B}\right) \cdot wt_{\%} = 0.254$$

Taking the partial derivatives of the above equation with MW_B as a constant and the other terms being variables, gives the following values:

$$a = \left(\frac{\partial N_B}{\partial W_1}\right) = \left(\frac{1}{MW_B}\right) \cdot wt_{\%} = 0.0128$$

$$b = \left(\frac{\partial N_B}{\partial W_2}\right) = \left(\frac{-1}{MW_B}\right) \cdot wt_{\%} = -0.0128$$

$$c = \left(\frac{\partial N_B}{\partial wt}\right) = \left(\frac{W_1 - W_2}{MW_B}\right) = 0.254$$

There are two s terms corresponding to the two weights (W_1 and W_2) above. These two terms are labelled s_{W_1} and s_{W_2} where W_1 is the weight before the transfer and the W_2 is the weight after. A third s term, corresponding to the percentage purity (wt%), is labelled $s_{wt\%}$. Substituting into the equations above leads to the accumulated error equation.

$$\begin{aligned} s^2 &= \left(\frac{1}{MW_B}\right)^2 wt_{\%}^2 \cdot s_{W_1}^2 + \left(\frac{-1}{MW_B}\right)^2 wt_{\%}^2 \cdot s_{W_2}^2 + \left(\frac{W_1 - W_2}{MW_B}\right)^2 s_{wt\%}^2 \\ &= a^2 s_{W_1}^2 + b^2 s_{W_2}^2 + c^2 s^2_{wt\%} \\ &= 1.6384\times10^{-10} + 1.6384\times10^{-10} + 1.6129\times10^{-10} \\ &= 4.8897\times10^{-10} \end{aligned}$$

The values for s_{W_1} and s_{W2} used in this calculation are both 0.001 g, which is the manufacturer's specified "error" for the analytical balance used in this transfer; this error is often reported as the *repeatability or reproducibility* of the balance. The wt% term may be taken as 0.99990 ± 0.00005, so that $s_{wt\%}$ equals to 0.00005. Calculating s yields.

$$s = 2.2112\times10^{-5}$$

so that

$$N_B = 0.254 \pm 2.2112\times10^{-5}\ \text{gmol}$$

A comparison of the value of s with the value of N_B, indicates that the transfer is accurate to five significant figures. On a percentage basis, s/N_B is accurate within 0.009%; clearly this very "good" data.

Illustrative Example 1.4 [4]

In a rotary dryer experiment, samples of partially dried cornmeal are collected in aluminum weighing pans and weighed (W_1). The pans are placed in the oven overnight to dry the remaining water from the cornmeal and reweighed (W_2). The dry cornmeal is discarded and the empty pan is weighed (W_3). Data obtained from one experiment appear below in Table 1.4.

Table 1.4 Rotary dryer experiment data.

	Jill and Joe's data	Analytical balance
Weight of weighing pan + Cornmeal before drying (W_1)	4.0 ± 0.1g	4.012 ± 0.001g
Weight of weighing pan + Cornmeal after drying (W_2)	3.7 ± 0.1g	3.638 ± 0.001g
Weight of empty pan (W_3)	1.3 ± 0.1g	1.394 ± 0.001g

(This example demonstrates a case where very "bad" data is obtained because the students, Joe Jasper and his partner Jill Joker, decide to use a measuring device that they liked instead of the best measuring device available. Jill and Joe were instructed to use an analytical balance accurate to 0.001 g. However, Jill and Joe didn't like the analytical balance, so they decided to use a balance that is only accurate to 0.1 g.)

Compare Jill and Joe's data with the correct analytical data, and determine whether using a less accurate balance significantly affects experimental error.

Solution

The weight of water removed from the cornmeal by drying is (W_1 – W_2). The weight of the dry cornmeal is (W_2 – W_3). First, use the student data in Table 1.4 to calculate the weight fraction, W, of water in the partially dried cornmeal, as defined by the following equation:

$$W = \left(\frac{W_1 - W_2}{W_2 - W_3}\right) = 0.125 \frac{\text{g water}}{\text{g dry solids}}$$

Taking the partial derivative of the above equation and following the procedure in the previous illustrated example yields the following equation for the accumulated error.

$$s^2 = \left(\frac{1}{W_2 - W_3}\right)^2 \cdot s_{W_1}^2 + \left(\frac{-W_1 - W_3}{(W_2 - W_3)^2}\right)^2 \cdot s_{W_2}^2 + \left(\frac{W_1 - W_3}{(W_2 - W_3)^2}\right)^2 \cdot s_{W_3}^2$$

The values for s_{W_1}, s_{W_2} and s_{W_3} used in this calculation are all 0.1 g, which is the manufacturer's specified "error" for the scale used by the students. Calculating s from their data yields,

$$s^2 = 6.1305 \times 10^{-3} \frac{\text{g}^2\text{water}}{\text{g}^2 \text{ dry solids}}$$

and

$$s = 0.07830\frac{\text{g water}}{\text{g dry solids}}$$

so that

$$W = 0.125 \pm 0.07830\frac{\text{g water}}{\text{g dry solids}}$$

A comparison of the value of s with the value of W shows that the data obtained with this balance is very "bad." On a percentage basis, s/W is accurate to within 50%. This value of W is then used in successive calculations with the accumulated errors getting even worse.

Much better results would have been obtained if the students used an analytical balance with 0.001 g accuracy. A value for s of 0.00086 g/g would result in a percentage error of only 0.52% on redoing the calculation with the correct weights. Details are left as an exercise for the reader.

References

1. L. Theodore and F. Taylor, *Probability and Statistics,* A Theodore Tutorial, Theodore Tutorials, East Williston, NY, originally published by the USEPA/APTI, RTP, NC, 1995.
2. J. Happel, *Chemical Process Economics,* John Wiley & Sons, Hoboken, NJ, 1958.
3. M. McHugh, doctoral presentation, Department of Chemical Engineering, University of Delaware, 1981.
4. Author(s) unknown, Unit Operation Experiment, Manhattan College, Bronx, NY, date unknown.

2

Series Analysis

An infinite series is an infinite sum of the form

$$a_1 + a_2 + \ldots + a_n + \ldots \tag{2.1}$$

going on to infinitely many terms. It can further be defined that the terms in some series are non-ending. Such series are familiar even in the simplest operations. For example, one may write

$$\frac{2}{3} = 0.66666\ldots \tag{2.2}$$

This is equivalent to

$$\frac{2}{3} = \frac{4}{6} = \frac{6}{10} + \frac{6}{100} + \frac{6}{1{,}000} + \frac{6}{10{,}000} + \ldots \tag{2.3}$$

One might also "truncate," or cut the infinite series after a certain number of decimal places and use the resulting *rational* number

$$0.6666 = \frac{6666}{10{,}000} \tag{2.4}$$

as a sufficiently good *approximation* to the number $2/3$. One may also "round off" the above number to 0.6667.

The procedure just used applies to the general series $a_1 + a_2 + \ldots + a_n + \ldots$. To evaluate it, one rounds off after m terms and replaces the series by a finite sum.

$$a_1 + a_2 + \ldots + a_m \tag{2.5}$$

However, the rounding off procedure must be justified, i.e., one must be sure that taking more than m terms would not significantly affect the result. For the series

$$3 + 3 + 3 + 3 + \ldots \tag{2.6}$$

such a justification would be inappropriate, since the first term gives a sum of 3, two terms give 6, three terms give 9, etc. Truncating is of no help here. This series is an example of a *divergent series*. On the other hand, for the series

$$S = 1 + \frac{1}{4} + \frac{1}{9} + \frac{1}{16} + \ldots + \frac{1}{n^2} + \ldots \tag{2.7}$$

it seems reasonable to truncate. Thus, one has as sums of the first four of the n terms:

$$S = 1 + \frac{1}{4} + \frac{1}{9} + \frac{1}{16} = \frac{205}{144} \tag{2.8}$$

In addition, for

$$n = 1,\ S = 1;\ n = 2,\ S = 1 + \frac{1}{4} = \frac{5}{4};\ n = 3,\ S = 1 + \frac{1}{4} + \frac{1}{5} = \frac{49}{36}$$

Three sections compliment the presentation of this chapter. Section numbers and subject titles follow:

2.1: Other Infinite Series
2.2: Tests for Convergence and Divergence
2.3: Infinite Series Equations

2.1 Other Infinite Series

A portion of differential equations (see Part II, Chapter 11) result in solutions that can be expressed in a closed or analytical form. In such cases, many of these solutions are obtained in terms of functions which actually represent an infinite series. Logarithmic, trigonometric, and hyperbolic functions are cases in point (see last section in this chapter). It is not too surprising, then, to find that the solutions to certain classes of differential equations are obtained in the form of an infinite series of terms, i.e., an infinite series. Certain equations of this class appear frequently enough that the particular forms of infinite series which represent their solutions have been given specific names and symbols and the numerical values of the series are tabulated in reference books. Bessel functions, Legendre polynomials, etc., are typical examples [1]. Power series and Taylor series are briefly addressed below.

An expression of the type

$$a_0 + a_1(x - x_0) + \ldots + a_n(x - x_0)^n + \ldots \tag{2.9}$$

is termed a "power series." Such a series is said to "converge" if it approaches a finite value as n approaches infinity. The simplest test for convergence is the *ratio test*; if the *absolute* value of the ratio of the $(n + 1)$ term to the nth term in any infinite series approaches a limit j as $n \to \infty$, then the series converges for $j < 1$, diverges for $j > 1$, and fails if $j = 1$.

A "Taylor series" expansion of y as a function of x (in the vicinity of x_0) may be written as

$$y(x) = y(x_0) + (x - x_0)y'(x_0) + \frac{(x - x_0)^2}{2!}y''(x_0) + \ldots \tag{2.10}$$

where

$$y' = \text{the first derivative}; \quad y'' = \text{the second derivative, etc.} \tag{2.11}$$

By setting the interval from the original point of expansion x_0 to a general point x as h, Equation (2.10) can be rewritten as

$$y(x_0 + h) = \sum_{n=0} \frac{(h)^n}{n!}\left(\frac{d^n y}{dx^n}\right)_{x=x_0} \tag{2.12}$$

where $0! = 1$. If the values of the function and its derivatives at the point x_0 are given, the solution for other values of x can be determined by simply adjusting h.

2.2 Tests for Convergence and Divergence [2]

In general, the problem of determining whether a given series will or will not converge can occasionally require a great deal of ingenuity and resourcefulness. The convergence or divergence of an infinite series is complicated by the removal of a finite number of terms. There is no all-inclusive test which can be applied to all series. As the only alternative, it is necessary to apply one or more of the theorems below in an attempt to ascertain the convergence or divergence of the series under study. The following 10 tests are given in relative order of effectiveness.

1. A series will converge if the absolute value of each term (with or without a finite number of terms) is less than the corresponding term of a known convergent series.
2. A positive series is divergent if it is term-wise larger than a known divergent series of positive terms.
3. A series is divergent if the nth term of the series does not approach zero as n becomes increasingly large.
4. If the absolute ratio of the $(n + 1)$ term divided by the nth term as n becomes unbounded approaches
   a. a number less than 1, the series is convergent.
   b. a number greater than 1, the series is divergent.
   c. a number equal to 1, the test for convergence or divergence is inconclusive.
5. The series converges if the partial summation of a series converges as n becomes unbounded.
6. The series diverges if the partial summation of a series diverges as n becomes unbounded.
7. If the terms of a series are alternately positive and negative and never increase in numerical value, the series will converge, provided that the terms tend to zero as a limit.
8. If the nth root of the absolute value of the nth term, as n becomes unbounded, approaches
   a. a number less than 1, the series is convergent.
   b. a number greater than 1, the series is divergent.
   c. a number equal to 1, the test for convergence or divergence is inconclusive.
9. If a series has no sum, it is divergent.
10. Multiplying every term of an infinite series by the same quantity will not affect the convergence or divergence of the series.

2.3 Infinite Series Equations

This section opens with 3 arithmetic series which are expressed in terms of n, followed by over 30 special cases involving infinite series.

$$1 + 2 + 3 + \ldots + n = \frac{n(n+1)}{2} \tag{2.13}$$

$$1 + 3 + 5 + \ldots + (2n-1) = n^2 \tag{2.14}$$

$$(1)^2 + (2)^2 + (3)^2 + \ldots + n^2 = \frac{n(n+1)(2n+1)}{6} \tag{2.15}$$

Special cases follow.

$$1 - \frac{1}{2} + \frac{1}{3} - \frac{1}{4} + \frac{1}{5} - \ldots = \ln(2) \tag{2.16}$$

$$1 - \frac{1}{3} + \frac{1}{5} - \frac{1}{7} + \frac{1}{9} - \ldots = \frac{\pi}{4} \tag{2.17}$$

$$\frac{1}{1^2} + \frac{1}{2^2} + \frac{1}{3^2} + \frac{1}{4^2} + \ldots = \frac{\pi^2}{6} \tag{2.18}$$

$$\frac{1}{1^4} + \frac{1}{2^4} + \frac{1}{3^4} + \frac{1}{4^4} + \ldots = \frac{\pi^4}{90} \tag{2.19}$$

$$\frac{1}{1^2} - \frac{1}{2^2} + \frac{1}{3^2} - \frac{1}{4^2} + \ldots = \frac{\pi^2}{12} \tag{2.20}$$

$$\frac{1}{1^4} - \frac{1}{2^4} + \frac{1}{3^4} - \frac{1}{4^4} + \ldots = \frac{7\pi^4}{720} \tag{2.21}$$

$$\frac{1}{1^2} + \frac{1}{3^2} + \frac{1}{5^2} + \frac{1}{7^2} + \ldots = \frac{\pi^2}{8} \tag{2.22}$$

$$\frac{1}{1^4} + \frac{1}{3^4} + \frac{1}{5^4} + \frac{1}{7^4} + \ldots = \frac{\pi^4}{96} \tag{2.23}$$

$$\frac{1}{1^3} - \frac{1}{3^3} + \frac{1}{5^3} - \frac{1}{7^3} + \ldots = \frac{\pi^3}{32} \tag{2.24}$$

$$\frac{1}{1\cdot 3} + \frac{1}{3\cdot 5} + \frac{1}{5\cdot 7} + \frac{1}{7\cdot 9} + \ldots = \frac{1}{2} \tag{2.25}$$

$$\frac{1}{1\cdot 3} + \frac{1}{2\cdot 4} + \frac{1}{3\cdot 5} + \frac{1}{4\cdot 6} + \ldots = \frac{3}{4} \tag{2.26}$$

$$\frac{1}{1^2\cdot 3^2}+\frac{1}{3^2\cdot 5^2}+\frac{1}{5^2\cdot 7^2}+\frac{1}{7^2\cdot 9^2}+\ldots=\frac{\pi^2-8}{16} \tag{2.27}$$

$$\frac{1}{1^2\cdot 2^2\cdot 3^2}+\frac{1}{2^2\cdot 3^2\cdot 4^2}+\frac{1}{3^2\cdot 4^2\cdot 5^2}+\ldots=\frac{4\pi^2-39}{16} \tag{2.28}$$

$$(1+x)^{-1}=1-x+x^2-x^3+x^4-\ldots \quad -1<x<1 \tag{2.29}$$

$$(1+x)^{-2}=1-2x+3x^2-4x^3+5x^4-\ldots \quad -1<x<1 \tag{2.30}$$

$$(1+x)^{-3}=1-3x+6x^2-10x^3+15x^4-\ldots \quad -1<x<1 \tag{2.31}$$

$$(1+x)^{-\frac{1}{2}}=1-\frac{1}{2}x+\frac{1\cdot 3}{2\cdot 4}x^2-\frac{1\cdot 3\cdot 5}{2\cdot 4\cdot 6}x^3+\ldots \quad -1<x<1 \tag{2.32}$$

$$(1+x)^{\frac{1}{2}}=1+\frac{1}{2}x-\frac{1}{2\cdot 4}x^2+\frac{1\cdot 3}{2\cdot 4\cdot 6}x^3-\ldots \quad -\infty<x<\infty \tag{2.33}$$

$$e^x=1+x+\frac{x^2}{2!}+\frac{x^3}{3!}+\ldots \quad -\infty<x<\infty \tag{2.34}$$

$$a^x=e^{x\ln a}=1+x\ln a+\frac{(x\ln a)^2}{2!}+\frac{(x\ln a)^3}{3!}+\ldots \quad -\infty<x<\infty \tag{2.35}$$

$$\ln x=2\left\{\left(\frac{x-1}{x+1}\right)+\frac{1}{3}\left(\frac{x-1}{x+1}\right)^3+\frac{1}{5}\left(\frac{x-1}{x+1}\right)^5+\ldots\right\} \quad x>0 \tag{2.36}$$

$$\ln x=\left(\frac{x-1}{x}\right)+\frac{1}{2}\left(\frac{x-1}{x}\right)^2+\frac{1}{3}\left(\frac{x-1}{x}\right)^3+\ldots \quad x\geq\frac{1}{2} \tag{2.37}$$

$$\ln(1+x)=x-\frac{x^2}{2}+\frac{x^3}{3}-\frac{x^4}{4}+\ldots \quad -1<x\leq 1 \tag{2.38}$$

$$\frac{1}{2}\ln\left(\frac{1+x}{1-x}\right)=x+\frac{x^3}{3!}+\frac{x^5}{5!}+\frac{x^7}{7}+\ldots \quad -1<x\leq 1 \tag{2.39}$$

$$\sin x = x - \frac{x^3}{3!} + \frac{x^5}{5!} - \frac{x^7}{7!} + \ldots \quad -\infty < x < \infty \tag{2.40}$$

$$\cos x = 1 - \frac{x^2}{2!} + \frac{x^4}{4!} - \frac{x^6}{6!} + \ldots; \quad -\infty < x < \infty \tag{2.41}$$

References

1. H. Mickley, T. Sherwood, and C. Reed, (adapted from) *Applied Mathematics in Chemical Engineering*, McGraw-Hill, New York City, NY, 1957.
2. D. Green and R. Perry, editors, (adapted from) *Perry's Chemical Engineers' Handbook, 8th ed.*, McGraw-Hill, New York City, NY, 2008.

3

Graphical Analysis

Engineering data are often best understood when presented in the form of graphs or mathematical equations. The preferred method of obtaining the mathematical relations between variables is to plot the data as straight lines and use the *slope-intercept* method to obtain coefficients and exponents; therefore, it behooves the environmental engineer and scientist to be aware of the methods of obtaining straight lines on various types of graph paper, and of determining the equations of such lines (see Part IV, Chapter 28). One always strives to plot data as straight lines because of the simplicity of the curve, ease of both interpolation and extrapolation. Graphical methods proved invaluable in the past in the analysis of relatively complex relationships. Much of the basic physical and chemical data are still best represented graphically, and graphical methods are often employed in the analytical treatment of processes.

One or more of the many types of graphical representations may be employed for the following purposes:

1. as an aid in visualizing a process for the representation of qualitative data
2. for the representation of quantitative data
3. for the representation of a theoretical equation
4. for the representation of an empirical equation

5. for the comparison of experimental data with a theoretical expression
6. for the comparison of experimental data with an empirical expression

The relation between two quantities y - the dependent variable and x - the independent variable is commonly obtained as a tabulation of values of y for several different values of x. The relation between y and x may not be easy to visualize by studying the tabulated results and is often best seen by plotting y vs. x. If the conditions are such that y is known to be a function of x only, the functional relation will be indicated by the fact that the points may be represented graphically by a smooth curve. Deviations of the points from a smooth curve generally indicates unreliability in the data. If y is a function of two variables x_1 and x_2, a series of results of y in terms of x_1 may be obtained in terms of x_2. When plotted, the data will be represented by a family of curves, each curve representing the relation between y and x_1 for a definite *constant* value of x_2. If another variable x_3 is involved, one may have separate graphs for constant values of x_3, each showing a family of curves of y vs. x_1. One may expand this method of representing data to express the relationship between more than four independent variables (This method is explored in more depth in Chapter 32, Part V).

Graphs can also be used for interpolation and extrapolation. In addition, graphs may be employed to perform common mathematical manipulations such as integration and differentiation. The data may be of almost any type encountered in environmental engineering and science practice. They may be physical property data, such as the variation in density and viscosity with temperature. They may be process data, such as the variation in temperature along a tube in a heat exchanger with respect to the flow rate of the fluid. One of the things that should be kept in mind whenever working with plotted data is the loss of accuracy arising due to the use of graphs. In general, the number of significant figures may be only as great as the size of the divisions on the graph.

Five sections compliment the presentation of this chapter. Section numbers and subject titles follow:

3.1: Rectangular Coordinates
3.2: Logarithmic-Logarithmic (Log-Log) Coordinates
3.3: Semi- Logarithmic (Semi-Log) Coordinates
3.4: Other Graphical Coordinates
3.5: Methods of Plotting Data

The reader should note that a good part of this material will be revisited in the Optimization section (Chapter 32, Part V).

3.1 Rectangular Coordinates

Rectangular (sometimes referred to as Cartesian) coordinates on graph paper are most generally used to represent equations of the form

$$y = mx + b \tag{3.1}$$

where y = dependent variable represented on the ordinate
x = independent variable represented on the abscissa
m = slope of the line
b = y-intercept at $x = 0$

The bulk of the development in this chapter will be based on rectangular coordinates. For general engineering use, the most common form of graph paper is the 8 ½ × 11-inch sheet, having 20 lines per inch, with every fifth line accented, and every tenth line heavily accented. This type of coordinate graph may be obtained on drawing or tracing paper, with lines in black, orange, or green, and with or without accented and heavy lines; it is also available in other sizes.

All graphs are created by drawing a line, or lines, about one or more axes. For the purpose of this chapter, the presentation will be primarily concerned with graphs built around two axes – the x axis and the y axis. The graph data may be presented as coordinates of the x and y axis in the form (x, y), or they may be presented as the abscissa (distance from the y axis or the x coordinate) and the ordinate (distance from the x axis or the y coordinate). Referring to Figure 3.1, point L5 has an abscissa (x coordinate) of +60 and an ordinate of +40 (y coordinate). Using coordinate notation, the point can be described as (60, 40) where the points are listed as (x, y). Point L2 can be described as (–6.7, 0).

All coordinates are produced as the result of solving an equation by assigning different values to x or y, and solving for the value of the other. By convention, problems are usually stated in the following form: y = some value(s) of x and other constant(s) (e.g., $y = mx + b$).

The x and y axes divide a graph into four quadrants. The quadrants are usually numbered in a counter-clockwise sequence, beginning with the upper right

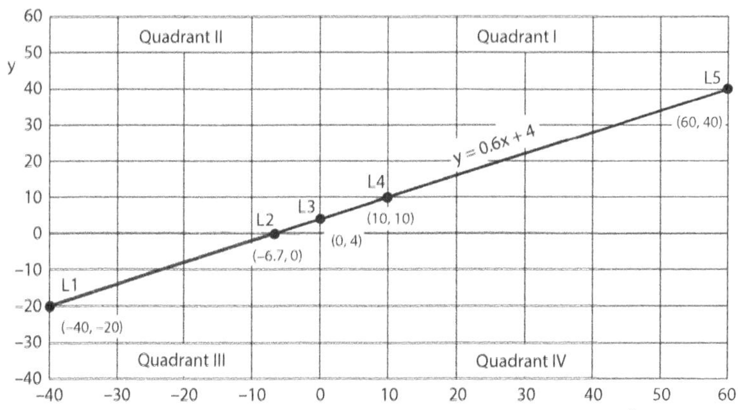

Figure 3.1 Linear coordinates.

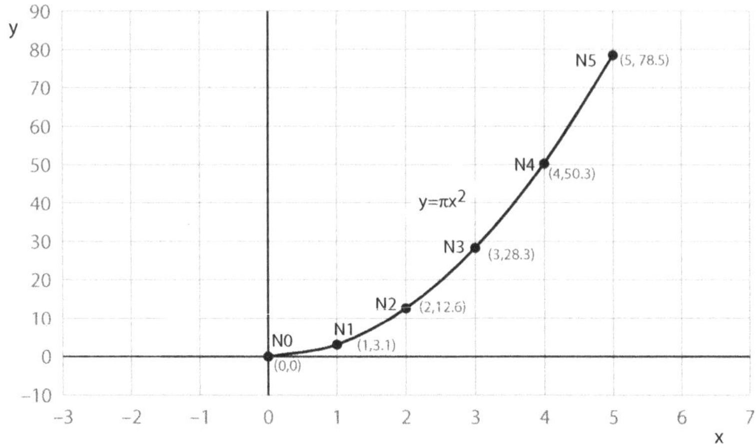

Figure 3.2 Non-linear equation.

quadrant (see Figure 3.1). Note that in quadrants I and IV, the x values are positive while the x values are negative in quadrants II and III. The y values are positive in quadrants I and II and negative in quadrants III and IV. The quadrant identifications are only of value when information is provided that a point or result lies in a certain quadrant. Without knowledge of the precise values of x or y one can still determine whether the x and y values are positive or negative.

Equations are identified by the type of line (or curve) they produce when plotted on a graph. A linear equation, as the name implies, will result in a plot of a straight line on a graph. With the linear equation, each incremental change of x (or y) will result in an incremental change of a fixed ratio in y (or x). The ratio of change between y and x is known as the *slope* of the line. A linear equation is also referred to as a *first-degree equation*. The line plotted between L1 and L5 in Figure 3.1, is linear. All solutions to the equation describing the line ($y = 0.6x + 4$) will fall somewhere on the line.

When plotted, a *non-linear equation* will result in a line having one or more curves, and the ratio of change between x and y is *not* constant. The line between points N0 and N5 in Figure 3.2 represents the plot of a non-linear equation. Non-linear equations come in a variety of forms and may be quickly identified since they contain functions or operators that cause non-linearity, i.e., sin, cos, ln, log, exponents. The equation $y = \pi x^2$ (see Figure 3.2) is representative of a non-linear equation.

Reference to the x and y intercepts are often made when working with graphs. An *intercept* is the point at which a line crosses the x or y axis. The *x-intercept* is the point where a line crosses the x axis ($y = 0$), and the *y-intercept* is the point where a line crosses the y axis ($x = 0$). In Figure 3.1, the x-intercept occurs at (–6.7, 0) and the y-intercept occurs at (0,4).

In a linear equation, there can be only one x and one y intercept. With non-linear equations, there may be no x or y intercept *or* there may be multiple intercepts.

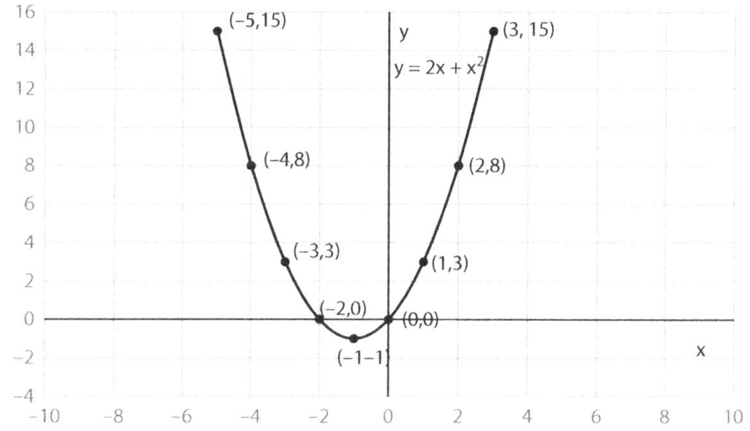

Figure 3.3 Intercepts analysis.

Note that in Figure 3.2, an intercept occurs when $x = 0$ or $y = 0$. In Figure 3.3, there are two x-intercepts [(−2, 0) and (0, 0)] and one y-intercept (0, 0).

The reader should note that a good part of the above material will be revisited in Part IV, Chapter 32.

3.2 Logarithmic-Logarithmic (Log-Log) Coordinates

A logarithmic scale is easily constructed by plotting (on rectangular paper) numbers from one to ten on the ordinate versus the logarithm of the number on the abscissa. If the points representing the numbers on the ordinate are projected to the resultant curve and then upward, the scale formed by the vertical lines will be a logarithmic scale. The construction is illustrated in Figure 3.4. The utility of the logarithmic scale lies in the fact that one can use actual numbers on the scale instead of the logarithms. When two logarithmic scales are placed perpendicular to each other and lines are drawn vertically and horizontally to represent major divisions, a full logarithmic graph results, i.e., a log-log graph.

The term *log-log* or *full logarithmic* referred to above is used to distinguish between another kind yet to be discussed, e.g., semi-logarithmic (see next section). The full logarithmic graphs may be obtained in many types depending on the scale length, sheet size, and number of cycles desired. The paper is specified by the number of cycles on the ordinate and abscissa; for example, *logarithmic 2 × 3 cycles* means two cycles on the ordinate and three on the abscissa. Cycle lengths are generally the same on both axes of a particular sheet. In addition, the distance between numbers differing by a factor of 10 is constant on a logarithmic scale.

When plotting lines on logarithmic paper, equations in the general form

$$y = bx^m \tag{3.2}$$

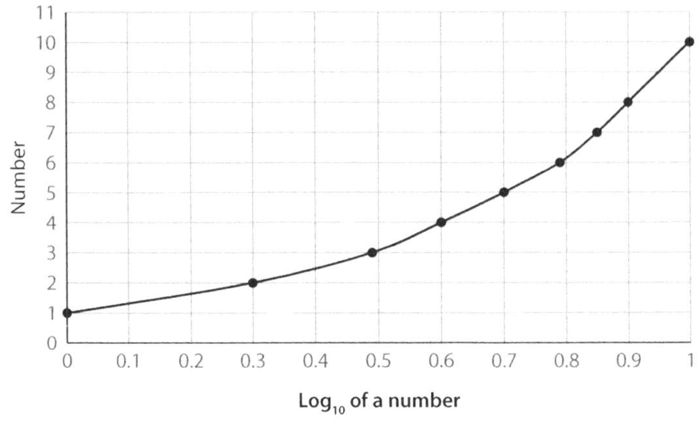

Figure 3.4 Log_{10} of a number.

where y = a dependent variable
x = an independent variable
b = a constant
m = a constant

will plot as straight lines on (full) logarithmic paper. A form of the equation analogous to the slope-intercept equation for a straight line is obtained by plotting the equation in logarithmic form, i.e.,

$$\log y = \log b + m \log x \tag{3.3}$$

If log y versus log x were plotted on rectangular coordinates, a straight line of slope m and y-intercept log b would result. An expression for the slope is obtained by differentiating Equation (3.3).

$$\frac{d(\log y)}{d(\log x)} = m \tag{3.4}$$

When $x = 1$, $m \log x = 0$, and the equation reduces to

$$\log y = \log b \tag{3.5}$$

Hence, log b is the y-intercept.

To illustrate the above, a linear plot of the equation

$$y = 2x^{1.3} \tag{3.6}$$

is given in Figure 3.5 on rectangular coordinate paper. It should be noted that although the slope was obtained from a ratio of logarithmic differences, the same result could have been obtained using measured differences if the scale of the ordinate and abscissa were the same (in this case they were).

Inspection of Figure 3.6 reveals that equal measured distances on the logarithmic scale are equivalent to equal logarithmic differences on the abscissa of that plot. It therefore follows that since the slope of Equation (3.3) is a ratio of logarithmic differences, the slope on full logarithmic paper could be obtained using measured distances. The plot of Equation (3.6) on 2 × 2 cycle logarithmic paper is given in Figure 3.6. The slope using logarithmic differences from Figure 3.5 is

$$m = \frac{y_2 - y_1}{x_2 - x_1} = \frac{1.08 - 0.43}{0.60 - 0.10} = 1.3 \tag{3.7}$$

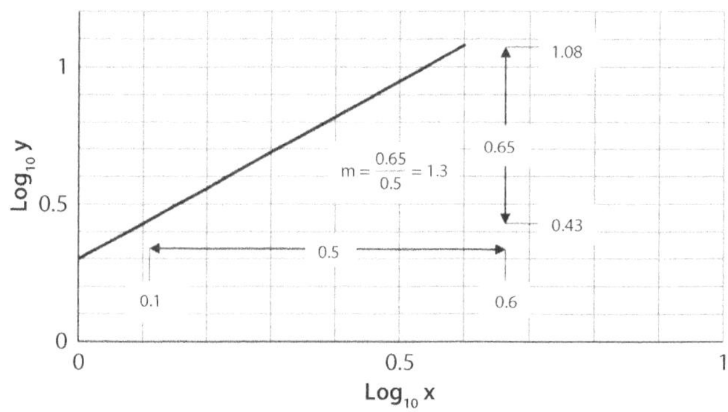

Figure 3.5 Linear plot of $y = 2x^{1.3}$.

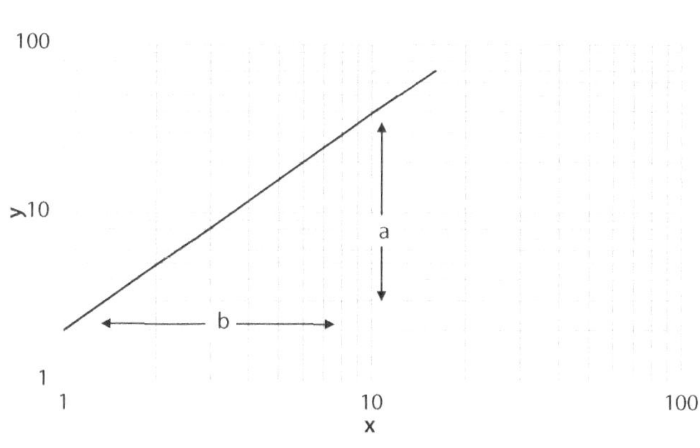

Figure 3.6 The equation $y = 2x^{1.3}$ on logarithmic-logarithmic paper.

Table 3.1 Log data.

x	y
2.0	3.6
3.9	6.0
7.0	9.4
14.0	16.0
26.5	26.0
62.0	50.0

and the slope using measured differences in Figure 3.6 is

$$m = \frac{a}{b} = 1.3 \tag{3.8}$$

To further illustrate the use of full logarithmic paper, consider the determination of a mathematical equation to represent data which are known to plot as a straight line on log-log paper. If data in Table 3.1 are to be plotted, it is evident that more than one cycle will be needed in both directions. For x, the data covers the range from

$$2\times10^0 \text{ (or } 0.2\times10^1\text{) to } 6.2\times10^1 \tag{3.9}$$

and for y the range is

$$3.6\times10^0 \text{(or } 0.36\times10^1\text{) to } 5.0\times10^1 \tag{3.10}$$

In both cases two cycles are needed to cover the ranges from 10^0 to 10^1 and from 10^1 to 10^2. A general rule would be that the number of cycles should be equal to the number of exponents of ten involved in the range of data – in this case, two for each coordinate. The data are plotted in Figure 3.7. The slope may be once again calculated in either of two ways:

$$m = \frac{\log 50 - \log 3.6}{\log 62 - \log 2.0} = 0.765 \tag{3.11}$$

or

$$m = \frac{c}{d} = 0.765 \tag{3.12}$$

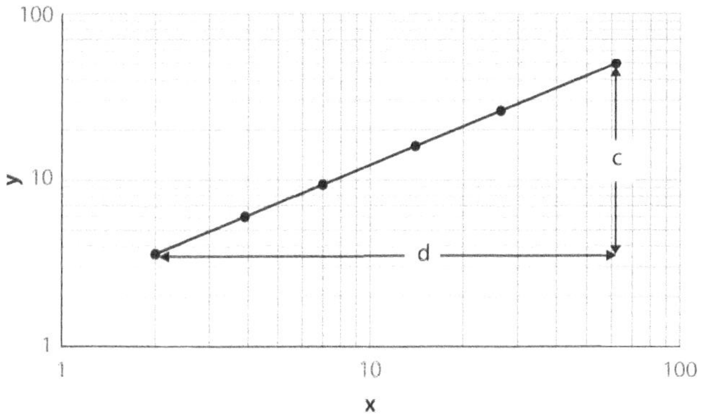

Figure 3.7 Determination of a mathematical equation from raw data.

The y-intercept is 2.13; hence, the equation of the line is

$$\log y = 2.13 + 0.765 \log x \tag{3.13}$$

In those cases where the line $x = 1$ is several cycles removed from the data range it may be inconvenient to extrapolate; the intercept can then be is determined by substituting the value of the slope and one data point and solving for b.

3.3 Semilogarithmic (Semi-Log) Coordinates

Graph paper made with one logarithmic scale and one arithmetic scale is termed *semilogarithmic*. The logarithmic scale is normally the ordinate and the arithmetic scale is the abscissa on semilogarithmic paper. The paper is available in many styles depending on the sheet size, type of paper, color of lines, number of cycles, length of cycles, and the divisions on the uniform scale. The designation *semilogarithmic, 7 × 5 to the ½ inch* refers to semilogarithmic paper whose logarithmic scale contains seven cycles whose uniform scale contains five divisions to the half inch. The designation *10 division per inch (70 divisions) by two 5-inch cycles* refers to paper whose arithmetic scale is seven inches long and contains ten division per inch, and whose logarithmic scale contains two-cycles for each five inches. Thus, a semilogarithmic graph uses a standard scale for one axis and a logarithmic scale for the other axis. The reason for this use of scales is that a logarithmic (exponential) plot would quickly exceed the physical boundaries of the graph.

Semilogarithmic paper can be used to represent equations of the form

$$y = ne^{mx} \tag{3.14}$$

or

$$y = n10^{mx} \tag{3.15}$$

where y = dependent variable
x = independent variable
n = a constant
m = a constant
e = the natural logarithm

Here again a form of the equation analogous to the slope-intercept equation can be obtained by placing one of the above equations in logarithmic form. For example, Equation (3.14) may be written as

$$\log y = b + mx \tag{3.16}$$

where $b = \log n$
An expression for slope is obtained from the differential form of Equation (3.16):

$$\frac{d(\log y)}{dx} = m \tag{3.17}$$

so that the y-intercept is log n or b. Expressions for the slope and intercept of Equation (3.15) are similar to those of Equation (3.14) except for the fact that natural logarithms are employed.

One can illustrate the use of semilogarithmic paper by plotting a segment of the curve representing the equation

$$\log y = 2.2 + \frac{x}{46.3} \tag{3.18}$$

The line may be constructed (see Figure 3.8) by choosing values of x, solving for y, and plotting the results. Note that values of y *and not* log y are plotted on the ordinate. The slope is a logarithmic difference divided by an arithmetic difference and, in this case, may be calculated from the two indicated points as follows:

$$m = \frac{\log 1585 - \log 158.5}{46.3 - 0} = \frac{1}{46.3} \tag{3.19}$$

The slope of a line on semilogarithmic coordinates may be calculated from any two points, but the simplest method is the one used above in which the slope is the logarithmic difference on one cycle (equal to log 10 or 1.0) divided by

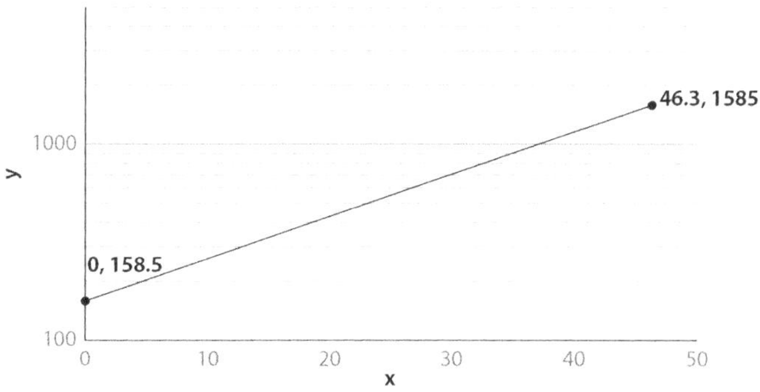

Figure 3.8 The equation $\log y = 2.2 + x/46.3$ plotted on semilogarithmic paper.

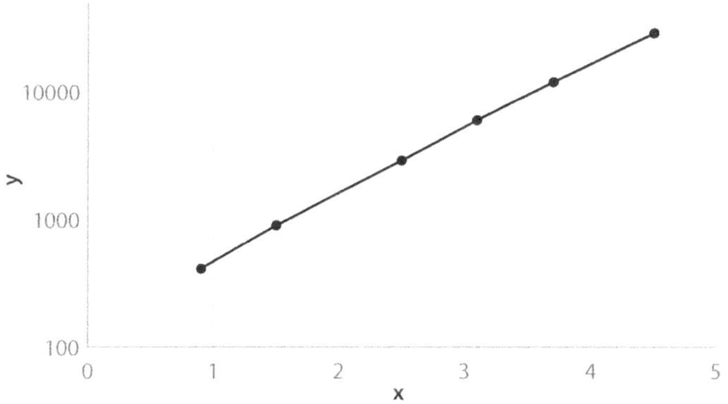

Figure 3.9 The determination of a mathematical equation from raw data.

the arithmetic difference cut by one cycle. Note that if the cycle had been taken between $y = 300$ and $y = 3000$, the slope would be

$$m = \frac{\log 3000 - \log 300}{59.1 - 12.8} = \frac{1}{46.3} \tag{3.20}$$

It should also be noted that the y-intercept is 2.2, as required by Equation (3.18).

The method of obtaining a mathematical equation from data that are known to plot as a straight line on semilogarithmic paper may be illustrated by the use of the following data provided in Table 3.2.

Table 3.2 Log data.

x	y
0.93	432
1.52	886
2.56	2,830
3.20	6,000
3.80	12,000
4.50	27,000

The data are plotted in Figure 3.9. Three cycles on the logarithmic scale were necessary because values of y ranged from

$$4.32\times10^{2} \text{ to } 2.7\times10^{4} \tag{3.21}$$

and three exponents of ten are involved, e.g., 2, 3, and 4. The slope may be calculated as follows:

$$m=\frac{\log(27{,}000)-\log(432)}{4.5-0.93}=\frac{1}{2} \tag{3.22}$$

The line is extrapolated to $x = 0$ to give a y-intercept of 2.175; hence the equation representing the data is

$$\log y=\log 2.175+\log\left(\frac{x}{2}\right) \tag{3.23}$$

or

$$y=2.175+(x/2) \tag{3.24}$$

3.4 Other Graphical Coordinates

Triangular (or trilinear) coordinates are used in some mass transfer studies to represent systems defined by three variables [1]. The graph can be in the form of an equilateral triangle, each side of which is usually divided into 100 equal parts (see Figure 3.10).

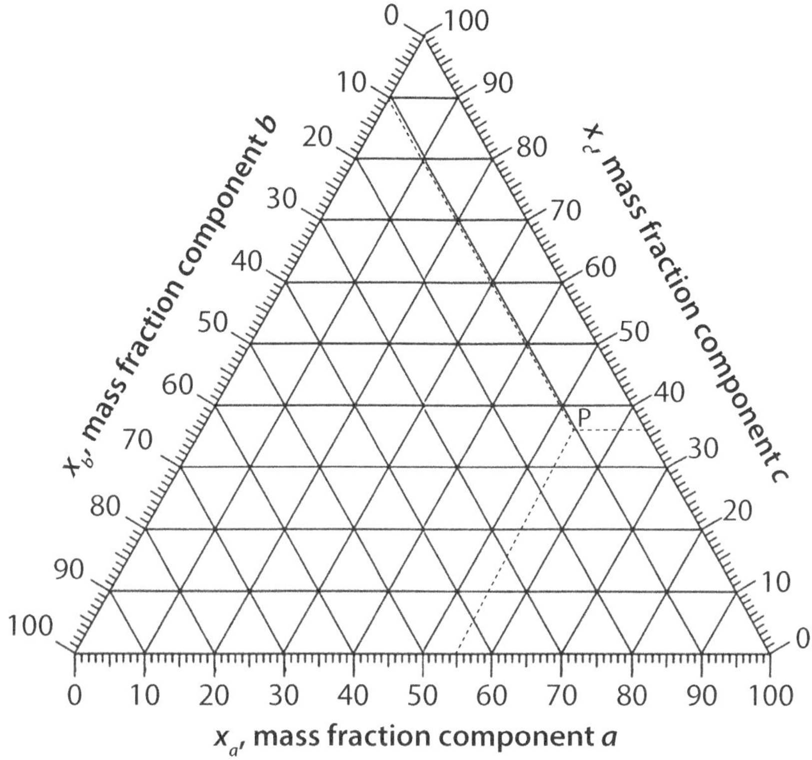

Figure 3.10 Equilateral triangular coordinates for a three-component mixture.

In addition to the more common types of graph paper previously discussed, the practicing environmental engineer and scientist may also have need for probability versus arithmetic, probability versus logarithmic, logarithmic versus reciprocals, and other special types of graphs. Non-standard scales, if desired, may be constructed in a similar manner to the logarithmic scale shown earlier. For example, logarithmic-probability (log-normal) graphics find applications in air pollution studies to describe particle size distribution [2].

3.5 Methods of Plotting Data

The simplest procedure to employ in plotting equations of various forms is detailed in Table 3.3 Various additional forms are available in the literature. Details on statistical methods for calculating the coefficients in the seven equations provided are reviewed in Part IV, Chapter 28.

Table 3.3 Procedures for plotting equations.

$y = a + bx$	plot y vs. x
$y = ax^n$	plot log y vs. log x or y vs. x on logarithmic crdinates
$y = c + ax^n$	first obtain c as an intercept on a plot of y vs. x; then plot log $(y - c)$ vs. x on logarithmic coordinates
$y = ae^{bx}$	plot log y vs. x on semilogarithmic coordinates
$y = ab^x$	plot log y vs. x on semilogarithmic coordinates
$y = a + \frac{b}{x}$	plot y vs $1/x$
$y = \frac{x}{a + bx}$	plot x/y vs. x or $1/y$ vs. $1/x$

References

1. L. Theodore and F. Ricci, *Mass Transfer Operations for the Practicing Engineer*, John Wiley & Sons, Hoboken, NJ, 2010.
2. L. Theodore, *Air Pollution Control Equipment Calculations*, John Wiley & Sons, Hoboken, NJ, 2008.

4

Flow Diagrams

The complete design specification for a medium-sized chemical process could cover several hundred pages. It would include diagrams, tables, and discussion of all aspects of the plant, including chemical, environmental, mechanical, structural, electrical, metallurgical, and civil engineering considerations. To understand a complete design, an environmental engineer or scientist must have training and experience in this area (Note: Computer flow diagrams are not addressed in this chapter.)

Any manufacturing plant, whether for the production of canned fruit, vacuum cleaners, sulfuric acid, rubber tires, or any environmental process, may be visualized as a box into which raw materials and energy are fed and from which useful products, waste, and energy emerge. The manufacturing process ordinarily involves several consecutive operations or steps through which the materials in the process pass.

Figure 4.1 is a schematic representation of materials passing into and out of an unidentified process for which there are environmental concerns. In this hypothetical three-step process, raw materials *A* and *B* are fed into step 1; material *C* is drawn off while material *D* is passed to step 2 for further processing. One cannot tell whether *C* is a useful product or waste. From the sketch in step 2, it is necessary to combine raw material *E* with *D* in order to produce *F*. In step 3, material *F*

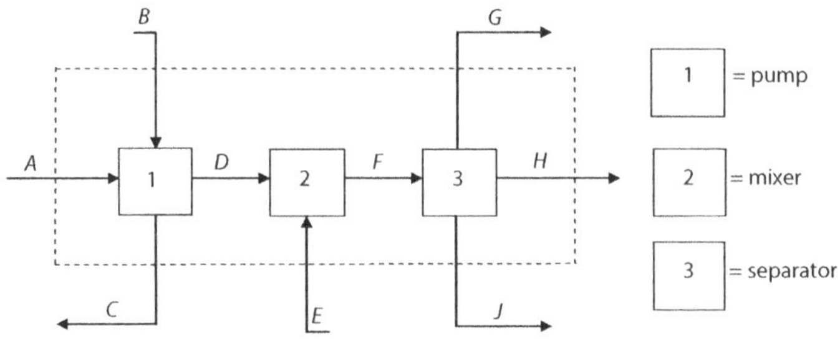

Figure 4.1 Flow diagram for a three-step manufacturing process.

is separated into *G, H,* and *J,* ending the process. Again, the figure does not show which of the last three items are useful products. The arrows lettered *A, B, C,* etc. represented *material streams* and the sketch is known as a *flow diagram.* Although nothing has been said about the specific nature of the process, this flow diagram nevertheless conveys a great deal of information; namely, that in this three-step process, three raw materials *A, B,* and *E* are required to produce four products *C, G, H,* and *J,* and that these seven materials enter and leave the manufacturing process at the points shown on the diagram [1].

As one might expect, a process flow diagram for a chemical or petroleum plant is usually significantly more complex than that for a simple environmental facility. For the latter case, the flow sequence and other determinations often reduce to an approach that employs a "railroad" or sequential type of calculation that does not require iterative calculations [1].

Five sections compliment the presentation of this chapter. Section numbers and subject titles follow:

4.1: Process Schematics
4.2: Flow Chart Symbols
4.3: Preparing Flow Diagrams
4.4: Simplified Flow Diagrams
4.5: Hazard Risk Assessment Flow Chart

4.1 Process Schematics

To the environmental engineer, but particularly the chemical engineer, the process flowchart is the key instrument for defining, refining, and documenting a chemical process. The process flow diagram is the authorized process blueprint, the framework for specifications used in equipment designation and design; it is the single, authoritative document employed to define, construct, and operate a process [1].

There are several essential constituents to a detailed process flowchart beyond equipment symbols and process stream flow lines. These include equipment identification numbers and names; temperature and pressure designations; utility designations; mass, molar, and volumetric flow rates for each process stream; and (usually), a material balance table pertaining to process flow lines. The process flow diagram may also contain additional information such as energy requirements, major instrumentation, environmental equipment (and concerns), and physical properties of the process streams. When properly assembled and employed, a process schematic provides a coherent picture of the overall process; it can pinpoint some deficiencies in the process that may have been overlooked earlier in a study, e.g., instrumentation overkill, by-products (undesirable or otherwise), and recycle needs. Basically, the flowchart symbolically and pictorially represents the interrelation between the various flow streams and equipment, and permits easy calculation of material and energy balances.

4.2 Flow Chart Symbols

Various symbols are universally employed to represent equipment, equipment parts, valves, piping, etc. Some of these are depicted in the schematic in Figure 4.2. Although a significant number of these symbols are used to describe some of the chemical and petrochemical processes, only a few are needed for simpler facilities. These symbols obviously reduce, and in some instances, replace detailed written descriptions of the process. Note that many of the symbols are pictorial, which helps in better describing process components, units and equipment [2].

The degree of sophistication and details of a flowchart usually vary with both the preparer and time. It may initially consist of a simple freehand block diagram with limited information that includes only the equipment; later versions may

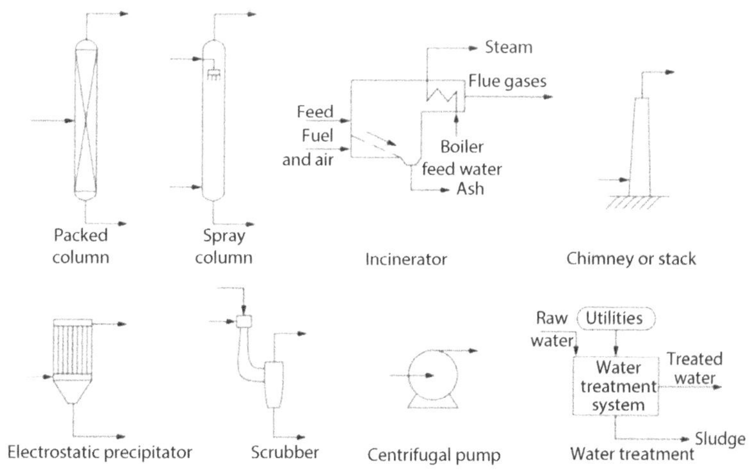

Figure 4.2 Flowchart symbols (*Cont.*).

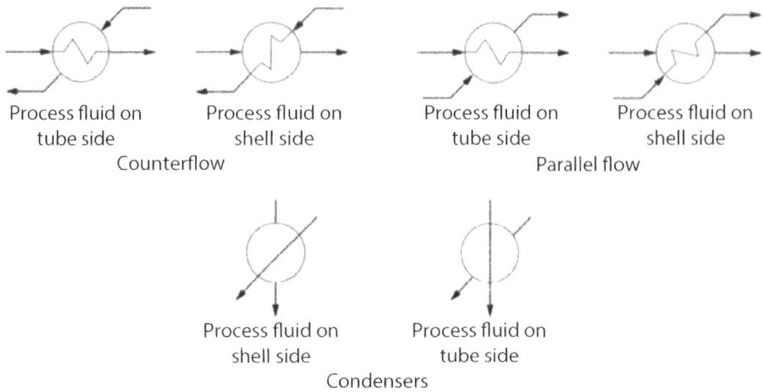

Figure 4.2 Flowchart symbols (*Continued*).

include line drawings with pertinent process data such as overall and componential flow rates, utility and energy requirements, environmental equipment, and instrumentation. During the later stages of a design project, the flowchart will be a highly-detailed P&I (piping and instrumentation) diagram; this aspect of the design procedure is beyond the scope of this text. The reader is referred to the literature for information on P&I diagrams [3].

In a sense, flowcharts are the international language of the engineer, particularly the practicing engineer. Chemical engineers conceptually view a (chemical) plant as consisting of a series of interrelated building blocks that are defined as *units* or *unit operations* [4–7]. The plant ties together the various pieces of equipment that make up the process. Flow schematics follow the successive steps of a process by indicating where the pieces of equipment are located and the material streams entering and leaving each unit [8–9].

4.3 Preparing Flow Diagrams

Before attempting to calculate the material or energy requirements of a process it is desirable to attain a clear picture of the process. The best way to do this is to draw the aforementioned flow diagram where the flow diagram is defined as a line diagram showing the succession of steps in the process. Flow diagrams are very important for saving time and eliminating mistakes. The beginner should learn how to draw them properly and cultivate the habit of sketching them on the slightest excuse. The following rules should be observed.

1. Show operating units by simple neat rectangles. Do not waste time in elaborate art work since it is without advantage or even meaning (see Figure 4.3).
2. Make each material stream line represent *an actual stream of material* passing along a pipe, duct, chute, belt, or other conveying device. Refer to Figure 4.4(a). Show the gaseous mixture of carbon dioxide,

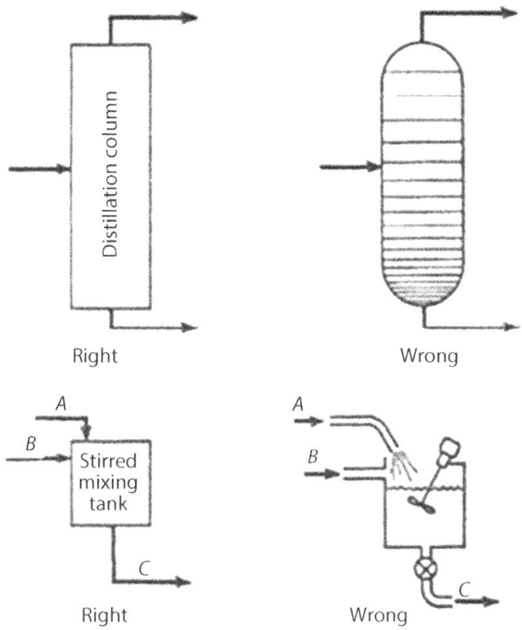

Figure 4.3 Simple flowchart symbols.

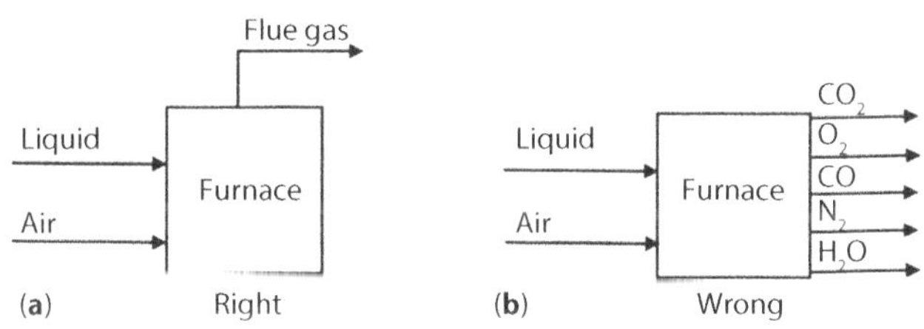

Figure 4.4 Right-and-wrong representation of a flue gas containing CO_2, O_2, CO, N_2, and water vapor.

oxygen, carbon monoxide, nitrogen, and water vapor obtained from the combustion of a hydrocarbon fuel and exiting from the stack of a furnace as a single material stream of *flue gas*, *not* to be confused as five separate streams. Figure 4.4(b) shows a sophisticated "device" that sorts components of the flue gas and delivers them as five separate products through five separate pipes.

3. Distinguish between "open" and "closed" material streams. In Figure 4.5(a), open steam is being blown into the tank, mixing with the other materials in it. In Figure 4.5(b), steam is passing through a coil, being kept separate from the tank contents, and heating them by the transfer of heat through the walls of the coil.

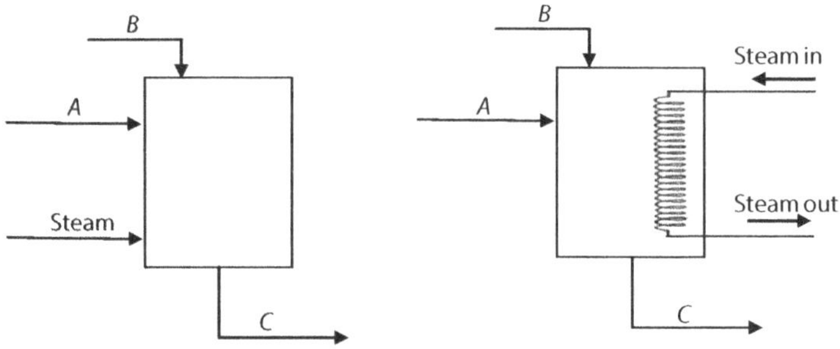

Figure 4.5 Open and closed material streams.

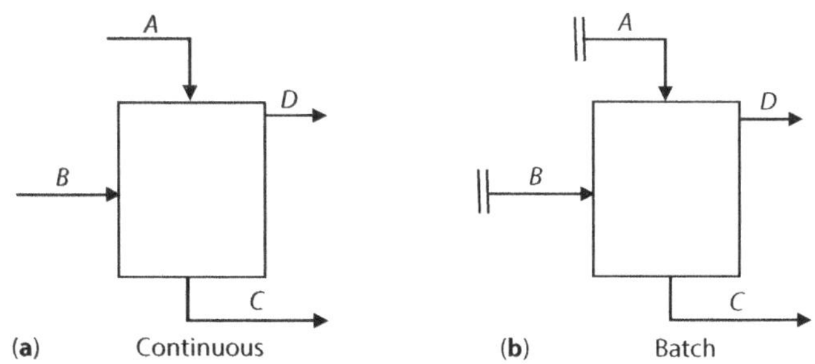

Figure 4.6 Continuous vs. batch operation.

4. Distinguish between a *continuous operation*, in which raw material is fed into the equipment in an uninterrupted stream and a *batch operation* in which a fixed charge of material is introduced, processed, and removed, followed by the feeding of a new charge with repetition of the cycle, etc. Batch operation is indicated conveniently by putting a double bar on the material stream lines of entering materials as shown in Figure 4.6(b).
5. Except for a few unusual situations which will be discussed later, keep flow diagrams free of data regarding the material streams. An over-burden of data clutters up the diagram and robs it of its main function, which is to present a clear picture of the materials as they move into, through, and out of the process. This is illustrated in Figure 4.7 with the air and coal entering a furnace.

4.4 Simplified Flow Diagrams

In many cases *simplified* flow sheets are prepared to illustrate a process. These are often for a special purpose and do not show all the details of the process.

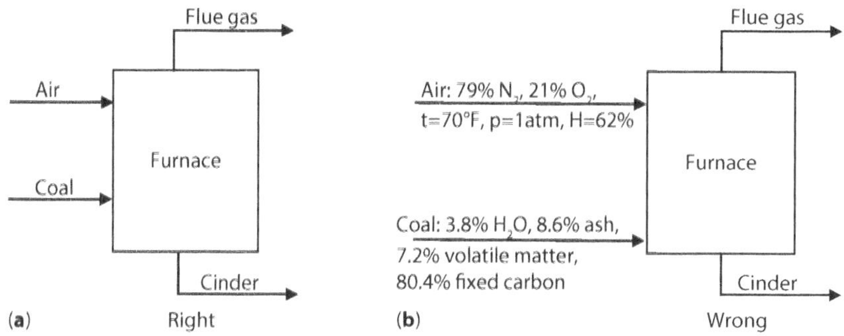

Figure 4.7 Inclusion versus exclusion of experimental data on flow diagrams.

A common type of flow sheet shows the major unit operations and chemical reactors with their interconnecting piping and an identification of the materials being processed. The units are not shown to scale, but the drawing may resemble the equipment used for the operation. Such a diagram would be called a *graphic* process flow sheet. For a small process of a few units, a *pictorial* flow sheet with drawings of equipment approximately to scale may be used. Photographs are usually unsatisfactory because the arrangement and connections are seldom clear. On a process flow sheet, the equipment is arranged logically to show the flow of materials through the process, but a photograph shows the final physical arrangement determined by structural requirements without regard to the process flow. All tall absorber columns may be grouped together for structural support, and large heat exchangers may be grouped together for ease of maintenance. The successful individual must learn to read flow sheets and to relate them to the actual plant layout.

If sufficient data are available, a *quantitative* process flowsheet may be prepared showing the flow rates, compositions, temperatures, and pressures throughout the process. Such flow sheets may be very complicated. Flow sheets showing instrumentation and controls are also occasionally prepared.

4.5 Hazard Risk Assessment Flow Chart

An *acceptable risk* is a risk whose probability is unlikely to occur during the lifetime of a plant or process. An acceptable hazard risk can also be defined as an accident that has a high probability of occurring, but with negligible consequences. Risks can be ranked qualitatively in categories of high, medium, and low. Risks can also be ranked quantitatively as the annual number of fatalities per million affect individuals.

Hazard risk assessment (HZRA) calculations are based on an algorithm proposed over 30 years ago. This algorithm is based on a simple flowchart (see also

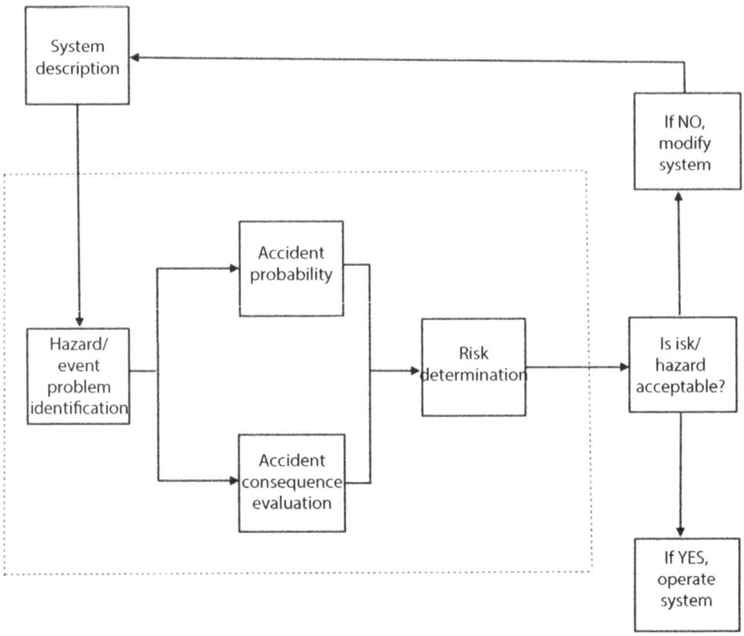

Figure 4.8 HZRA flowchart for a chemical plant.

Figure 4.8). There are several steps in evaluating the risk of an accident; if the system in question is a chemical plant, the following guidelines apply:

1. A brief description of the equipment and chemicals used in the plant is needed.
2. Any hazard in the system has to be identified. Hazards that may occur in a chemical plant include:
   a. Corrosion
   b. Explosions
   c. Fires
   d. Rupture of a pressurized vessel
   e. Runaway reactions
   f. Slippage
   g. Unexpected leaks
3. The event or series of events that will initiate an accident must be identified. An event could be a failure to follow correct safety procedures, to improperly repair equipment, or the failure of a safely mechanism.
4. The probability that the accident will occur often has to be determined. For example, if a nuclear power plant has a 10-year life, what is the probability that the temperature in a reactor will exceed

the specified temperature range? The probability can be ranked qualitatively from low to high. A low probability means that it is unlikely for the event to occur in the life of the plant. A medium probability suggests that there is a possibility that the event will occur. A high probability means that the event will probably occur during the life of the plant.
5. The severity of the consequence of the accident must be determined.
6. If the probability of the accident and the severity of its consequences are low, then the risk is usually deemed acceptable and the plant should be allowed to operate. If the probability of occurrence is too high or the damage to the surroundings is too great, then the risk is usually unacceptable and the system needs to be modified to minimize these effects.

The heart of the hazard risk assessment algorithm provided is enclosed in the dashed-line box in Figure 4.8. This algorithm allows for reevaluation of the process if the risk is deemed unacceptable (the process is repeated starting with step 1 or 2).

The reader should note that both health risk assessment and hazard risk assessment plus accompanying calculations have received extensive treatment in the literature [10–11].

References

1. L. Theodore, personal notes, East Williston, NY, 1967.
2. G. Ulrich, *A Guide to Chemical Engineering Process Design and Economics*, John Wiley & Sons, NY, 1984.
3. T. Shen, Y. Choi, and L. Theodore. *Hazardous Waste Incineration Manual*, A Theodore Tutorial, Theodore Tutorials, East Williston, NY, originally published by USEPA/APTI, RTP, NC, 1985.
4. P. Abulencia and L. Theodore, *Fluid Flow for the Practicing Chemical Engineer*, John Wiley & Sons, Hoboken, NJ, 2007.
5. L. Theodore, *Heat Transfer Applications for the Practicing Engineer*, John Wiley & Sons, Hoboken, NJ, 2011.
6. L. Theodore and F. Ricci, *Mass Transfer Operations for the Practicing Engineer*, John Wiley & Sons, Hoboken, NJ, 2010.
7. L. Theodore, R. Dupont, K. Ganesan, *Unit Operations in Environmental Engineering*, Scrivener-Wiley, Salem, MA, 2017.
8. R. Perry and D. Green, editors, *Perry's Chemical Engineers' Handbook*, 8th ed., McGraw-Hill, New York City, NY, 2008.
9. W. McCabe, J. Smith, and P. Harriot, *Unit Operations of Chemical Engineering*, 5th ed., McGraw-Hill, New York City, NY, 1993.
10. L. Theodore, J. Reynolds, and K. Morris, *Accident and Emergency Management*, A Theodore Tutorial, Theodore Tutorials, East Williston, NY, originally published by USEPA/APTI, RTP, NC, 1996.
11. L. Theodore and R. Dupont, *Environmental Health Risk and Hazard Risk Assessment: Principles and Calculations*, CRC Press/Taylor & Francis Group, Boca Raton, Fla., 2012.

5

Dimensional Analysis

A surprising amount of a practicing environmental engineer's time (and scientist's as well) is spent in converting data and equations from one set of units to another. Keep in mind that a unit is defined as a *measure* of a physical extent while a dimension is a *description* of the physical extent. As long as most physicists and chemist make measurement in grams and centimeters while many engineers in the United States employ pounds and feet, confusion in terminology will continue to exist.

Units, unlike physical laws, can be considered as either *derived* or *basic*. There is a certain latitude in choosing the basic units and, unfortunately, this free choice has resulted in the aforementioned mild form of confusion. Two systems of units have arisen: metric-the cgs, or centimeters-gram-second system *and* the English the fps, or foot-pound-second system of engineering. The metric system has come to be defined as the System of International Units, or more commonly for some as the SI system.

Six sections compliment the presentation of this chapter. Section numbers and subject titles follow:

5.1: The Metric System
5.2: The SI System
5.3: Conversion of Units

5.4: Select Common Abbreviations
5.5: Dimensionless Numbers
5.6: Buckingham Pi (π) System

5.1 The Metric System [1]

The need for a single worldwide coordinated measurement system was recognized nearly 350 years ago. In 1670, Gabriel Mouton, Vicar of St. Paul's Church in Lyon, France, proposed a comprehensive decimal measurement system based on the length of one minute of arc of a great circle of the Earth. On 1671, Jean Picard, a French astronomer, proposed the length of a pendulum arc beating seconds as the unit of length. (Such a pendulum would be fairly easy to reproduce, thus facilitating the widespread distribution of uniform standards.) Other proposals were made, but over a century elapsed before any action was taken.

In 1790, in the midst of the French Revolution, the National Assembly of France requested the French Academy of Scientists to "deduce an invariable standard for all the measures and weights." The Commission appointed by the Academy created a system that was, at once, simple and scientific. The unit of length was to be a portion of the Earth's circumference. Measures for capacity (volume) and mass (weight) were to be derived from the unit of length, thus relating the basic units of the system to each other and to nature. Furthermore, the larger and smaller versions of each unit were to be created by multiplying or dividing the basic units by 10 and its multiples. This feature provided a great convenience to users of the system by eliminating the need for calculating and dividing by 16 (to convert ounces to pounds) or by 12 (to convert inches to feet). Similar calculations in the metric system could be performed simply by shifting the decimal point. Thus, the metric system is a *base-10* or *decimal* system.

The Commission assigned the name *metre* (which is now spelled *meter*) to the unit of length. This name was derived from the Greek word *metron* meaning "a measure." The physical standard representing the meter was to be constructed so that it would equal one ten-millionth of the distance from the north pole to the equator along the meridian of the Earth running near Dunkirk in France and Barcelona in Spain.

The metric unit of mass, called the *gram*, was defined as the mass of one cubic centimeter (a cube that is 1/100 of a meter on each side) of water at its temperature of maximum density. The cubic decimeter (a cube 1/10 of a meter on each side) was chosen as the unit of fluid capacity. This measure was given the name *liter.*

Although the metric system was not accepted with enthusiasm at first, adoption by other nations occurred steadily after France made its use compulsory in 1840. The standardized character and decimal features of the metric system made it well suited to scientific and engineering work. Consequently, it is not surprising that the rapid spread of the system coincided with an age of rapid technological development. In the United States, by Act of Congress in 1866, it was made "lawful throughout the United States of America to employ the weights and measures of the metric system in all contracts, dealings, or court proceedings."

By the late 1860's, even better metric standards were needed to keep pace with scientific advances. In 1875, an international treaty, the "Treaty of the Meter," set up well-defined metric standards for length and mass, and established permanent machinery to recommend and adopt further refinements in the metric system. This treaty, known as the *Metric Convention*, was signed by 17 countries, including the United States.

As a result of the treaty, metric standards were constructed and distributed to each nation that ratified the Convention. The internationally agreed metric standards have served as the fundamental weights and measures standards of the United States since 1893.

A total of 35 nations- including the major nations of continental Europe and most of South America- had officially accepted the metric system by 1900. Today, with the exception of the United States and a few small countries, the entire world is predominately using the metric system, or is committed to such use. The Secretary of Commerce, in transmitting to Congress the results of a 3-year study authorized by the Metric Study Act of 1968, recommended in 1971 that the US change to the predominant use of the metric system through a coordinated national program.

The International Bureau of Weights and Measures located at Sevres, France, serves as a permanent secretariat for the Metric Convention, coordinating the exchange of information about the use and refinement of the metric system. As measurement science develops, more precise and easily reproducible ways of defining the measurement units will emerge. The General Conference of Weights and Measures- the diplomatic organization made up of adherents to the Convention-meets periodically to ratify improvements in the system and the standards.

5.2 The SI System

In 1960, the General Conference adopted an extensive revision and simplification of the system. The name *Le Systeme International d'Unites* (International System of Units), with the international abbreviation SI, was adopted for this modernized metric system. Further improvements in and additions to SI were made by the General Conference in 1964, 1968, and 1971.

The basic units in the SI system are the *kilogram* (mass), *meter* (length), *second* (time), *Kelvin* (temperature), *ampere* (electric current), *candela* (the unit of luminous intensity), and *radian* (angular measure), all of which are commonly used by the practicing environmental engineer and scientist. The Celsius scale of temperature (0 °C = 273.15 K) is commonly used with the absolute Kelvin scale. The important derived units are the *newton* (SI unit of force), the *joule* (SI unit of energy), the *watt* (SI unit of power), the *pascal* (SI unit of pressure), and the *hertz* (SI unit of frequency). There are a number of electrical units: *coulomb* (charge), *farad* (capacitance), *henry* (inductance), *volt* (potential), and *weber* (magnetic flux). As noted above, one of the major advantages of the metric system is that larger and smaller units are given in powers of ten. In the SI system, a further simplification is introduced by recommending only the use of units with multipliers of 10^3. Thus, for lengths in engineering, the *micrometer* (previously referred to as *micron*), *millimeter*, and the *kilometer* are recommended, and the *centimeter* is generally avoided. A further simplification is that the decimal point may be substituted by a comma (as in France, Germany, and South Africa), while the other number, before and after the comma, is separated by spaces between groups of three.

The aforementioned seven base units have been associated with the SI.

1. Length – meter (m)
2. Mass – kilogram (kg)
3. Time – second (s)
4. Electric current – ampere (A)
5. Temperature – Kelvin (K)
6. Amount of substance – mole (mol) (aka gram · mole)
7. Luminous intensity – candela (cd)

Note that two supplementary units are occasionally employed in geometric studies:

1. Phase angle – radian (rad)
2. Solid angle – steradian (sr)

5.3 Conversion of Units

Momentum and rate of momentum come into play in both fluid flow and the conservation law in momentum studies. The latter term is defined (in terms of units) as

$$\text{Rate of Momentum} = \frac{\text{lb}\cdot\text{ft}}{\text{s}^2} \tag{5.1}$$

The above units can be converted to lb_f, if multiplied by an appropriate constant. A *conversion constant* is a term that is used to obtain units in a more convenient form. All conversion constants have magnitude and units in the term, but can also be shown to be equal to 1.0 (unity) with *no* units.

A conversion constant used often is

$$12\ \text{in/ft} \tag{5.2}$$

This term is obtained from the following defining equation:

$$12\ \text{in} = 1\ \text{ft} \tag{5.3}$$

If both sides of this equation are divided by 1 ft, one obtains:

$$12 \text{ in/ft} = 1 \tag{5.4}$$

Note that this conversion constant, like all others, is also equal to unity without any units.

Another defining equation is:

$$1\,\text{lb}_f = 32.2\frac{\text{lb}\cdot\text{ft}}{\text{s}^2} \tag{5.5}$$

If this equation is divided by lb_f, Equation (5.6) results.

$$1.0 = 32.2\frac{\text{lb}\cdot\text{ft}}{\text{lb}_f\cdot\text{s}^2} \tag{5.6}$$

This serves to define the conversion constant g_c. If the rate of momentum is divided by g_c as $32.2\ \text{lb}\cdot\text{ft}/\text{lb}_f\cdot\text{s}^2$ – this operation being equivalent to dividing by 1 – the following units result:

$$\text{Rate of Momentum} = \left(\frac{\text{lb}\cdot\text{ft}}{\text{s}^2}\right)\left(\frac{\text{lb}_f\cdot\text{s}^2}{\text{lb}\cdot\text{ft}}\right) = \text{lb}_f \tag{5.7}$$

One can conclude from the above dimensional analysis that a force is equivalent to a rate of momentum.

There are hundreds of conversion constants employed by engineers and scientists. Some of the more common "conversion constants" are provided in Table 5.1 [1].

Conversion of units can be accomplished by the multiplication of the quantity to be converted by appropriate unit ratios, i.e., the conversion constants. For example, suppose an energy of 50 Btu must be converted to units of $(\text{ft}\cdot\text{lb}_f)$. From the energy section in Table 5.1 one notes that to convert from Btu to $(\text{ft}\cdot\text{lb}_f)$, multiply by 778. Therefore,

$$778(\text{ft}\cdot\text{lb}_f) = 1\,\text{Btu} \tag{5.8}$$

and the conversion constant or unit ratio is similarly,

$$\frac{778(\text{ft}\cdot\text{lb}_f)}{1\,\text{Btu}} = 1.0 \tag{5.9}$$

Table 5.1 Conversion constants.

To convert from	To	Multiply by
	Length	
m	cm	100
m	mm	1000
m	microns (μm)	10^6
m	angstroms (Å)	10^{10}
m	in	39.37
m	ft	3.281
m	mi	6.214×10^{-4}
ft	in	12
ft	m	0.3048
ft	cm	30.48
ft	mi	1.894×10^{-4}
	Mass	
kg	g	1000
kg	lb	2.205
kg	oz	35.24
kg	ton	2.268×10^{-4}
kg	grains	1.543×10^4
lb	oz	16
lb	ton	5×10^{-4}
lb	g	453.6
lb	kg	0.4536
lb	grains	7000
	Time	
s	min	0.01667
s	h	2.78×10^{-4}
s	day	1.157×10^{-7}
s	week	1.653×10^{-6}
s	yr	3.171×10^{-8}
	Force	
N	kg • m/s^2	1
N	dynes	10^5
N	g • cm/s^2	10^5

(Continued)

Table 5.1 Cont.

To convert from	To	Multiply by
N	lb_f	0.2248
N	lb • ft/s^2	7.233
lb_f	N	4.448
lb_f	dynes	4.448×10^5
lb_f	g • cm/s^2	4.448×10^5
lb_f	lb • ft/s^2	32.17
Pressure		
atm	N/m^2 (Pa)	1.013×10^5
atm	kPa	101.3
atm	bars	1.013
atm	dynes/cm^2	1.013×10^6
atm	lb_f/in^2 (psi)	14.696
atm	mm Hg at 0 °C (torr)	760
atm	in Hg at 0 °C	29.92
atm	ft H_2O at 4 °C	33.9
atm	in H_2O at 4 °C	406.8
psi	atm	6.80×10^{-2}
psi	mm Hg at 0 °C (torr)	51.71
psi	in H_2O at 4 °C	27.70
in H_2O at 4 °C	atm	2.458×10^{-3}
in H_2O at 4 °C	psi	0.0361
in H_2O at 4 °C	mm Hg at 0 °C (torr)	1.868
Volume		
m^3	L	1000
m^3	cm^3 (cc, mL)	10^6
m^3	ft^3	35.31
m^3	gal (U.S.)	264.2
m^3	qt	1057
ft^3	in^3	1728
ft^3	gal (U.S.)	7.48
ft^3	m^3	0.02832
ft^3	L	28.32

(Continued)

Table 5.1 Cont.

To convert from	To	Multiply by
	Energy	
J	N • m	1
J	erg	10^7
J	dyne • cm	10^7
J	kW· h	2.778×10^{-7}
J	cal	0.2390
J	ft • lb_f	0.7376
J	Btu	9.486×10^{-4}
cal	J	4.186
cal	Btu	3.974×10^{-3}
cal	ft • lb_f	3.088
Btu	ft • lb_f	778
Btu	hp • h	3.929×10^{-4}
Btu	cal	252
Btu	kW • h	2.93×10^{-4}
ft • lb_f	cal	0.3239
ft • lb_f	J	1.356
ft • lb_f	Btu	1.285×10^{-3}
	Power	
W	J/s	1
W	cal/s	0.2390
W	ft • lb_f/s	0.7376
W	kW	10^{-3}
kW	Btu/s	0.949
kW	hp	1.341
hp	ft • lb_f/s	550
hp	kW	0.7457
hp	cal/s	178.2
hp	Btu/s	0.707
	Concentration	
$\mu g/m^3$	lb/ft^3	6.243×10^{-11}
$\mu g/m^3$	lb/gal	8.346×10^{-12}
$\mu g/m^3$	gr (grain)/ft^3	4.370×10^{-7}

(*Continued*)

Table 5.1 Cont.

To convert from	To	Multiply by
gr/ft^3	μg/m^3	2.288×10^6
gr/ft^3	g/m^3	2.288
lb/ft^3	μg/m^3	1.602×10^{10}
lb/ft^3	μg/L	1.602×10^{-8}
lb/ft^3	lb/gal	7.48
	Viscosity	
P (poise)	g/cm • s	1
P	cP (centipoise)	100
P	kg/m • h	360
P	lb/ft • s	6.72×10^{-2}
P	lb/ft • h	241.9
P	lb/m • s	5.6×10^{-3}
lb/ft • s	P	14.88
lb/ft • s	g/cm • s	14.88
lb/ft • s	kg/m • h	5.357×10^3
lb/ft • s	lb/ft • h	3600
	Heat Capacity	
cal/g • °C	Btu/lb • °F	1
cal/g • °C	kcal/kg • °C	1
cal/g • °C	cal/gmol • °C	Molecular weight
cal/gmol • °C	Btu/lbmol • °F	1
J/g • °C	Btu/lb • °F	0.2389
Btu/lb • °F	cal/g • °C	1
Btu/lb • °F	J/g • °C	4.186
Btu/lb • °F	Btu/lbmol • °F	Molecular weight

The 50 Btu may be multiplied by the above conversion constant without changing its value. Therefore,

$$(50\ \text{Btu})\left(\frac{778(\text{ft}\cdot\text{lb}_f)}{1\ \text{Btu}}\right) = 38{,}900(\text{lb}_f\cdot\text{ft}) \tag{5.10}$$

with the Btu units canceling, just like numbers.

Illustrative Example 5.1

Convert:

1. 150 miles/h to yards/h
2. 100.0 m/s^2 to ft/min^2
3. 0.03 g/m^3 to lb/ft^3

Solution

Employ Table 5.1.

1. $$\left(\frac{150\text{ mile}}{\text{h}}\right)\left(\frac{5,280\text{ ft}}{\text{mile}}\right)\left(\frac{\text{yd}}{3\text{ ft}}\right)=2.6\times10^5\text{ yd/h}$$

2. $$100.0\text{ m/s}^2\left(\frac{100\text{ cm}}{\text{m}}\right)\left(\frac{\text{ft}}{30.48\text{ cm}}\right)\left(\frac{60\text{s}}{\text{min}}\right)^2=1.181\cdot10^6\text{ ft/min}^2$$

3. $$(0.06\text{ g/cm}^3)\left(\frac{\text{lb}}{454\text{ g}}\right)\left(\frac{30.48\text{ cm}}{\text{ft}}\right)^3=2.0\text{ lb/ft}^3$$

5.4 Select Common Abbreviations [1]

Some typical abbreviations that are employed in environmental engineering and science practice are provided in Table 5.2.

5.5 Dimensionless Numbers

Problems are frequently encountered in environmental studies and other engineering work that involve several variables. Engineers and scientists are generally interested in developing functional relationships (equations) between these variables. When these variables can be grouped together in such a manner that they can be used to predict the performance of similar pieces of equipment, independent of the scale or size of the operation, something very valuable has been accomplished.

Consider, for example, the problem of establishing a method of calculating power requirements for mixing liquids in an open tank at a water treatment facility. The obvious variables would be the depth of liquid in the tank, the density and viscosity of the liquid, the speed of the agitator, the geometry of the agitator, and the diameter of the tank. There are therefore six variables that affect the power, or a total of seven terms that must be considered. To generate a general equation to describe power variation with these variables, a series of tanks having different diameters would have to be set up in order to gather data for various values of each

Table 5.2 Selected common abbreviations.

Å, A	angstrom unit of length
abs	absolute
amb	ambient
app. MW, M	apparent molecular weight, molecular weight
atm	atmosphere
at. wt.	atomic weight
b.p.	boiling point
bbl	barrel
Btu	British thermal unit
cal	calorie
cg	centigram
cm	centimeter
cgs system	centimeter gram second system
conc	concentration
cc, cm^3	cubic centimeter
cu ft, ft^3	cubic feet
cfh	cubic feet per hour
cfm	cubic feet per minute
cfs	cubic feet per second
m^3, M^3 (rarely)	cubic meter
°	degree
°C	degree Celsius, degree Centigrade
°F	degree Fahrenheit
°R	degree Rankine, degree Reamur (rarely)
ft	foot
ft • lb	foot pound mass
ft • lb_f	foot pound force
fpm	feet per minute
fps	feet per second
fps system	foot • pound • second system
f.p.	freezing point
gr	grain
g, gm	gram
h	hour

(*Continued*)

Table 5.2 Cont.

in	inch
kcal	kilocalorie
kg	kilogram
km	kilometer
liq	liquid
L	liter
log	logarithm (common)
ln	logarithm (natural)
m.p.	melting point
m, M (rarely)	meter
μm	micrometer (micron)
mks system	meter • kilogram • second system
mph	miles per hour
mg	milligram
ml	milliliter
mm	millimeter
mμ	millimicron
min	minute
mol wt, MW, M	molecular weight
oz	ounce
ppb	parts per billion
pphm	parts per hundred million
ppm	parts per million
lb	pound
psi	pound force per square inch
psia	pound force per square inch absolute
psig	pound force per square inch gage
rpm	revolutions per minute
s, sec	second
sp gr	specific gravity
sp ht	specific heat
sp wt	specific weight
sq	square
scf	standard cubic foot
STP	standard temperature and pressure

(*Continued*)

Table 5.2 Cont.

t	time, temperature
T, temp.	temperature
wt	weight

variable. Assuming that the 10 different values for each of the six variables were imposed on the process, 10^6 runs would be required. Obviously, a mathematical method for handling several variables that requires considerably less than one million runs to establish a *design method* must be available. In fact, such a method is available and it is defined as *dimensional analysis* [1].

Dimensional analysis is a powerful tool that is employed by environmental engineers and scientists in planning experiments, presenting data compactly, and making practical predictions from models without detailed mathematical analysis. The first step in an analysis of this nature is to write down units of each variable. The end result of a dimensional analysis is a list of pertinent dimensionless numbers. A partial list of common dimensionless numbers used in fluid flow analyses is given in Table 5.3 [2]. The reader is referred to the literature for definitions of these terms [2].

Dimensional analysis is a relatively "compact" technique for reducing the number and the complexity of the variables affecting a given phenomenon, process or calculation. It can help obtain not only the most out of experimental data but also scaleup data from a model to a prototype. To do this, one must achieve similarity between the prototype and the model. This similarity may be achieved through dimensional analysis by determining the important dimensionless numbers, and then designing the model and prototype such that the important dimensionless numbers are the same in both.

There are three steps in dimensional analysis. These are:

1. List all parameters and their primary units.
2. Formulate dimensionless numbers (or ratios).
3. Develop the relation between the dimensionless numbers experimentally.

Further details on this approach are provided in the next section.

5.6 Buckingham Pi (π) Theorem

This theorem provides a simple method to obtain the aforementioned dimensionless numbers (or ratios) termed π parameters. The steps employed in obtaining the dimensionless π parameters are given below [2]:

1. List all parameters. Define the number of parameters as n.
2. Select a set of primary dimensions, e.g., kg, m, s, K (English units may also be employed). Let r = the number of primary dimensions.

Table 5.3 Dimensionless numbers [2].

Parameter	Definition	Area of importance	Qualitative ratio
Cavitation number	$Ca = \frac{P - p^{!}}{\rho v^2 / 2}$	Cavitation	$\frac{\text{Pressure}}{\text{Inertia}}$
Eckert number	$Ec = \frac{v^2}{C_p \cdot T}$	Dissipation	$\frac{\text{Kinetic energy}}{\text{Inertia}}$
Euler number	$Eu = \frac{\Delta P}{\rho v^2 / 2}$	Pressure drop	$\frac{\text{Pressure}}{\text{Inertia}}$
Froude number	$Fr = \frac{v^2}{gL}$	Free surface flow	$\frac{\text{Inertia}}{\text{Gravity}}$
Mach number	$Ma = \frac{v}{c}$	Compressible flow	$\frac{\text{Flow speed}}{\text{Sound speed}}$
Poiseuille number	$P_0 = \frac{D^2 \cdot \Delta P}{\mu L v}$	Laminar flow in pipes	$\frac{\text{Pressure}}{\text{Viscous forces}}$
Relative roughness	$\frac{k}{D}$	Turbulent flow, rough walls	$\frac{\text{Wall roughness}}{\text{Body length}}$
Reynolds number	$\mathrm{Re} = \frac{\rho v D}{\mu} = \frac{vD}{\nu}$	Various uses	$\frac{\text{Inertia forces}}{\text{Viscous forces}}$
Strouhal number	$St = \frac{wL}{v}$	Oscillating flow	$\frac{\text{Oscillation speed}}{\text{Mean speed}}$
Weber number	$We = \frac{\rho v^2 L}{\sigma}$	Surface forces effect	$\frac{\text{Inertia}}{\text{Surface tension}}$

3. List the units of all parameters in terms of the primary dimensions, e.g., L [=] m, where "[=]" means "has the units of." This is a critical step and often requires some creativity and ingenuity on the part of the individual performing the analysis.
4. Select a number of variables from the list of parameters (equal to r). These are called repeating variables. The selected repeating parameters must include all r independent primary dimensions. The remaining parameters are called "non-repeating" variables.
5. Set up dimensional equations by combining the repeating parameters with each of the non-repeating parameters in turn to form the dimensionless parameters. π. There will be $(n - r)$ dimensionless groups of (πs).
6. Check that each resulting π group is in fact dimensionless.

Note that it is permissible to form a different π group from the product or division of other πs, e.g.,

$$\pi_5 = \frac{2\pi_1\pi_2}{\pi_3^2} \quad \text{or} \quad \pi_6 = \frac{1}{\pi_4} \tag{5.11}$$

Note, however, that a dimensional analysis approach will fail if the fundamental variables are not correctly chosen. The Buckingham Pi theorem approach to dimensionless numbers is given in the Illustrative Example that follows.

Illustrative Example 5.2

When a fluid flows through a horizontal circular pipe, it undergoes a pressure drop. $\Delta P = (P_2 - P_1)$. For a rough pipe, ΔP will be higher than a smooth pipe. The extent of non-smoothness of a material is expressed in terms of the roughness, k [2]. For steady state incompressible Newtonian fluid flow, the pressure drop is believed to be a function of the fluid average velocity v, viscosity μ, density ρ, pipe diameter D, pipe length L, pipe roughness k, and the speed of sound in a fluid (an important variable if the flow is compressible) c, i.e.,

$$\Delta P = f(v, \mu, \rho, D, L, k, c)$$

Determine the dimensionless numbers of importance for this flow system [1, 2].

Solution

A pictorial representation of the system in question is provided in Figure (5.1). List all the parameters and find the value of n:

$$\Delta P, v, \mu, \rho, D, L, k, c$$

There are 8 parameters therefore, $n = 8$. Choose primary units (employ SI).

$$\text{m, s, kg, K}$$

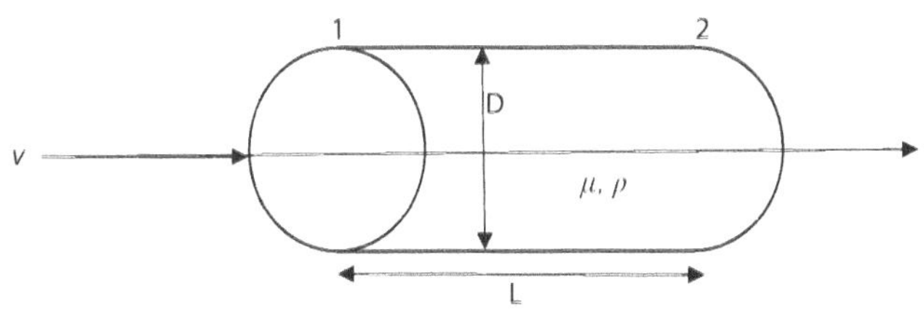

Figure 5.1 Pipe.

List the primary units of each parameter:

$$\Delta P\,[=]\,\text{Pa} = \text{kg}\,\text{m}^{-1}\,\text{s}^{-2}$$

$$v\,[=]\,\text{ms}^{-1}$$

$$\mu[=]\,\text{kg}\,\text{m}^{-1}\,\text{s}^{-1}$$

$$D\,[=]\,\text{m}$$

$$L\,[=]\,\text{m}$$

$$\rho\,[=]\,\text{kg}\,\text{m}^{-3}$$

$$k\,[=]\,\text{m}$$

$$c\,[=]\,\text{ms}^{-1}$$

Therefore, $r = 3$ with primary units m, s, kg.

Select three parameters from the list of eight parameters. These are the repeating variables:

$$D\,[=]\,\text{m}$$

$$\rho[=]\,\text{kg}\,\text{m}^{-3}$$

$$v\,[\,=]\,\text{ms}^{-1}$$

The non-repeating parameters are then ΔP, μ, k, c and L. Determine the number of πs:

$$n - 3 = 8 - 3 = 5$$

Formulate the first π, i.e., π_1, employing ΔP as the non-repeating parameter.

$$\pi_1 = \Delta P v^a \rho^b D^f$$

Determine a, b, and f by comparing the units on both sides of the following equation:

$$0[=](\text{kg}\cdot\text{m}^{-1}\cdot\text{s}^{-2})(\text{m}\cdot\text{s}^{-1})^a(\text{kg}\cdot\text{m}^{-3})^b(\text{m})^f$$

Compare kg:

$$0 = 1 + b. \text{ Therefore, } b = -1$$

Compare s:

$$0 = -2 - a. \text{ Therefore, } a = -2$$

Compare m:

$$0 = -1 + a - 3b + f. \text{ Therefore, } f = 0$$

Substituting back into π_1 leads to:

$$\pi_1 = \Delta P v^a \rho^b = \frac{\Delta P}{\rho v^2}$$

This represents the Euler number (see Table 5.3).

Formulate the second π, i.e., π_2, as

$$\pi_2 = \mu v^a \rho^b D^f$$

Determine a, b, and f by comparing the units on both sides:

$$0[=](\text{kg}\cdot\text{m}^{-1}\cdot\text{s}^{-2})(\text{m}\cdot\text{s}^{-1})^a(\text{kg}\cdot\text{m}^{-3})^b(\text{m})^f$$

Compare kg:

$$0 = 1 + b. \text{ Therefore, } b = -1$$

Compare s:

$$0 = -1 - a. \text{ Therefore, } a = -1$$

Compare m:

$$0 = -1 + a - 3b + f. \text{ Therefore, } f = -1$$

Substituting back into π_2 yields:

$$\pi_2 = \mu v^{-1} \rho^{-1} D^{-1} = \frac{\mu}{v \rho D}$$

Replace π_2 by its reciprocal:

$$\pi_2 = \frac{v\rho D}{\mu} = \text{Re}$$

where Re = Reynolds number.

Similarly, the remaining non-repeating variables lead to

$$\pi_3 = kv^a\rho^b D^f \rightarrow \frac{k}{D}$$

and from

$$\pi_4 = cv^a\rho^b D^f$$

one obtains

$$\pi_4 = \frac{v}{c} = \text{the Mach Number}$$

Similarly,

$$\pi_5 = \frac{L}{D}$$

Combine the πs into an equation, expressing π_1 as a function of π_2, π_3, π_4 and π_5.

$$Eu = \frac{\Delta P}{\frac{\rho v^2}{2}} = f\left(\text{Re}, \frac{k}{D}, Ma, \frac{L}{D}\right) = \text{the Euler number}$$

Consider the case of incompressible flow.

$$Eu = \frac{\Delta P}{\frac{\rho v^2}{2}} = f\left(\text{Re}, \frac{k}{D}, \frac{L}{D}\right)$$

The result indicates that to achieve similarity between a model (m) and a prototype (p), one must have the following:

$$\text{Re}_m = \text{Re}_p$$

$$\left(\frac{k}{D}\right)_{\mathrm{m}} = \left(\frac{k}{D}\right)_{\mathrm{p}}$$

$$\left(\frac{L}{D}\right)_{\mathrm{m}} = \left(\frac{L}{D}\right)_{\mathrm{p}}$$

Since $\mathrm{Eu} = f\left(\mathrm{Re}, \frac{k}{D}, \frac{L}{D}\right)$, then it follows that $\mathrm{Eu}_{\mathrm{m}} = \mathrm{Eu}_{\mathrm{p}}$ (see also Table 5.3).

References

1. I. Farag and J. Reynolds, *Fluid Flow,* A Theodore Tutorial, Theodore Tutorials, East Williston, NY, originally published by the USEPA/APTI, RTP, NC, 1995.
2. P. Abulencia and L. Theodore, *Fluid Flow for the Practicing Chemical Engineer,* John Wiley & Sons, Hoboken, NJ, 2009.

6

Economics

Regardless of the type of society or government within which a business is operated, it must either show a profit or perish. What applies to an individual business applies equally to a community or an entire economy. Without profit, the structure is weak; with profit, it *may* be strong. This conclusion is not debatable.

Because of the necessity for combining economic considerations with environmental fundamentals and principles, the environmental engineer or scientist must be aware of the many different types of costs involved in environmental processes and issues. Money must be paid out for different plant expenses such as those for raw materials, labor, and equipment. In addition, many other indirect expenses are incurred, and these must be included if a complete analysis of the total cost is to be obtained. Some examples of these are addressed later in the chapter.

Since this chapter was written primarily for the beginning student it would be almost impossible at this level to cover all of the various aspects of engineering economics. This chapter therefore introduces the philosophy of economics in engineering and shows, by means of a few illustrative examples, the use of economics as a tool. It is realized that the broad subject of engineering economics cannot be fitted to any rigid set of formulas. For the interested reader, there are many texts available in the literature which provide a much broader treatment of the subject. Both the qualitative and quantitative viewpoints are emphasized in the material to follow.

Four sections compliment the presentation of this chapter. Section numbers and subject titles follow:

6.1: Definitions
6.2: The Need for an Economic Analysis
6.3: Capital Investment and Risk
6.4: Applications

6.1 Definitions

Before proceeding to the applications later in the chapter, it would be wise to provide the reader with certain key definitions used in the field. Fourteen concepts that often come into play in an economic analysis are given below. The definitions have been drawn from the literature [1].

Simple Interest

The term interest can be defined as the money paid for the use of money. It is also referred to as the value or worth of money. Two terms of concern are simple interest and compound interest. Simple interest is always computed on the original principal. The basic formula to employ in simple interest calculations is:

$$S = P(1 + ni) \tag{6.1}$$

where P = original principle
n = time in years
i = annual interest rate
S = sum of interest and principal after n years

Normally, the interest period is one year, in which case i is referred to as the effective annual interest rate.

Compound Interest

Unlike simple interest, with compound interest, interest is added periodically to the original principle. The term conversion or compounding of interest simply refers to the addition of interest to the principal. The interest period or conversion period in compound interest calculations is the time interval between successive conversions of interest. The interest rate is the ratio of the stated annual rate to the number of interest periods in one year. Thus, if the given interest rate is 10% compounded semiannually, the interest period is six months and the interest rate per interest period is 5%. Alternatively, if the given interest rate is 10% compounded quarterly, then the interest period is 3 months and the interest rate per interest period is 2.5%. One should always assume the interest is compounded annually unless otherwise stated. The basic formula to employ for compound interest is:

$$S = P(1+i)^n \tag{6.2}$$

If interest payments become due m times per year at compound interest, $(m)(n)$ payments are required in n years. An effective annual interest rate, i', may be defined by

$$i' = \left(1+\frac{i}{m}\right)^m - 1 \tag{6.3}$$

In this case, the new equation for S becomes:

$$S = P\left(1+i'\right)^n \tag{6.4}$$

In the limit (as m approaches infinity), such payments may be considered to be required at infinitesimally short intervals, in which case, the interest is said to be compounded *continuously*. Numerically, the difference between compounding may be significant when applied to vary large sums of money.

Present Worth

The present worth is the current value of a sum of money due at time n and at interest rate i. The following equation is the compound interest equation solved for the present worth term P.

$$P = S(1+i)^{-n} \tag{6.5}$$

Evaluation of Sums of Money

The value of a sum of money changes with time because of interest considerations. \$1,000 today, \$1,000 ten years from now, and \$1,000 ten years ago all have different meanings when interest is taken into account. \$1,000 today would be worth more than \$1,000 ten years from now because of the interest that could be accumulated in the interim. On the other hand, \$1,000 today would have been worth less ten years ago because a smaller sum of money could have been invested then so as to yield \$1,000 today. Therefore, one must refer to the date as well as the sum of money when evaluating sums of money.

Summarizing, evaluating single sums of money requires multiplying by $(1 + i)^n$ if the required date of evaluation is *after* the date associated with the obligation or multiplying by $(1 + i)^{-n}$ if the required date of evaluation is *before* the date associated with the obligation. The term n is always the time in periods between the date associated with the obligation and the evaluation date.

The evaluation of sums of money may be applied to the evaluation of a uniform series of payments. A uniform series is a series of equal payments made at equal intervals. Suppose R is invested at the end of every interest period for n periods. The total value of all these payments, S, as of the date of the last payment, may be calculated from the equation

$$S = \frac{R[(1+i)^n - 1]}{i} \tag{6.6}$$

The term S is then called the amount of the uniform series.

Depreciation

The term depreciation refers to the decrease in the value of an asset. Two approaches that can be employed are the straight line and sinking fund method. In the straight line method of depreciation, the value of the asset is decreased each year by a constant amount. The annual depreciation amount, D, is given by:

$$D = \frac{\text{Original cost} - \text{Salvage value}}{\text{Estimated life in years}} \tag{6.7}$$

In the sinking fund method of depreciation, the value of the asset is determined by first assuming that a "sinking" fund consisting of uniform annual payments had been set up for the purpose of replacing the asset at the end of its estimated useful life. The uniform annual payment (UAP) may be calculated from the equation

$$\text{UAP} = (\text{Original cost} - \text{Salvage value})(\text{SFDF}) \tag{6.8}$$

where SFDF is the sinking fund deposit factor and is given by

$$SFDF = \frac{i}{[(1+i)^n - 1]} \tag{6.9}$$

The value of the asset at any time is estimated to be the difference between the original cost and the amount that would have accumulated in the sinking fund. The amount accumulated in the sinking fund is obtained by multiplying the SFDF by the compound amount factor (CAF) where

$$CAF = \frac{[(1+i)^n - 1]}{i} \tag{6.10}$$

Fabricated Equipment Cost Index

A simple process is available to estimate the equipment cost from past cost data. The method consists of adjusting the earlier cost data to present values using factors that correct for inflation. A number of such indices are available; one of the most commonly used is the fabricated equipment cost index (FECI).

$$\text{Cost}_{\text{year B}} = \text{Cost}_{\text{year A}} \left(\frac{\text{FECI}_{\text{year B}}}{\text{FECI}_{\text{year A}}} \right) \quad (6.11)$$

Given the cost and FECI for year A, as well as the FECI for year B, the cost of the equipment in year B can be estimated by employing Equation (6.11).

Capital Recovery Factor

In comparing alternative processes or different options for a particular process from an economic point-of-view, one recommended procedure to follow is that the total capital cost can be converted to an annual basis by distributing it over the projected lifetime of the facility (or the equivalent). The sum of both the annualized capital cost (ACC), including installation, and the annual operating cost (AOC), is called the total annualized cost (TAC) for the project or facility. The economic merit of the proposed facility process, or scheme can be examined once the total annual cost is available.

The conversion of the total capital cost (TCC) to an ACC requires the determination of an economic parameter known as the capital recovery factor (CRF). This parameter can be found in any standard economics textbook or calculated directly from the following equation:

$$\text{CRF} = \frac{i(1+i)^n}{[(1+i)^n - 1]} \quad (6.12)$$

where n = projected lifetime of the system
i = annual interest rate (as a fraction)

The CRF is a positive, fractional number. Once this factor has been determined, the ACC can be calculated from the following equation:

$$\text{ACC} = (\text{TCC})(\text{CRF}) \quad (6.13)$$

The annualized capital cost can be viewed as the cost associated with recovering the initial capital expenditure over the depreciable life of the system.

Present Net Worth

There are various approaches that may be employed in the economic selection of the best of several alternatives. For each alternative in the present net worth (PNW) method of economic selection, a single sum is calculated that would provide for all expenditures over a common time period. The alternative having the least PNW of expenditures is selected as the most economical. The equation to employ is

$$PNW = CC + PN + PWD - PWS \tag{6.14}$$

where

CC = Capital cost
PN = Future renewals
PWD = Other disbursements
PWS = Salvage value

If the estimate lifetimes differ for the various alternatives, one should employ a period of time equal to the least common multiple of the different lifetimes for renewal purposes.

Perpetual Life

Capitalized cost can be viewed as present worth under the assumption of perpetual life. Computing capitalized cost involves, in a very real sense, finding the present worth of an infinite series of payments. To obtain the present worth of an infinite series of payments of $R at the end of each interest period forever, one needs simply to divide R by i, where i is the interest rate per interest period. Thus, to determine what sum of money, P, would have to be invested at 8.0% to provide payments of $100,000 at the end of each year forever, P would have to be such that the interest on it each period would be $100,000. Withdrawal of the interest at the end of each period would leave the original sum intact to again draw $100,000 interest at the end of the next period. For this example,

$$P = \frac{100{,}000}{0.08} = \$1{,}250{,}000 \tag{6.15}$$

The $1,250,00 would be the present worth of an infinite series of payments of $100,000 at the end of each year forever, assuming money is worth 8%.

To determine the present worth of an infinite series of payments of $R at the end of each n periods forever, first multiply by the SFDF to convert to an equivalent single period payment and then divide by i (fractional value) to obtain the present worth.

Break-Even Point

From an economic point-of-view, the break-even point of a process operation is defined as that condition when the costs (C) exactly balance the income (I). The profit (P) is therefore:

$$P = I - C \tag{6.16}$$

At break-even, the profit is zero.

Approximate Rate of Return

Rate of return can be viewed as the interest that will make the present worth of net receipts equal to the investment. The approximate rate of return (ARR), denoted by some as p, may be estimated from the equation below:

$$p = ARR = \frac{\text{Average annual profit or earnings}}{\text{Initial total investment}} \tag{6.17}$$

To determine the average annual profit, simply divide the difference between the total money receipts (income) and the total money disbursements (expenses) by the number of years in the period of the investment.

Exact Rate of Return

Using the approximate rate of return as a guide, one can generate the exact rate of return (ERR). This is usually obtained by trial-and-error and interpolation calculations of the rate of interest that makes the present worth of the net receipts equal to the investments. The approximate rate of return will tend to overestimate the exact rate of return when all or a large part of the receipts occur at the end of a period of investment. The approximate rate will tend to underestimate the exact rate when the salvage value is zero and also when the salvage value is a high percentage of the investment.

Bonds

A bond is a written promise to pay both a certain sum of money (redemption price) at a future date (redemption date) and equal interest payments at equal intervals in the interim. The holder of a $1,000, 5% bond, redeemable at 105 (bond prices are listed without the last zero) in 10 years, with interest payable semiannually would be entitled to semiannual payments of ($1,000) (0.025) or $25 for 10 years and 105% of $1,000, that is $1,050, at the end of 10 years when the bond is redeemed.

The interest payment on a bond is found by multiplying the face value of the bond by the bond interest rate per period. From the above, the face value is $1,000 and the bond interest rate per period is 0.025. Therefore, the periodic interest payment is the aforementioned \$25. Redeemable at 105 means that the redemption price is 105% of the face value of the bond.

The purchase price of a bond depends on the yield rate, i.e., the actual rate of return on the investment represented by the bond purchase. Therefore, the purchase price of a bond is the present worth of the redemption price plus the present worth of future interest payments, all computed at the yield rate. The bond purchase price formula is:

$$V = \frac{C(1+i)^n + R}{i} \tag{6.18}$$

where

V = purchase price
C = redemption price
R = periodic interest payment
n = time in periods to maturity
i = yield rate

Incremental Cost

By definition, the average unit increment cost is the increase in cost divided by the increase in production. Only those cost factors which vary with the production can affect the average unit incremental cost. In problems involving decisions as to whether to stay in production or (temporarily) shut down, the average unit increment cost may be compared with the unit increment cost or the unit selling price.

6.2 The Need for an Economic Analysis [1]

Before an environmental engineer or scientist can successfully apply the principles of economics to operational expenses, he/she must first become familiarized with the existing economics with which he/she is associated, as well as with the policies, if applicable, of the company itself. A company or individual hoping to increase its profitability must carefully assess a range of investment opportunities and select the most profitable options from those available. Increasing competitiveness also requires that efforts be expended to reducing costs of existing processes. In order to accomplish this, environmental engineers and scientists should be fully aware of not only technical but also economic factors, particularly those that have the largest effect on profitability.

In earlier years, environmental engineers (and scientists) concentrated on the technical side of projects and left the financial studies to the economist. In effect, all those engineers involved in making estimates of the capital and operating costs had often left the overall economic analysis and investment decision-making to others. This approach is no longer acceptable.

Some engineers are not equipped to perform a financial and/or economic analysis. Furthermore, many environmental engineers and scientists already working for companies have never taken courses in this area. This shortsighted attitude is surprising in a group of people who normally go to great lengths to get all the available technical data before making an assessment of a project or study. The attitude is even more surprising when one notes that data are readily available to enable an engineer to assess the prospects of both his/her own company and those of his/her particular industry [2].

Bridging the gap between theory and practice is often a matter of experience acquired over a number of years. Even then, methods developed from experience all too often must be re-evaluated in the light of changing economic conditions if optimum profits are to result. The approach presented here therefore represents an attempt to provide a consistent and reasonably concise method for the solution of these problems involving economic alternatives [3].

As noted above, the purpose of this chapter is to provide a working tool to better assist the reader in not only understanding economics and finance but also in applying technical information to the economic design and operation of processes and plants. The material to follow will often focus on environmental and/or plant applications. Hopefully, this approach will provide the reader with a better understanding of some of the fundamentals and principles.

The term "economic analysis" in engineering problems generally refers to calculations made to determine the conditions for realizing maximum financial return for a design or operation. The same general principles apply, whether one is interested in the choice of alternatives for competing projects, in the design of plants or projects so that the various components are economically proportioned, or in the optimization for economical operation of existing plants. General considerations that form the framework on which sound decisions must be made are often simple. Sometimes their application in the problems encountered in the development of a commercial enterprise involves too many intangibles to allow exact analysis, in which case judgments must be intuitive. Often, however, such calculations may be made with a considerable degree of exactness. This chapter attempts to develop a relatively concise method for applying these principles.

Finally, concern with the maximum financial returns implies that the criterion for judging projects involves profit. While this is usually true, there are many important objectives which, though aimed at ultimate profit increase, cannot be immediately evaluated on quantitative terms. Perhaps the most significant of these has been the tendency in recent years to regard management of commercial organizations as a profession with environmental obligations and responsibilities; considerations other than the profit motive *may* govern business decisions. However, these additional social objectives are for the most part often consistent with the economic goal of satisfying human wants with the minimum cost and effort. In fact, even in the operation of primarily nonprofit organizations, it is still important to determine the effect of various policies on profit [1].

6.3 Capital Investment and Risk

A capital investment is required for any process, and the determination of the necessary investment is an important part of a project. The total investment for any process consists of the fixed capital investment for the physical equipment and facilities plus the working capital for money which must be available to pay salaries, keep raw materials and products on hand, and handle other special items requiring a direct cash outlay. Thus, in an analysis of costs in industrial processes, investment costs, factory manufacturing costs, and general expenses that include income taxes must be taken into consideration.

The following material primarily addresses chemical plants. The following sequence of steps must be taken before a chemical plant can be put into operation:

1. A process must be planned and evaluated, or even invented.
2. The equipment is designed and the operational procedure worked out and analyzed.
3. Land is acquired for the site and services made available through roads, lines for gas, water, and electricity, sewage disposal systems, etc.
4. Requisite materials are purchased, and the buildings and all equipment erected.
5. Funds must be made constantly available to keep the venture in operation once the plant has been put "on stream."

All five of the above steps require money. The total money involved is called the *capital investment*. This consists of two parts; that required for steps (1), (2), (3), and (4), known as *fixed capital* and that required for step (5) called *working capital*. In more graphic terms, the money required to bring the process from the moment of conception to the moment of when the "button is pushed" and the plant is first set in operation is *fixed capital*. Typical items of the fixed capital are salary paid, e.g., to researchers and engineers for conception and design. All money required for the venture thereafter (with the exception of capital improvements or expansions of the plant) is *working capital* that can include the cost of raw materials; salaries and wages paid to production engineers, foremen, and plant workers; packaging and shipping costs; the costs of water, gas, fuel, electricity, etc., and, the salaries paid to clerical personnel involved with the process [4].

Many a process that looked good in an early stage of research or development proved to be economically unattractive after closer and more intensive examination. Often, the process does not go beyond step (2) listed above. Money invested up to this point is partly lost and may be completely lost. In other cases, the process gets all the way to the operational stage before proving to be unprofitable, in which case the loss may be significant.

Certain things should be obvious. All investment of capital in chemical processing involves a risk. Manufacturing profit cannot start unless and until the process is put into actual operation, and since for every process that proves profitable there are others that do not, the profitability of the successful processes should be sufficient to absorb the losses incurred by the unsuccessful ventures. What potential profit level should seem probable before capital is invested in a manufacturing process? A partial answer is provided in the next paragraph.

Money put into a savings account or invested in bonds may draw a certain return without risk, effort, or worry. It follows that the investment of capital in a manufacturing process that would yield as little as this percent return would be unsound since this would represent the assumption of business risk *with no profit at all.* How far in excess above the potential rate of return should be to justify the investment of capital depends upon the nature of both the industry and the individual process. When new frontiers are being broached, a potential profit of 30 percent or more may be essential to make a plant design attractive. On the other hand, starting a plant to produce a heavy chemical by a process with a long record of successful performance for which detailed economic data are available is another matter and may be attractive at a lower profitability potential.

6.4 Applications

The remainder of the chapter is devoted to illustrative examples, many of which contain technical development material. A good number of these environmental applications have been drawn from the National Science Foundation (NSF) literature [5–8] and one other key source [9].

Illustrative Example 6.1

List the major fixed capital costs for the chemical process industry.

Solution

1. Major process equipment (i.e., reactors, tanks, pumps, filters, distillation columns, etc.)
2. Installation of major process equipment
3. Process pumping
4. Insulation
5. Instrumentation
6. Auxiliary facilities (i.e., power substations, transformers, boiler houses, fire-control equipment, etc.)
7. Outside lines (i.e., piping external to the building, supports and posts for overhead piping, electric feeders from the power substations, etc.)

8. Land and site improvements
9. Building and structures
10. Consultant fees
11. Engineering and construction (design and engineering fees plus supervision of plant erection)
12. Contractors' fees (administrative)

Illustrative Example 6.2

List the major working capital costs for the chemical process industry.

Solution

1. Raw materials for plant startup
2. Raw materials, intermediate and finished product inventories
3. Cost of handling and transportation of materials to and from sites
4. Cost inventory control, warehouses, associated insurance, security arrangements, etc.
5. Money to carry accounts receivable (i.e., credit extended to customers) less accounts payable (i.e., credit extended by suppliers)
6. Money to meet payrolls when starting up
7. Readily available cash for emergencies
8. Any additional cash required to operate the process or business
9. Expenses associated with new hires
10. Startup consultant fees

Illustrative Example 6.3

A fluid is to be transported 4 miles under turbulent fluid flow conditions. An engineer is confronted with two choices in designing the system:

a. Employ a 2-inch ID pipe at a cost of $1/foot.
b. Employ a 4-inch ID pipe at a cost of $6/foot.

Pressure drop costs for the 2-inch ID pipe are $20,000/year. Assume that the only operating cost is the pressure drop and the only capital cost is the pipe. The capital recovery factor (CRF) for either pipe system is 0.1. Estimate the operating cost for the 4-inch ID pipe. Also, determine which is the more economical pipe system?

Solution

The pressure drop is (approximately) proportional to the velocity squared for turbulent flow [10]. The velocity of the fluid is lower for the 4-inch ID pipe. Since the area ratio is 4 (diameter squared), the 4-inch ID pipe velocity is one quarter of the velocity of the fluid in the 2-inch ID pipe. Therefore, the pressure drop for the 4-inch ID pipe is approximately one sixteenth of the pressure drop for the 2-inch ID pipe. The operating cost associated with this pressure drop is equal to:

$$\text{Operating Cost} = \frac{\frac{\$20{,}000}{\text{yr}}}{16} = \$1250/\text{yr}$$

To select the more economic pipe system, calculate the total cost

$$\text{Total Cost} = \text{Operating Cost} + \text{Capital Cost}$$

The operating cost for a 2-inch pipe system and a 4-inch pipe system are given in the problem statement and calculated above, respectively. The annual capital cost is calculated as follows:

$$\text{Capital Cost} = (\text{Distance})(\text{Cost})(\text{Capital recovery factor})$$

$$\text{Capital cost}(2\text{ inch}) = (4\text{ miles})(5280\text{ ft/mile})(\$1/\text{ft})(0.1) = \$2{,}110$$

$$\text{Capital cost}(4\text{ inch}) = (4\text{ miles})(5280\text{ ft/mile})(\$6/\text{ft})(0.1) = \$12{,}700$$

Annual cost data is summarized in Table 6.1. Obviously, the 4-inch pipe is more economical.

Illustrative Example 6.4

A process emits 50,000 acfm of gas containing dust (it may be considered metal) at a loading of 2.0 gr/ft^3. A particulate control device is employed for particle capture and the dust captured from the unit is worth \$0.03/lb of dust. Experimental data have shown that the collection efficiency, E, is related to the system pressure drop, ΔP, by the formula:

$$E = \frac{\Delta P}{\Delta P + 15.0}$$

Table 6.1 Costs results for two different sized pipes.

	2-inch	4-inch
Operating cost	\$20,000	\$1,250
Capital cost	\$2,110	\$12,700
Total Cost	\$22,110	\$13,950

where, E = fractional collection efficiency
ΔP = pressure drop, lb_f/ft^2

If the fan is 55% efficient (overall) and electric power costs \$0.18/kW · h, at what collection efficiency is the cost of power equal to the value of the recovered material? What is the pressure drop in inches of water (in H_2O) at this condition?

Solution

The value of the recovered material (*RV*) may be expressed in terms of the volumetric flowrate *q*, the inlet dust loading *c*, the value of the dust (*DV*), and the fractional collection efficiency *E*:

$$RV = (q)(c)(DV)(E)$$

Substituting yields

$$RV = \left(\frac{50{,}000\ \text{ft}^3}{\text{min}}\right)\left(\frac{2.0\ \text{gr}}{\text{ft}^3}\right)\left(\frac{1\ \text{lb}}{7{,}000\ \text{gr}}\right)\left(\frac{0.03\$}{\text{lb}}\right)(E)$$

$$= \frac{0.429E\$}{\text{min}};\ \$/\text{min}$$

The recovered value can be expressed in terms of pressure drop, i.e., replace *E* by ΔP:

$$RV = \frac{(0.429)(\Delta P)}{\Delta P + 15.0};\$/min$$

The cost of power (*CP*) in terms of ΔP, *q*, the cost of electricity (*CE*), and the fan fractional efficiency, E_f, is

$$CP = \frac{(q)(\Delta P)(CE)}{E_f}$$

Substitution yields

$$CP = \left(\frac{50{,}000\ \text{ft}^3}{\text{min}}\right)\left(\frac{\Delta P \text{lb}_f}{\text{ft}^2}\right)\left(\frac{0.18\$}{\text{kW}\cdot\text{h}}\right)\left(\frac{1\ \text{min}\cdot\text{kW}}{44{,}200\ \text{ft}\cdot lb_f}\right)\left(\frac{1}{0.55}\right)\left(\frac{1\ \text{h}}{60\ \text{min}}\right)$$

$$= \frac{0.006\Delta P\$}{\text{min}};\ \$/\text{min}$$

The pressure drop at which the cost of power is equal to the value of the recovered material is found by equating RV with CP:

$$RV = CP$$

Solving yields

$$\Delta P = 56.5 \text{lb}_f/\text{ft}^2 = 10.9 \text{inH}_2\text{O}$$

The collection efficiency corresponding to the above calculated ΔP is

$$E = \frac{\Delta P}{\Delta P + 15.0} = \frac{56.5}{56.5 + 15} = 0.79 = 79\%$$

The reader should note that operating below this efficiency (or the corresponding pressure drop) will produce a profit; operating above this value leads to a loss.

Calculating the maximum profit is left as an exercise for the reader. [*Hint:* Set the first derivative of the profit (i.e., RV – CP) with respect to ΔP equal to zero.] This problem will be revisited several times in Part V – Optimization.

Illustrative Example 6.5

The annual operating cost of an outdated environmental control device is $75,000. Under a proposed emission reduction plan, the installation of a new fan system will require an initial cost of $150,000 and an annual operating cost of $15,000 for the first 5 years. Determine the annualized cost for the new processing system by assuming the system has only 5 years (n) of operational life. The interest rate (i) is 7%. The capital recovery factor (CRF) or annual payment of a capital investment can be calculated from Equation (6.12) as follows:

$$CRF = \left(\frac{A}{P}\right)_{i,n} = \frac{i(1+1)^n}{[(1+i)^n - 1]}$$

where A is the annual cost and P is the present worth.

Compare the costs for both the outdated and proposed operations,

Solution

The annualized cost for the new fan is determined based on the following input data:

Capital cost $= \$150{,}000$
Interest, i $= 7\%$
Term, n $= 5$ yr

For $i = 0.07$ and $n = 5$, the CRF is

$$CRF = \frac{0.07(1+0.07)^5}{(1+0.07)^5 - 1} = 0.2439$$

The total annualized cost for the fan is then

$$\begin{aligned} \text{Annualized cost} &= \text{Installation cost} + \text{Operation cost} \\ &= (0.2439)(\$150{,}000) + \$15{,}000 = \$51{,}585 \end{aligned}$$

Since this cost is lower than the annual cost of $75,000 for the old process, the proposed plan should be instituted.

References

1. L. Theodore and K. Neuser, *Engineering Economics and Finance,* a Theodore Tutorial, Theodore Tutorials, East Williston, NY, originally published by the USEPA/APTI, RTP, NC, 1996.
2. F. Holland, F. Watson, and J. Wilkinson, *Financing Assets by Equity and Debt,* Chem. Eng. New York City, NY, September 2, 1974.
3. J. Happel, *Chemical Process Economics,* John Wiley & Sons, Hoboken, NJ, 1958.
4. A. Schmidt and H. List (adapted from), *Material and Energy Balances*, Prentice-Hall, Upper Saddle River, NJ, 1962.
5. R. Dupont, L. Theodore, and J. Reynolds, *Accident and Emergency Management: Problems and Solutions*, VCH Publishers, New York City, NY, 1991.
6. L. Theodore, R. Dupont, and J. Reynolds, *Pollution Prevention: Problems and Solutions*, Gordon and Breach Publishers, Amsterdam, Holland, 1994.
7. K. Ganeson, L. Theodore, and J. Reynolds, *Air Toxics: Problems and Solutions*, Gordon and Breach Publishers, Amsterdam, Holland, 1996.
8. R. Dupont, T. Baxter, and L. Theodore, *Environmental Management: Problems and Solutions*, CRC Press, Boca Raton, FL, 1998.
9. L. Theodore, R. Dupont, K. Ganesan, "*Unit Operations in Environmental Engineering,*" Scrivener-Wiley, Salem, MA, 2017.
10. L. Theodore, "Chemical Engineering: The Essential Reference," McGraw Hill, New York City, NY, 2014.

7

Problem Solving

The engineer (and to a lesser degree the scientist) is known for his/her problem-solving ability. It is probably this ability more than any other that has enabled many engineers to rise to positions of leadership and top management within their companies.

In problem-solving, both in school and in industry, considerable importance is attached to a proper analysis of the problem, to a logical recording of the problem solution, and to the overall professional appearance of the finished produced calculations. Neatness and clarity of presentation should be distinguishing marks of the work. Environmental engineers and scientists should always strive to practice professional habits of problem solving analysis and to make a conscious effort to improve the appearance of each document whether it is submitted for grading or is included in a notebook.

The value of an environmental engineer or scientist is determined by his/her ability to apply basic principles, facts, and methods in order to accomplish some useful purpose. In this modern age of industrial competition, the ultimate definition of a useful purpose is usually based on a tangible profit of dollars and cents. It is not sufficient, therefore, to have a knowledge and understanding of physics, chemistry, mathematics, mechanics, stoichiometry, thermodynamics, the unit operations, chemical technology, engineering, and other related scientific subjects; he/she must also have the ability to apply this knowledge to practical situations, and, in making these applications, recognize the importance of the dollar sign (see also the previous chapter).

Environmental engineers and scientists who have mastered the method of problem solving are considerably more successful in their work than are people who have not been trained in this technique. Many engineering problems in the past were of such a routine nature that a resort to deductive reasoning would suffice, and premises of deduction could be taken from handbooks. However, many of the engineering problems of today cannot be solved by mere "handbook techniques." Experimentation, research, and development have indeed become more significant activities in today's world.

Four sections compliment the presentation of this chapter. Section numbers and subject titles follow:

7.1: Sources of Information
7.2: Generic Problem-Solving Techniques
7.3: An Approach
7.4: Some Generic Comments

7.1 Sources of Information [1]

The field of environmental engineering and science encompasses many diverse fields of activity. It is therefore important that means be available to place widely scattered information in the hands of the reader. One need only examine the wide variety of problems seen in practice to appreciate the need for a readily available source of a wide variety of information. Problems should almost always be approached by checking on and reviewing all sources of information. Most of this information can be classified in the following source categories:

Traditional Sources
Engineering and Science Sources
New Sources
Personal Experience

Each topic is discussed below.

Traditional Sources

The library is the major repository of traditional information. To efficiently use the library, one must become familiar with its classification system. In general, environmental engineers and scientists use the technical section of a library to seek information on a certain subject and not material by a particular author or in a specific book. For this reason, classification systems are based on subject matter. The question confronting the user then is how to find and/or obtain information on a particular subject. Two basic subject-matter-related classification systems are employed in U.S. libraries: the Dewey Decimal System and the Library of Congress System. (The systems have changed somewhat with the advent of computers and the internet.) A library (particularly a technical one) usually contains all or some of the following:

1. General books (nonfiction, etc.)
2. Reference books
3. Handbooks
4. Journals
5. Transactions
6. Encyclopedias
7. Periodicals
8. Dictionaries
9. Trade literature
10. Catalogs

Engineering and Science Sources

Engineering and science technical literature includes textbooks, handbooks, periodicals, magazines, journals, dictionaries, encyclopedias, and industrial catalogs. There are a host of textbooks, including some of the classic works. The three major engineering handbooks include those by Perry and Green [2], Kirk and Othmer [3], and Albright [4]. Journals and magazines abound in the literature.

New Sources

The internet has a variety of resources available within it where information regarding just about anything, including precise technical information, can be found. Search engines such as Google, Bing, Yahoo!, and DuckDuckGo allow one to search public webpages, one of the largest repositories of information in history, free of charge and in a fraction of a second.

Wikipedia (derived from the phrase "What I Know Is...") is essentially an online site that provides free and open information. It is a collaborative encyclopedia where one can obtain or apply information to a host of topics. It is a great tool for fast reference but information can be provided by users and as such may contain errors so sources must be checked to confirm their validity. Sources are listed at the bottom of the Wikipedia pages usually in the form of clickable links.

Virtually every university, business, government (national, state, local) organization has a website. Individuals can also have one or more websites. Even one of your humble authors has two websites. One site, *www.theodorenewsletter.com*, provides information on education, sports, economics, politics, and a host of other topics. Try *Basketball Coaching 101* if the reader is interested in basketball [5].

Engineering Toolbox (http://www.engineeringtoolbox.com) provides most of the tables and reference material one would find in engineering handbooks organized by category. This site, along with many others, includes imbedded calculators that can perform many different calculations ranging from simple unit conversion to complex thermodynamic and optimization calculations.

There are also third party computer programs such as Aspen HYSYS and CHEMCAD that specialize in chemical process simulations. With proper training, these programs can be used to make extensive and complicated process related calculations in seconds. This ability is only possible with their vast databases. However, these programs come at great monetary cost, with licenses being anywhere from the tens to hundreds of thousands of dollars. This high cost limits access of these programs to large institution's such as universities, colleges, and large companies.

Problem solving will usually require a combination of different sources and websites. It is rare that one source will contain all the information necessary to complete a problem. Note also that some of the above will almost certainly have changed between the times when this section was initially written (March 2016) and when it reaches the reading audience.

Personal Experience

Another information source – and a frequently underrated one – is one's personal experience, personal files, and company experience files. Although traditional engineering and science problem-solving methods and the new sources previously described are important, other fields can also provide useful information. In fact, most equipment designs, process changes, and new-plant designs in the chemical process industries are based on either prior experience, company files, or both.

7.2 Generic Problem-Solving Techniques [6]

Certain methods of logic and techniques of calculation are fundamental to the solution of many problems, and there is a near infinite number of methods. Words such as creative, ingenuity, original, etc., etc., etc., appear in all these approaches. What do they all have in common? They usually provide a systematic, logical approach to solving problems, and what follows is the authors' definition of a generic approach.

The methodology of solving problems has been discussed by most mathematicians and logicians since the days of Aristotle. *Heuristic* ("serving to discover") is the term often given to this study of the methods and rules of solving problems.

Nearly always, a stepwise approach to the solution is desirable. The broad steps are:

1. Understanding the problem
2. Devising a plan

3. Carrying out the plan
4. Looking back

Details on each of the above four points follows.

Understanding the Problem

1. Get acquainted with the problem. Read it and attempt to visualize it in its entirety without concerning yourself with detail. Be particularly sure that you understand what is known and what is unknown.
2. Organize the available data. Engineering and science data are often insufficient, redundant, or contradictory. The units may have to be converted. The accuracy of the data must also be verified.
3. Sketch a flow sheet. This step will usually facilitate the "get acquainted" operation. If the problem is a simple one, this sketch may consist of nothing more than a small box with arrows on the back side of an envelope. For chemical processes, the diagrams could indicate the amount and composition of all streams entering or leaving the process.
4. Introduce suitable and consistent symbols for the unknowns. It is usually helpful to adopt a notation which will serve to identify the unknowns. This is particularly important when a problem has many unknowns.

Devising a Plan

Memory plays an important part in even the most creative thinking processes. It is almost impossible to generate ideas or plans with little or no background knowledge. Although mere memory is not enough to create new ideas, such ideas should be based on experience and acquired knowledge.

Here are some questions to consider:

1. Have I seen a similar or slightly different problem before?
2. Is there a very general equation or theorem which applies to this type of problem?
3. Can the problem be restated?
4. If I cannot solve the problem, can I solve a related problem?
5. Can part of the problem be solved?
6. What data are lacking to solve the rest of problem?
7. Did I use all the data given?

Carrying out the Plan

Choose a basis of calculation, if applicable. During various stages in the solution of a complex problem, it may be necessary to use more than one basis. Also, remember that the most brilliant work is useless if it cannot be conveyed clearly to other persons by written or oral presentation(s).

Looking Back

The amount of time it takes to check a problem (if applicable) is usually only a small fraction of the time it takes to solve it. Nearly all solutions to physical problems are amendable to verification. Thus, one should always:

1. Look for order-of-magnitude errors. Can the answer be checked by an approximate solution? Does the answer make sense?
2. Attempt to solve the problem by an alternate method.
3. Can the results be verified?

7.3 An Approach

Here is an approach drawn from the files of one of the authors [7]. This method involves six steps (see Figure 7.1).

1. The first step, and perhaps the most important one, is that of problem definition. It is not possible to establish rules for problem definition that are sufficiently general to be useful. Environmental problems are so diverse that it is up to the analyst to clearly state the nature of the problem. This will establish a definite objective for the analysis and is invaluable in outlining a path from the problem to the solution.
2. The next step is a definition of the theory that governs the phenomena of the environmental problem. This theory is usually available from a variety of sources, both published and unpublished, but for those isolated cases where there is no theory available, it is worthwhile to postulate one (or several) and to test its validity later by

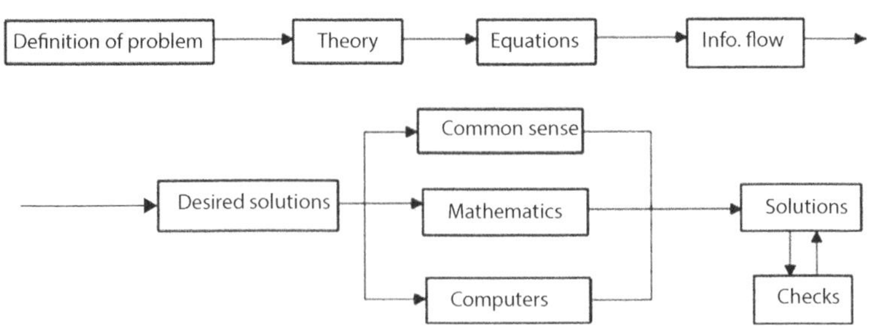

Figure 7.1 A problem-solving approach.

comparing (if available) the solution of the mathematical model with experimental results. One of the advantages of the computerized approach is the facility of rapidly obtaining solutions of various cases; this makes comparisons between alternate theories possible.

3. In the third step, the theory, as applied to the problem, is written in mathematical symbology. This necessary step forces the analyst to a clear unambiguous definition of the problem. The physical system may be described by a set of simultaneous algebraic and/or differential equations. These equations must be written in the most direct form possible and no manipulations are required at this stage. It is, however, worthwhile at this point to simplify the equations (where possible) by omitting insignificant terms. Care is needed though to ensure that any terms omitted are indeed insignificant during the entire course of the problem run. It is often possible to eliminate entire equations by merely neglecting minor fluctuations in certain intermediate variables. For example, if the heat capacity calculation of a component mixture (to be purified) required for a heat balance varies by only 1% of its value due to expected variations in composition, an average constant number could be substituted rather than include an equation in the model to compute a value (perhaps continuously).
4. Having assembled the equations, a procedural method for the solution is required. A consideration of the solution required from the model is a necessary preliminary step to the computation phase. A list of the various cases to be studied and the information that is expected in each case will reveal possible redundant situations as well as assist the programming of the computation phase.
5. The computation phase that follows may offer several alternate routes to the solution. The method selected will depend on the complexity of the equations to be solved. There are three general levels.
   a. The most elementary is common sense, i.e., the solutions desired can be obtained from the model by inspection if the equations or the solutions required are sufficiently simple. It should be realized that this technique cannot be extrapolated to more complex cases without requiring increasing amounts of pure guesswork.
   b. The next level, again restricted to systems of modest complexity, solves the equation by analytical techniques (see Part II). As pointed out in the previous discussion, a considerable amount of skill is required to solve even some of the most simple ones. Selecting the most fruitful path may be the only feasible method for problems of even fair complexity.
   c. The solution may require a numerical method (see Part III).

6. The sixth and last phase is the study and verification of the solution(s) obtained from the mathematical model. Any unexpected solution(s) should be scrutinized so as to ensure that no errors have occurred in the computation; also, some of the computer logic should be specifically designed to check the validity of the mathematical model.

7.4 Some General Concerns

Here is what one of the authors [8] has stressed in terms of developing problem-solving skills and other creative thinking.

1. Carefully define the problem at hand.
2. Obtain all pertinent data and information.
3. Initially, generate an answer or solution.
4. Examine and evaluate as many alternatives as possible, employing "what if" scenarios [9].
5. Reflect on the above over time.
6. Consider returning to step 1 and repeat/expand the process.

As noted earlier, the traditional methodology of solving problems has been described for decades in the following broad stepwise manner:

1. Understand the problem.
2. Devise a plan.
3. Carry out the plan.
4. Look back and (possibly) revise.

Many now believe creative thinking should be part of every student's education. Here are some ways that have proven to nudge the creative process along:

1. Break out of the one-and-only answer rut.
2. Use creative thinking techniques and games.
3. Foster creativity with assignments and projects.
4. Be careful not to punish creativity.

The above-suggested activities will ultimately help develop a critical thinker that:

1. Raises important questions and problems, formulating them clearly and precisely.
2. Gathers and assesses relevant information, using abstract ideas to interpret it effectively.

3. Comes to well-rounded conclusions and solutions, testing them against relevant criteria and standards.
4. Thinks open mindedly with alternative systems of thought, recognizing and assessing, as need be, their assumptions, implications, and practical consequences.
5. Communicates effectively with others in figuring out solutions to complex problems.

The analysis aspect of a problem remains. It essentially has not changed. The analysis of a new problem in environmental engineering and science can still be divided into four steps.

1. Consideration of the process in question
2. Mathematical description of the process, if applicable
3. Solution of any mathematical relationships to provide a solution
4. Verification of the solution

References

1. L. Theodore, *Chemical Engineering: The Essential Reference*, McGraw-Hill, New York City, NY, 2014.
2. D. Green and R. Perry (editors), *Perry's Chemical Engineers' Handbook*, 8th ed., McGraw-Hill, New York City, NY, 2008.
3. R. Kirk and D. Othmer, *Encyclopedia of Chemical Technology*, 4th ed., John Wiley & Sons, Hoboken, NJ, 2001.
4. L. Albright, *Albright's Chemical Engineering Handbook*, CRC Press/Taylor & Francis Group, Boca Raton, FL, 2008.
5. L. Theodore, *Basketball Coaching 101*, Kourtakis - Tonry Publishers, East Williston, NY, (available from Amazon or Createspace), 2015.
6. L. Theodore, personal notes, original sources unknown, East Williston, NY, 1980-1984.
7. L. Theodore, personal notes, original sources unknown, East Williston, NY, 1962.
8. P. Abulencia and L. Theodore, *Open-Ended Problems: A Future Chemical Engineering Education Approach*, Scrivener - Wiley Publishing, Salem, MA, 2015.
9. L. Theodore and R. Dupont, *Environmental Health and Hazard Risk Assessment: Principles and Calculations*, CRC/Taylor &Francis Group, Boca Raton, FL, 2014.

Part II
ANALYTICAL ANALYSIS

Perhaps the most important step in many environmental engineering/science problems is the determination of the numerical value(s) of one or more equations. These solutions may be obtained analytically (in closed/exact form) or numerically (by some approximate method). A great number of methods have been developed for doing this. The purpose of this part is to introduce and provide some of the more important analytical methods. Part III addresses numerical methods.

There are seven chapters in Part II. The chapter numbers and accompanying titles are listed below:

Chapter 8: Analytical Geometry
Chapter 9: Differentiation
Chapter 10: Integration
Chapter 11: Differential Calculus
Chapter 12: Integral Calculus
Chapter 13: Matrix Algebra
Chapter 14: Laplace Transforms

8

Analytical Geometry

A host of subject matter receives treatment in this introductory chapter in Part II. The key topics will involve the three coordinate systems that the environmental engineer or scientist encounters in practice – rectangular, cylindrical, and spherical.

The location of a point or a system in space may be defined in various coordinate systems. Normally, the presentation keys on rectangular coordinates. However, many applications involve curvilinear-coordinate systems-thus, the inclusion of cylindrical and spherical coordinates in this chapter.

Five sections complement the presentation of this chapter. Section numbers and subject titles follow.

8.1: Rectangular Coordinates
8.2: Cylindrical Coordinates
8.3: Spherical Coordinates
8.4: Key Physical Equations
8.5: Applications

8.1 Rectangular Coordinates

In rectangular coordinates, the position of a point $P(x, y, z)$ is determined by the intersection of three mutually perpendicular planes $x =$ constant, $y =$ constant, and $z =$ constant. The Point P in Figure 8.1 (a) is defined by the rectangular

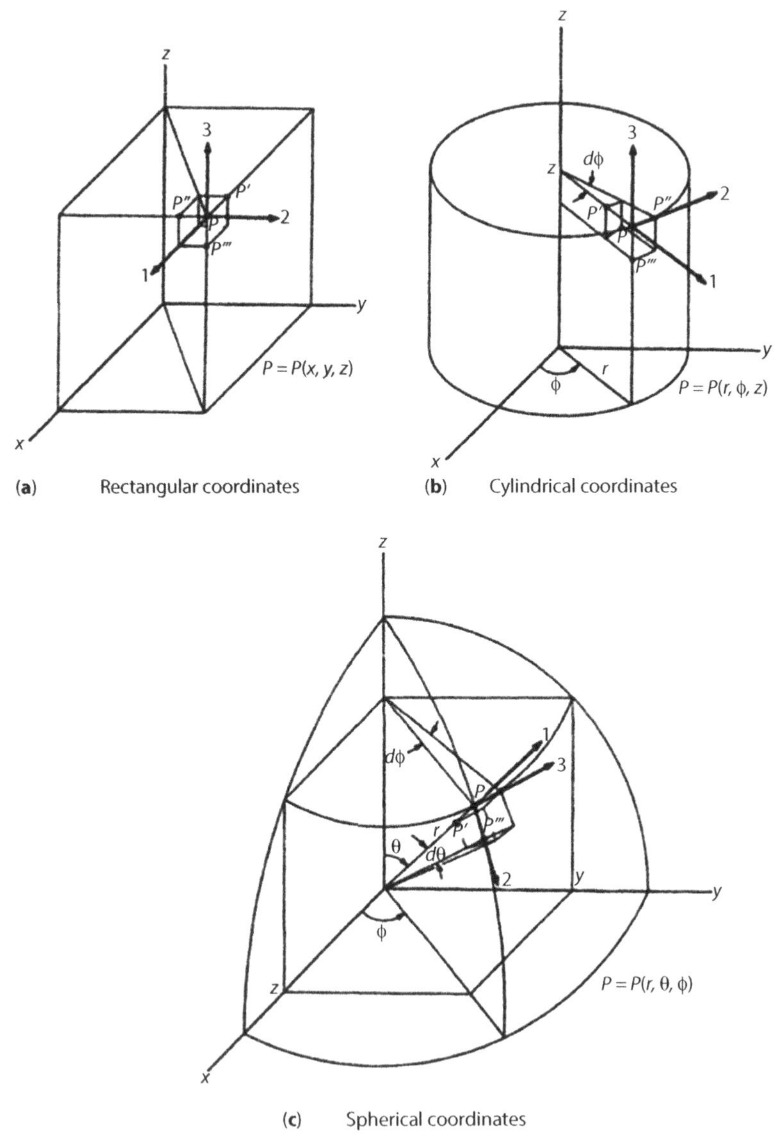

Figure 8.1 Coordinate systems.

coordinates x, y, and z. The differential lengths in the x, y, and z directions are given by

$$ds_1 = PP' = dx \tag{8.1}$$

$$ds_2 = PP'' = dy \tag{8.2}$$

$$ds_3 = PP''' = dz \tag{8.3}$$

Any differential in length ds in this coordinate system is given by

$$(ds)^2 = (PP')^2 + (PP'')^2 + (PP''')^2 = (dx)^2 + (dy)^2 + (dz)^2 \tag{8.4}$$

The differential area elements df in the planes of constant x, y, and z take the form

$$df_x = dy\ dz \tag{8.5}$$

$$df_y = dx\ dz \tag{8.6}$$

$$df_z = dx\ dy \tag{8.7}$$

The differential volume element $d\tau$ is

$$d\tau = dx\ dy\ dz \tag{8.8}$$

8.2 Cylindrical Coordinates

The same point P in space may also be represented by the cylindrical coordinates r, ϕ, and z. With reference to Figure 8.1(b) one can easily follow the transformation of x, y, and z into cylindrical coordinates.

$$x = r\cos\phi \tag{8.9}$$

$$y = r\sin\phi \tag{8.10}$$

$$z = z \tag{8.11}$$

The right-handed, mutually perpendicular (orthogonal) axes for this coordinate system are 1, 2, and 3. The differential lengths ds_1, ds_2, and ds_3 in the r, ϕ, and z, direction, respectively, are given by

$$ds_1 = PP' = dr \tag{8.12}$$

$$ds_2 = PP'' = r\, d\phi \tag{8.13}$$

$$ds_3 = PP''' = dz \tag{8.14}$$

The differential length ds now takes the form

$$\begin{aligned}(ds)^2 &= (PP')^2 + (PP'')^2 + (PP''')^2 \\ &= (dr)^2 + (r\, d\phi)^2 + (dz)^2\end{aligned} \tag{8.15}$$

The differential area element df on the curved surface is given by

$$\begin{aligned}df &= (PP'')(PP''') \\ &= (r\, d\phi)(dz) \\ &= r\, d\phi\, dz\end{aligned} \tag{8.16}$$

(The differential element on the top or bottom planes that bound the cylinder is $(PP')(PP'') = r\, dr\, d\phi$). The differential volume element $d\tau$ is

$$\begin{aligned}d\tau &= (PP')(PP'')(PP''') \\ &= (dr)(r\, d\phi)(dz) \\ &= r\, dr\, d\phi\, dz\end{aligned} \tag{8.17}$$

8.3 Spherical Coordinates

The point P is described by the spherical coordinates r, θ, and ϕ in Figure 8.1 (c). One can easily show that

$$x = r \sin\theta \cos\phi \tag{8.18}$$

$$y = r \sin\theta \sin\phi \tag{8.19}$$

$$z = r \cos\theta \tag{8.20}$$

The right-handed orthogonal axes for this coordinate system are once again 1, 2, and 3. The different lengths ds_1, ds_2, and ds_3 in the r, θ, and ϕ directions, respectively, are given by

$$ds_1 = PP' = dr \tag{8.21}$$

$$ds_2 = PP'' = r\, d\theta \tag{8.22}$$

$$ds_3 = PP''' = r \ \sin\theta d\phi \tag{8.23}$$

The differential length ds equation takes the form

$$\begin{aligned}(ds)^2 &= (PP')^2 + (PP'')^2 + (PP''')^2 \\ &= (dr)^2 + (r \ d\theta)^2 + (r\sin\theta d\phi)^2\end{aligned} \tag{8.24}$$

The differential area element df on the curved surface is given by

$$\begin{aligned}df &= (PP'')(PP''') \\ &= (r \ d\theta)(r\sin\theta d\phi) \\ &= r^2 \sin\theta \, d\theta \, d\phi\end{aligned} \tag{8.25}$$

while the differential volume element $d\tau$ is

$$\begin{aligned}d\tau &= (PP')(PP'')(PP''') \\ &= (dr)(r \ d\theta)(r\sin\theta \, d\phi) \\ &= r^2 \sin\theta \, dr \ d\theta \, d\phi\end{aligned} \tag{8.26}$$

8.4 Key Physical Equations

This section provides equations for calculating the area and/or volume and/or perimeter of a host of objects listed below.

a. Sphere, Cube, Rectangular Parallelepiped, and Cylinder
b. Parallelogram, Triangle, and Trapezoid
c. Polygon
d. Ellipse and Ellipsoid
e. Cone
f. Torus

a. Sphere, Cube, Rectangular Parallelepiped, and Cylinder

Describing equations for the area and volume of the following shaped particles are provided below.

1. Sphere of radius R
2. Cube of side A
3. Rectangular parallelepiped of sides A and B
4. Cylinder of radius R and height H

Equations

1. For a sphere of radius R:

$$\text{Volume} = \frac{4}{3}\pi R^3 \tag{8.27}$$

$$\text{Area} = 4\pi R^2 \tag{8.28}$$

2. For a cube of side A:

$$\text{Volume} = A^3 \tag{8.29}$$

$$\text{Area} = 6A^2 \tag{8.30}$$

3. For a rectangular parallelepiped of depth C:

$$\text{Volume} = ABC \tag{8.31}$$

$$\text{Area} = 2(AB + AC + BC) \tag{8.32}$$

4. For a cylinder of radius R and height H:

$$\text{Volume} = \pi R^2 H \tag{8.33}$$

$$\text{Area} = 2\pi RH \tag{8.34}$$

b. Parallelogram, Triangle, and Trapezoid

Describing equations for the area and perimeter of the following shaped particles are provided below.

1. Parallelogram of height H, side A, and base B
2. Triangle of height H, base B, and left side A; the angle between side A and base B is ϕ
3. Trapezoid of height H and parallel sides A and B; the lower left angle is ϕ and the lower right angle is θ

Equations

1. For the parallelogram:

$$\text{Area} = (H)(B) = (A)(B)\sin\phi \tag{8.35}$$

$$\text{Perimeter} = 2A + 2B \tag{8.36}$$

2. For the triangle:

$$\text{Area} = \frac{1}{2}(B)(H) = \frac{1}{2}(A)(B)\sin\phi \tag{8.37}$$

$$\text{Perimeter} = A + B + C \tag{8.38}$$

3. For the trapezoid:

$$\text{Area} = \frac{1}{2}H(A + B) \tag{8.39}$$

$$\begin{aligned}\text{Perimeter} &= A + B + H\left(\frac{1}{\sin\phi} + \frac{1}{\sin\theta}\right) \\ &= A + B + H(\csc\phi + \csc\theta)\end{aligned} \tag{8.40}$$

c. Polygon

Describing equations for the area and perimeter of the following shaped particles are provided below.

1. Regular polygon of N sides, each of length B
2. Regular polygon of N sides inscribed in circle of radius R

Equations

1. For case (1):

$$\text{Area} = \frac{1}{4}NB^2\cot\left(\frac{\pi}{N}\right) = \frac{1}{4}NB^2\frac{\cos(\pi/N)}{\sin(\pi/N)} \tag{8.41}$$

$$\text{Perimeter} = (N)(B) \tag{8.42}$$

2. For case (2):

$$\text{Area} = \frac{1}{2}NR^2\sin\left(\frac{2\pi}{N}\right) = \frac{1}{2}NR^2\sin\left(\frac{360^\circ}{N}\right) \tag{8.43}$$

$$\text{Perimeter} = 2(N)(R)\sin\left(\frac{\pi}{N}\right) = 2(N)(R)\sin\left(\frac{180}{N}\right) \tag{8.44}$$

d. Ellipse and Ellipsoid

Describing equations for the area, perimeter, or volume of the following shaped particles are provided below.

1. Ellipse of semimajor axis A and semiminor axis B
2. Ellipsoid of semi-axes A, B, C

Equations

1. For case (1):

$$\text{Area} = \pi(A)(B) \tag{8.45}$$

$$\text{Perimeter} = 4A\int\sqrt{1-k^2\sin^2\theta}\,d\theta \tag{8.46}$$

$$= 2\pi\sqrt{\frac{1}{2}(A^2+B^2)} \quad \text{(approximate)} \tag{8.47}$$

where $k = \sqrt{(A^2+B^2)/A}$

2. For case (2):

$$\text{Volume} = \frac{1}{2}\pi B^2 A$$

e. Cone

Describing equations for the volume and surface area of the following shaped particles are provided below.

1. Right circular cone of radius R, lateral length L, and height H
2. Frustum of right circular cone of radii A, B, lateral length L, and height H

Equations

1. For case (1):

$$\text{Volume} = \frac{1}{3}\pi R^2 H \tag{8.48}$$

$$\text{Surface area} = \pi R\sqrt{R^2+H^2} = \pi RL$$

2. For case (2):

$$\text{Volume} = \frac{1}{3}\pi H(A^2 + AB + B^2) \tag{8.50}$$

$$\text{Surface area} = \pi(A+B)\sqrt{H^2 + (B-A)^2} = \pi(A+B)L \tag{8.51}$$

f. Torus

Describing equations for the volume and surface area off a torus of inner radius A and outer radius B are provided below.

Equations

$$\text{Volume} = \frac{1}{4}\pi^2(A+B)(B-A)^2 \tag{8.52}$$

$$\text{Surface Area} = \pi^2(B^2 - A^2) \tag{8.53}$$

8.5 Applications

Illustrative Example 8.1

Develop the area to volume ratio equations for the shaped particles from sections 8.4A, 8.4E, and 8.4F.

Solution

For 8.4A:

$$\text{Area/Volume} = 3/R$$

$$\text{Area/Volume} = 6/A$$

$$\text{Area/Volume} = 2[(1/A) + (1/B) + (1/C)]$$

$$\text{Area/Volume} = 2/R$$

For 8.4E:

$$\text{Area/Volume} = 3L/RH$$

$$\text{Area/Volume} = [3(A+B)L]/[H(A^2 + AB + B^2)]$$

For 8.4 F:

$$\text{Area/Volume} = 4/(B - A)$$

Illustrative Example 8.2

Calculate the area to volume ratio of shaped particles from sections 8.4A, 8.4E, and 8.4F given the following data:

$$\text{A} = 1, \text{B} = 2, \text{C} = 2, \text{L} = 1, \text{H} = 1, \text{and R} = 1$$

Solution

For 8.4A:

$$\text{Area/Volume} = 3$$

$$\text{Area/Volume} = 6$$

$$\text{Area/Volume} = 4$$

$$\text{Area/Volume} = 2$$

For 8.4E:

$$\text{Area/Volume} = 3$$

$$\text{Area/Volume} = 9/7$$

For 8.4F:

$$\text{Area/Volume} = 4$$

Illustrative Example 8.3 [1]

Consider a sphere with a diameter of 1.0 μm (micrometer or micron). If this same mass of sphere is converted through a size reduction process to spheres with a diameter of 1.0 nm (nanometer), calculate the increase in surface area of the smaller sized spheres [1].

Solution

From 8.4A, the volume of a sphere is given by

$$V = \frac{4}{3}\pi R^3; \ R = D/2$$
$$= \pi D^3/6$$

The volume of a 1.0 μm sphere is therefore

$$V(1.0\ \mu\text{m}) = \pi(1.0)^3/6$$
$$= 0.5236(\mu\text{m})^3 \quad 1.0\ \mu\text{m} = 10^3\ \text{nm}$$
$$= 0.5236 \times 10^9\,(\text{nm})^3$$

The volume of a 1.0 nm sphere is correspondingly

$$V(1.0\ \text{nm}) = \pi(1.0)^3/6$$
$$= 0.5236(\text{nm})^3$$

Since the mass – as well as the total volume – remain the same, the number of smaller particles is

$$N = \frac{0.5236 \times 10^9}{0.5236} = 10^9$$

The surface area of the 1.0 μm particle is

$$SA(1.0\ \mu\text{m}) = 4\pi R^2 = \pi D^2$$
$$= (3.14)(1)^2$$
$$= 3.14(\mu\text{m})^2$$
$$= 3.14 \times 10^6\ (\text{nm})^2$$

and

$$SA(1.0\ \text{nm}) = (3.14)(1)^2$$
$$= 3.14(\text{nm})^2$$

However, there are 10^9 of the 1.0 nm particles. Therefore, the 10^9 smaller sized particles have a total surface area of 3.14×10^9 $(\text{nm})^2$. This reduction in size has increased the surface area by a factor of 1000 or 10^3. The reader should note that the particle diameter has decreased by the same factor.

Illustrative Example 8.4

Consider a cube with side of 1.0 μm. If the same mass of cube is converted to cubes with sides of 1.0 nm, calculate the increase in surface area of the smaller sized cubes.

Solution

From section 8.4A, the volume of a cube is given by

$$V = A^3$$

Therefore,

$$\begin{aligned} V(1.0\ \mu\text{m}) &= (1)^3 \\ &= 1.0(\mu\text{m})^3 \\ &= 1.0 \times 10^9 (\text{nm})^3 \end{aligned}$$

The volume of the 1.0 nm sided cube is

$$V(1.0\ \text{nm}) = 1.0\ (\text{nm})^3$$

Since the volume remains the same, the number of smaller cubes is

$$N = 10^9/1.0 = 10^9$$

The surface area of a 1.0 μm sided cube is

$$\begin{aligned} SA(1.0\ \mu\text{m}) &= 6A^2 \\ &= 6\ (\mu\text{m})^2 \\ &= 6 \times 10^6 (\text{nm})^2 \end{aligned}$$

and

$$\begin{aligned} SA\ (1.0\ \text{nm}) &= 6A^2 \\ &= 6(\text{nm})^2. \end{aligned}$$

Since there are 10^9 of the 1.0 nm sized cubes, their total surface area is 6×10^9 $(\text{nm})^2$. Once again, the reduction in size has increased the surface area by a factor of 1000 due to a similar corresponding decrease in side length of the cube.

Illustrative Example 8.5 [2]

A new baghouse [3–4] is structured as shown in Figure 8.2. (A baghouse contains multiple filter bags for removing particulate matter from a pollutant/latent gas stream.) The outside surface of the baghouse is to be covered with a new experimental outer insulating covering. Assume that there will be no waste in the covering process. The material is ordered to the nearest 0.01 yd^2 and the cost is \$87.54/

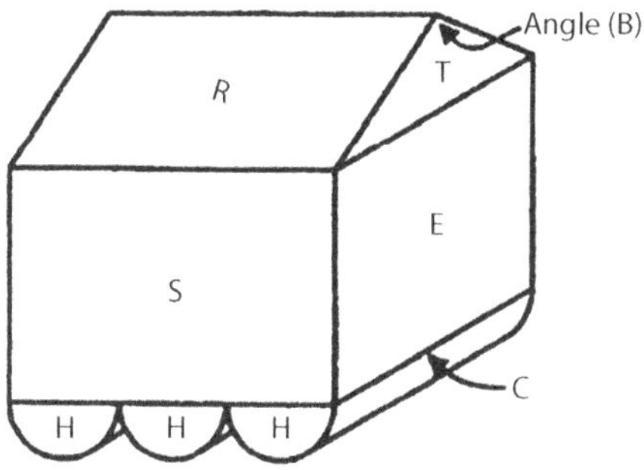

Figure 8.2 Physical configuration of a new baghouse.

yd^2. The minimum order that will be shipped is 5 yd^2. Your department budget must bear the cost. What will be the debit to your budget?

Data:

1. Surface E is a square with a measurement of 3 m per side.
2. The top roof angle (B) is 80°, the remaining two angles are equal.
3. The ends of the rounded hoppers (H) are semicircles (half circles) of equal size.
4. The length of the unit side (S) is 6.6 m.

Note: The problem requires several steps/processes. Therefore, one needs not only to know and use the appropriate mathematical principles but also needs to organize their work to avoid "oversight" errors.

Solution

There are two identical surfaces for E, R, S, and T. There are three surfaces of dimension C, and there are six ends (H). Therefore, the total area (A_t) to be covered is calculated as follows:

$$A_t = X = 2(E + R + S + T) + 3C + 6H$$

Note that the dimensions are provided in metric units and the insulating material is provided in English units. In any problem where the units of measure are mixed, terms must be converted to consistent units. In this problem, each of the measures provided in meters could be changed to yards, but that would be time consuming. Since all the linear measures are in meters, one would be better served by computing the areas in metric units and then converting the final number of square meters to square yards.

Note also that one square meter (1 m^2) is equal to 1.196 yd^2. Therefore

$$X(\text{yds}^2) = X(\text{m}^2) \times 1.196$$

For the interested reader, the solution, left as an exercise, is $14,300.

Illustrative Example 8.6

Determine the total particulate matter that can be collected in the hoppers, i.e., how many cubic feet will be occupied by the particulate matter when the three hoppers are full?

Solution

The calculation essentially requires determining the volume of the free hoppers. The interested reader is also left the exercise of verifying the solution provided below.

$$V = 17.1\ \text{m}^3$$
$$= 604\ \text{ft}^3$$

References

1. L. Theodore, *Nanotechnology: Basic Calculations for Engineers and Scientists*, John Wiley & Sons, Hoboken, NJ, 2006.
2. L. Theodore, personal notes (original source unknown), East Williston, NY, 1998.
3. L. Theodore, *Air Pollution Control Equipment Calculations*, John Wiley & Sons, Hoboken, NJ, 2008.
4. L. Theodore, R. Dupont, K. Ganesan, *Unit Operations in Environmental Engineering*, Scrivener-Wiley, Salem, MA, 2017.

9

Differentiation

The terms derivative and differentiation have not only come to mean different things to different people but have also been used interchangeably. The authors offer the following definitions. The former term provides information on how one variable, e.g., a dependent variable, changes with respect to another variable, e.g., an independent variable. The derivative dy/dx relates how a dependent variable y varies with an independent variable x. Differentiation is the operation that enables one to determine the value of the derivative.

One form of a derivative enables one to determine how fast something is moving at a given point or instant in time. To use a familiar example, if points A and B are 60 miles apart and a car travels from point A to point B in 60 minutes, one may say the car traveled at a rate of 60 mph (miles per hour). However, stating that the car traveled at 60 mph is an average that assumed the car was moving at a steady rate of 60 mph from the start to the finish. However, in reality, the car must accelerate from 0 to 60 mph beginning at point A and it must decelerate from 60 to 0 mph to stop at point B. Obviously the car traveled at varying rates of speed, and at some point had to be traveling faster than 60 mph. A derivative called velocity can be used to determine the speed of the car at any point along the route. The car's speedometer produces a derivative function – one can look at the speedometer at any time and determine the velocity (instantaneous speed) of the car. Many pollution control equipment and other environmental engineering calculations involving derivatives relate to velocity.

Two sections complement the presentation of this chapter. Section numbers and subject titles follow:

9.1: Graphical Methods
9.2: Finite Differencing

9.1 Graphical Methods

The fundamental laws on which engineering and science are based are frequently expressed in the form of differential equations providing the rate of change of an important dependent quantity with respect to an independent variable. It is therefore particularly necessary in the interpretation of experimental data to be able to carry out the process of differentiation analytically, graphically, or numerically. This chapter reviews the subject from a graphical perspective. A later chapter examines differentiation from an analytical point of view, while Part III provides material on numerical methods of analysis.

Since the derivative of a function y at any particular value of x is the slope of the curve of y vs. x at that value of x, it follows that any graphical means of measuring the slope of a curve may be employed to obtain the derivative. One rather obvious method is to plot the function and obtain the slope with the aid of a straightedge. The straightedge is swung about the point on the curve until it is judged "by eye" to best represent the slope of the curve at the point. A straight line is then drawn, and the slope of the straight line is calculated by any suitable means. In this last step, it is important to obtain the slope in terms of the variables, taking into account the scales used for the ordinate and abscissa. Only where these are equal is the geometric slope the correct value.

A better way to obtain the derivative is to place a protractor or triangle on a fixed straightedge or T square at a definite angle to the x axis and slide it along until it touches the curve representing the function. This procedure locates the point on the curve where the slope is equal to the predetermined value. As an aid in locating the exact point in tangency, several chords, such as AB of Figure 9.1, may be drawn parallel to the tangent. These are bisected, and a line QP is drawn through their mid-points to intersect the curve at the desired point P. Although the principle is quite similar to the direct measurement of the slope at a point, the results obtained are found to be somewhat better. With a graph of reasonable size, it is usually possible to obtain the derivative in this way with an error of less than 5 percent, although the accuracy (as one would expect) varies considerably with the curvature of the line representing the function.

A very convenient device for measuring slopes of curves in past years is a movable protractor fitted with a prism at the center. This is placed over the curve at the point in question and turned until the two branches of the curve seen through the two faces of the prism appear to join to form a smooth continuous curve. The slope

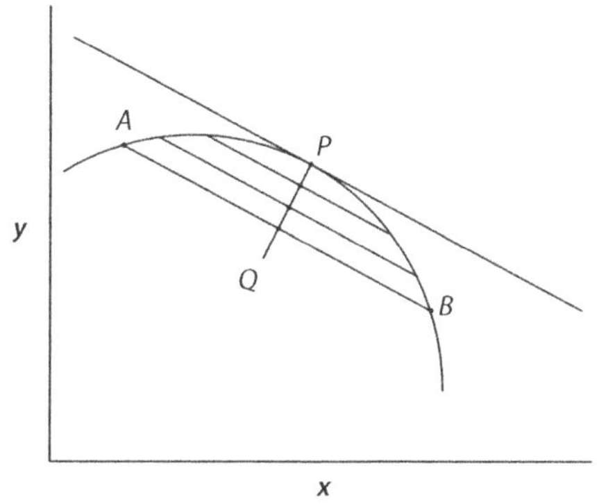

Figure 9.1 Graphical differentiation.

of the curve with reference to the base of the instrument may then be read. Some of the instruments of this type employ a protractor of such a small diameter that the accuracy is no better than, if as good as, that obtainable by the preceding methods.

It is important to note that the *first* derivative, dy/dx, represents the ratio of two distinct quantities, dy and dx. Consequently, this ratio may be handled by ordinary algebraic procedures. For example

$$dx\frac{dy}{dx} = dy \tag{9.1}$$

$$\frac{dy}{dx}\frac{dx}{dz} = \frac{dy}{dz} \tag{9.2}$$

On the other hand, in general, the higher derivatives cannot be treated in this manner.

As noted above, one frequently uses graphs to find the derivative of one variable with respect to another. The slope of a curve of distance vs. time (discussed earlier) would give the velocity at that point. In an another example, the derivative of the molal volume of a solution with respect to concentration is the partial molal volume [1, 2]. In cases where the relationship between the variables can be expressed by an equation, the computation of slopes and derivatives is best done either analytically or numerically. If empirical equations are not available, the aforementioned graphical techniques must be brought to bear.

Thus, graphical differentiation may be used to determine the slope of a curve relating two variables. If the analytical relationship is known, the differentiation may be performed analytically for a precise answer. If the function is not known, the data may be plotted and the slope may be determined by drawing a tangent as described above to the curve at a point. However, the determination of the proper slope of the tangent is often uncertain, especially if the experimental data are inaccurate. Graphical differentiation may then be used to obtain values of the slope. If tabulated data are available, the data are plotted as y as a function of x [1–3].

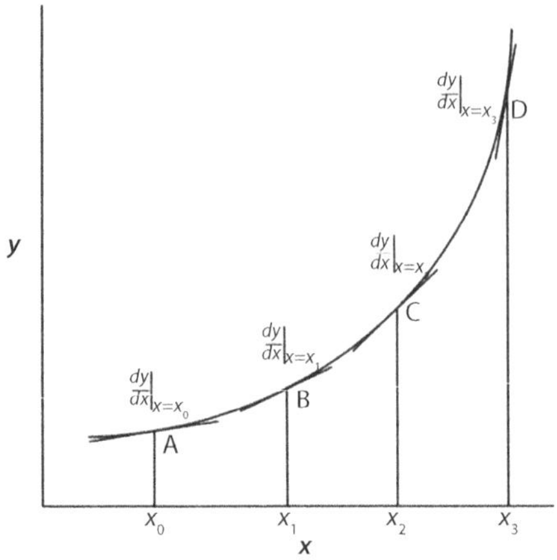

Figure 9.2 Derivatives of several points.

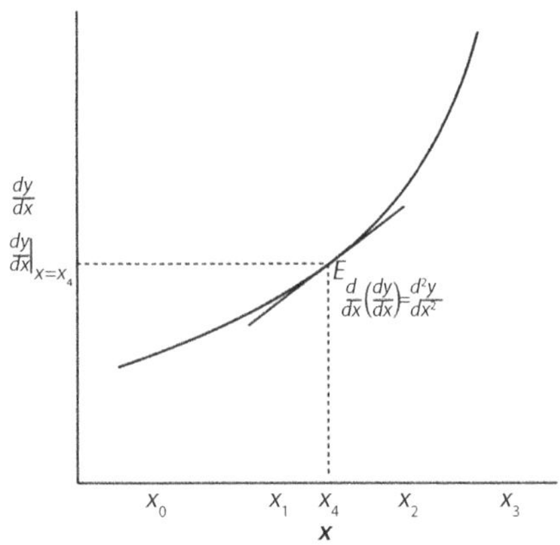

Figure 9.3 The second derivative.

The above graphical analysis provides what is referred to as the first derivative, i.e., dy/dx. The second derivative, d^2y/dx^2, may also be obtained in a similar manner. Refer to Figure 9.2. If the slopes obtained at $x = x_0$, $x = x_1$, $x = x_2$, and $x = x_3$, i.e., at points A, B, C, and D respectively, are plotted vs. x (see Figure 9.3) one may obtain the second derivative. The slope of point $x = x_4$ (Point E) represents d^2y/dx^2 at the point $x = x_4$.

9.2 Finite Differences

Suppose that the function $y = f(x)$ is given in tabular form for a series of equally spaced values of x. As an aid in performing necessary interpolations in such a

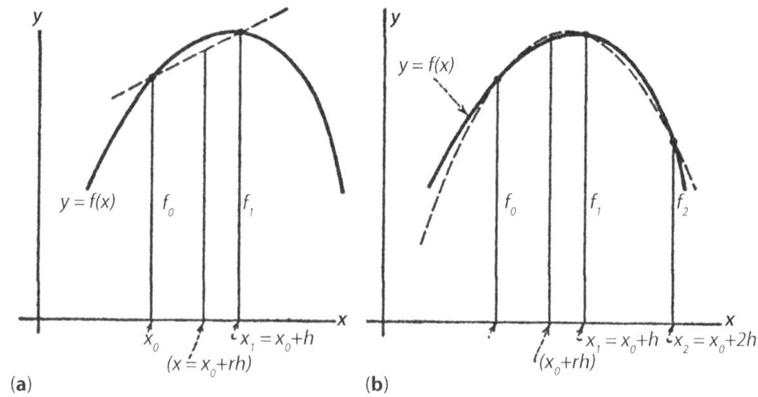

Figure 9.4 Approximation of derivatives by finite differences.

table one may, and in fact often does, list explicitly the differences between successive values of $f(x)$. In the same way, the differences between these differences may be calculated and listed, and so on indefinitely, unless perhaps at some stage the differences turn out to be zero. In the field of numerical mathematics these differences play an important role not only in interpolation, but also in such processes as the construction of tables, curve fitting, the differentiation and integration of tabular functions, and the approximate solution of differential equations. In this section, the fundamental properties of the differences of a function are developed, and their use in carrying out the operations just mentioned are detailed.

Begin by investigating from an elementary point of view the general problem of interpolation. In its simplest form, interpolation consists of substituting the chord of a curve for the curve itself, as in Figure 9.4 (a), and subsequently reading values of the function from the chord rather than from the graph of the function. Analytically, let h be the interval between successive values of x at which $f(x)$ is known, and let $x = x_0 + rh$, where r is some fraction, be a point intermediate between x_0 and $x_1 = x_0 + h$ at which the value of $f(x)$ is desired. Then manipulating the equation with the assumption that $x_0 = 0$, one has for the equation of the chord joining the points (x_0, f_0) [now $(0, f_0)$] and (x_1, f_1) [now (h, f_1)]

$$f(x) = y = f_0 + \frac{f_1 - f_0}{h} x \tag{9.3}$$

Since $x = x_0 + rh$ and $x_0 = 0$ in this manipulation, x can be replaced simply with rh. Substituting rh for x into Equation(9.3) one is left with

$$y = f_0 + (f_1 - f_0) r \tag{9.4}$$

One may also introduce the notation

$$\Delta f_0 = f_1 - f_0 \tag{9.5}$$

Finally, reverse the initial manipulation (x_0 = its original value and not 0) and one can write

$$f(x_0 + rh) = f_0 + r\Delta f_0 \tag{9.6}$$

This derivation was provided for an explanation of Equation(9.6) and is not required to calculate an interpolated value. One can simply place the original values into Equation(9.6).

On the other hand, it would be reasonable to attempt to approximate the graph of $y = f(x)$ not by a straight line through two successive points where $f(x)$ is known, but by a parabola through three such points, as in Figure 9.4 (b). In general, because of its curvature, a parabola will fit the graph of $f(x)$ more closely than will a straight line, and hence interpolation based on a parabolic approximating arc will be more accurate than ordinary linear interpolation.

A host of procedures are available for estimating the derivative in this and other ways. For example, to develop a formula for parabolic interpolation, one must first determine the coefficients a, b, c in the equation of the approximating arc

$$y = a + bx + cx^2 \tag{9.7}$$

This, as well as several other methods, receive treatment in Part III.

Naturally, by approximating $f(x)$ by polynomial functions of higher and higher order, and proceeding exactly as indicated above, increasingly refined interpolation formulas can be obtained [4]. This topic will be re-visited in Part III, Chapter 28. However, the labor involved in proceeding in this fashion soon becomes prohibitive, and many turn their attention to more efficient operational methods of deriving such results via numerical analysis.

References

1. H. Mickley, T. Sherwood, and C. Reed, *Applied Mathematics in Chemical Engineering*, 2nd Ed, McGraw-Hill, New York City, NY, 1957.
2. L. Theodore, *Chemical Engineering: The Essential Reference*, McGraw-Hill, New York City, NY, 2014.
3. R. Ketter and S. Prawel, *Modern Methods of Engineering Computation*, McGraw-Hill, New York City, NY, 1969.
4. S. Shaefer and L. Theodore, *Probability and Statistics for Environmental Science*, CRC Press/Taylor & Francis Group, Boca Raton, FL, 2007.

10

Integration

Three words that will appear regularly in this chapter are: integral, integrate, and integration. Most practicing environmental engineers and scientists use the terms interchangeably. But is there a difference? Here is what Webster [1] has to say:

1. Integral. "…the result of integrating a function."
2. Integrate. "…to calculate the integral or integrals of (a function, equation, etc.)."
3. Integration. "…the process of finding the quantity or function of which a given quantity or function is the derivative or differential."

Does this help? Add to this that an integral equation is one which contains the unknown function *behind* the integral sign. Its importance for physical problems lies in the fact that most *differential equations together with their boundary conditions* may be reformulated to give a *single integral equation*. If the latter can be solved, the mathematical difficulties are not appreciably greater even when the number of independent variables is increased.

As an example, assume that a source emits known quantities of hazardous particulate matter comprised of various sizes of particles into the atmosphere. As the particulate is dispersed into the surrounding area, the small particles travel a greater distance than the large particles, so that several concentric zones surrounding the source contain particles of predictable sizes. This particle size information can be used in integral calculations to compute the quantity of particulate matter dispersed to each zone [2].

Five sections complement the presentation of this chapter. Section numbers and subject titles follow:

10.1: Graphical Integration
10.2: The Rectangle Method
10.3: The Method of Rectangles
10.4: The Method of Trapezoids
10.5: Simple Batch (Differential) Distillation Example

10.1 Graphical Integration

This technique can be applied whenever it is arduous to fit an equation to experimental data, or when it is easier and faster to integrate graphically rather than analytically. Consider, for example, Figure 10.1. The data depicted are a plot of volumetric flow (ft^3/h) readings versus time (sec) as recorded at a small stack emitting

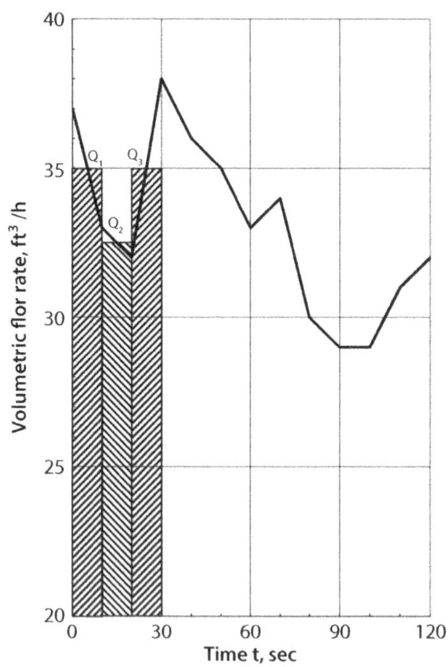

Figure 10.1 Flow readings.

a pollutant gas. It is obviously difficult, if not impossible, to describe these data by means of an equation. However, since the definite integral is simply a number representing the infinite sum

$$\lim_{\Delta x \to 0} \sum_{x=a}^{x=b} f(x)\Delta x \tag{10.1}$$

or

$$\int_{x=a}^{x=b} f(x)dx \tag{10.2}$$

where a and b are variable limits, the integration can be carried out graphically without recourse to an equation.

The average flow rate during the first 10 sec, Δt, is approximately Q_1 ft^3, in the next 10 sec the flow rate is Q_2 ft^3, etc. The total volumetric flow rate for the first two minutes Q_t is thus

$$Q_t = Q_1\left(\frac{10}{3,600}\right) + Q_2\left(\frac{10}{3,600}\right) + \ldots + Q_{12}\left(\frac{10}{3,600}\right) \tag{10.3}$$

$$\Sigma Q_i \, \Delta t = \int_0^{120/3,600} Q(t)dt$$

Hence, in order to calculate the total volume of gas emitted over a 2 minute period (120 seconds), one has only to add the entire area under the curve Q versus t. The total emission obtained in this way is approximately 1.10 ft^3.

There are many different methods for obtaining areas graphically. Counting areas or arbitrary blocks is often the quickest and surest way. A number of other techniques may be applied, such as:

1. Weighing: If the graph paper is of uniform thickness one can simply cut the figure out and weigh it. The weight is then compared to that of a standard rectangle and multiplied by the appropriate ratio of weights. This method is rarely, if ever, used today.
2. Planometers: Mechanical devices consisting of lever arms and a wheel which traces the curve are available commercially. This method is also rarely used today.
3. Analytical approximations: The methods of Simpson, Gauss, etc., as described in any appropriate mathematics text [3, 4], and in Part III of this text.

It is fair to say that graphical integration is useful when no analytical expression of the function exists. To evaluate the integral

$$y = \int_{x_1}^{x_2} I\,dx; \quad I = I(x) \tag{10.4}$$

analytically, it is necessary to know how I varies with x. If I is constant, the integral is

$$y = I(x_2 - x_1) \tag{10.5}$$

If I can be expressed as an analytical function of temperature T where $x = \mathrm{T}$ (see also Figure 10.2), for example, where I is the heat capacity c_p

$$I = 4.3 \times 10^{-3} x^2 \tag{10.6}$$

the integral could be evaluated analytically:

$$\begin{aligned} y &= \int_{x_1}^{x_2} 4.3 \times 10^{-3} x^2\,dx \\ &= \frac{4.3 \times 10^{-3}}{3}(x_2^3 - x_1^3) \end{aligned} \tag{10.7}$$

Substitution of numerical values of the limits x_1 and x_2 would enable quantitative evaluation of the integral. For example, if $x_1 = 200$ and $x_2 = 600$, then

$$y = \frac{4.3 \times 10^{-3}}{3}(600^3 - 200^3) = 298 \times 10^3 \tag{10.8}$$

However, the above integral can also be evaluated graphically by determining the area under the curve of I as a function of x. For example, Equation 10.6 is replotted in Figure 10.2. The limits $c_{p_1} = 200$, $c_{p_2} = 600$ are shown. The area under the curve and between the limits is equal to the integral. The area may be evaluated by counting areas and then multiplying by the value of each area. This is an exercise left for the reader.

There are several methods of counting areas. The three most often employed are:

1. One Rectangle
2. Several Rectangles
3. Several Trapezoids

The next three sections review these approaches to graphically evaluate an integral. The development is based on calculating the *average* heat capacity of a substance

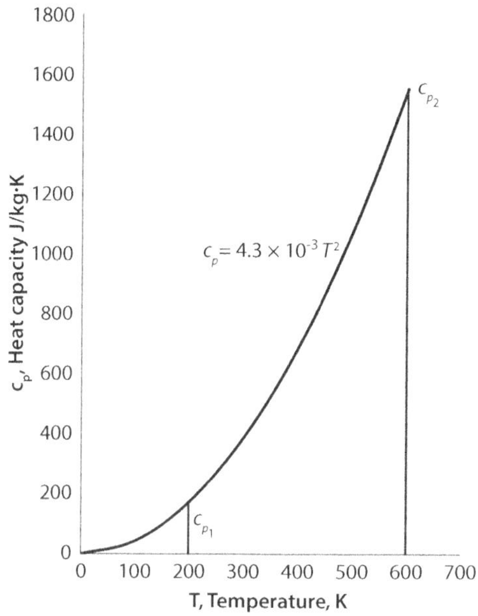

Figure 10.2 Graphical integration, area under curve.

over a specified temperature range. By definition, the average value $\bar{W}$ of a function W over the range of its independent variable x from x_1 to x_2 is given by

$$\bar{W} = \frac{\int_{x_1}^{x_2} W\, dx}{\int_{x_1}^{x_2} dx} \tag{10.9}$$

If this is applied to heat capacity, c_p, as a function of temperature, T

$$\bar{c}_p = \frac{\int_{T_1}^{T_2} c_p\, dT}{\int_{T_1}^{T_2} dT} \tag{10.10}$$

10.2 The Rectangle Method [5]

For the rectangle method, refer to Figure 10.3 once again. To calculate the average heat capacity over the 200–600K range, one draws a horizontal line such that approximately half that area lies above the line and half the area lies below the line. The value of c_p at that line is approximately 690. That area under the one rectangle is (690)(400) = 276,000, and the average heat capacity is 276,000/400 = 690 J/kg·K [5].

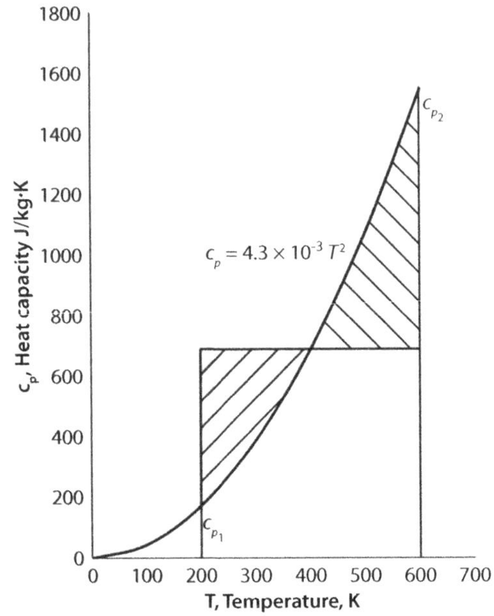

Figure 10.3 Graphical integration; area under curve.

10.3 The Method of Rectangles

The method of rectangles approximates the area by a series of vertical rectangles. It is convenient, but not essential, to take equal increments along the horizontal scale. The upper end of each rectangle is placed so that the area included below the curve but *excluded* from the rectangle is approximately equal to the area above the curve *included* in the rectangle. Figure 10.4 shows the general method applied to the previous example. The c_p ordinates are chosen for n increments in T, and the integral is

$$\int_{T_1}^{T_2} c_p\, dx = \sum_{i=1}^{i=n} c_p\, \Delta T \tag{10.11}$$

If the T increments are all equal,

$$\int_{T_1}^{T_2} c_p\, dT = \Delta T \sum_{i=1}^{i=n} c_{pi} \tag{10.12}$$

For the case at hand, $\Delta T = 100\text{K}$, and c_p values are obtained from Figure 10.4.

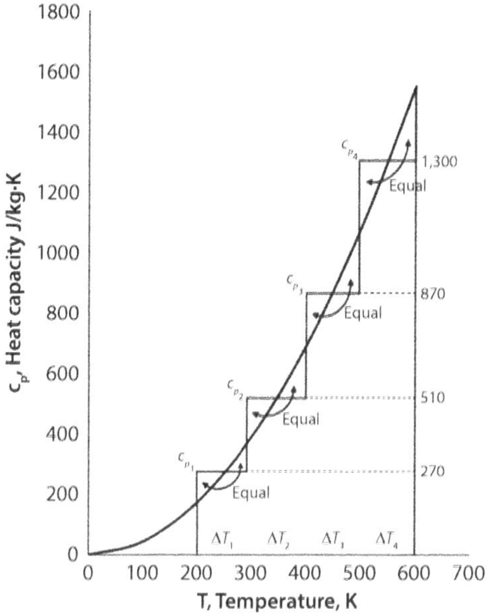

Figure 10.4 Graphical integration, method of rectangles.

$$\int_{T_1}^{T_2} c_p \, dT = (100)(270 + 510 + 870 + 1300) \tag{10.13}$$

$$= 295{,}000$$

The average heat capacity is then

$$\bar{c}_p = \frac{295{,}000}{600 - 200} = \frac{295{,}000}{400} = 773.8 \text{ J/kg} \cdot \text{K} \tag{10.14}$$

10.4 The Method of Trapezoids

The method of trapezoids replaces the smooth curve by a series of straight-line segments between data points. The area of each trapezoid is evaluated and the results are summed for the total area. With reference to Figure 10.5, the area of the first trapezoid A_1 is

$$A_1 = c_{p_1} \Delta T_1 + \frac{1}{2}(c_{p_2} - c_{p_1})\Delta T_1 = \left(\frac{c_{p_2} + c_{p_1}}{2} \right) \Delta T_1 \tag{10.15}$$

The total integral is

$$\int_{T_1}^{T_2} c_p \, dT = \sum_{i=1}^{i=n} \left(\frac{c_{p_{i+1}} + c_{p_i}}{2} \right) \Delta T_i; \quad T_1 = 200K, \; T_2 = 600K \tag{10.16}$$

If all T increments are equal,

$$\int_{T_1}^{T_2} c_p dT = \Delta T \sum_{i=1}^{i=n} \left(\frac{c_{p_1} + c_{p_n}}{2} + c_{p_{i+1}} + c_{p_{i+2}} \ldots + c_{p_{n-1}} \right) \tag{10.17}$$

For this calculation, Figure 10.5 indicates that with

$$\Delta T = 100, \; c_{p_1} = 172, \; c_{p_2} = 387, \; c_{p_3} = 688, \; c_{p_4} = 1075, \; c_{p_5} = 1548$$

Thus,

$$\begin{aligned} \int_{T_1}^{T_2} c_p \, dT &= (100) \left[\left(\frac{172 + 1548}{2} \right) + 387 + 688 + 1075 \right] \\ &= 301{,}000 \end{aligned} \tag{10.18}$$

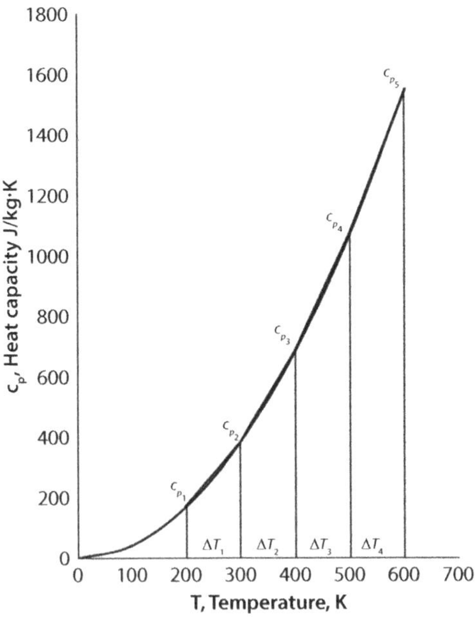

Figure 10.5 Graphical integrations; method of trapezoids.

The average heat capacity is then

$$\bar{c}_p = \frac{301{,}000}{400} = 752.5\ \text{J/kg}\cdot\text{K} \tag{10.19}$$

In this case, the area is somewhat high because each straight-line approximation includes slightly more area than that provided by the curve.

The method of a rectangle, the method of rectangles, and the method of trapezoids may be applied to a series of data points *without* plotting the data. Note also that the three methods produced results that were within reasonable agreement. The above equations are introductory examples of the general subject of *numerical methods*, to be discussed in Part III, and they may be applied to graphs or numerical data.

10.5 Rayleigh Equation for Simple Batch (Differential) Distillation

In a simple batch distillation (see Figure 10.6), the composition of the evolved vapors and the composition of the liquid in the still pot are continuously changing. Therefore, the mathematical analysis of the operation must be based on differential changes. Although batch distillations are generally more costly than their continuous counterparts, there are certain applications in which batch distillation is the method of choice. Batch distillation is typically chosen when it is not possible to run a continuous process due to limiting process constraints, the need to distill other process streams, or because the low frequency use of distillation does not warrant a unit devoted solely to a specific product [6–9].

A relatively efficient separation of two or more components may be accomplished through batch distillation in a pot or tank. Although the purity of the distilled product varies throughout the course of batch distillation, it still has its use in industry. As shown in Figure 10.6, a feed is initially charged to a tank, and the

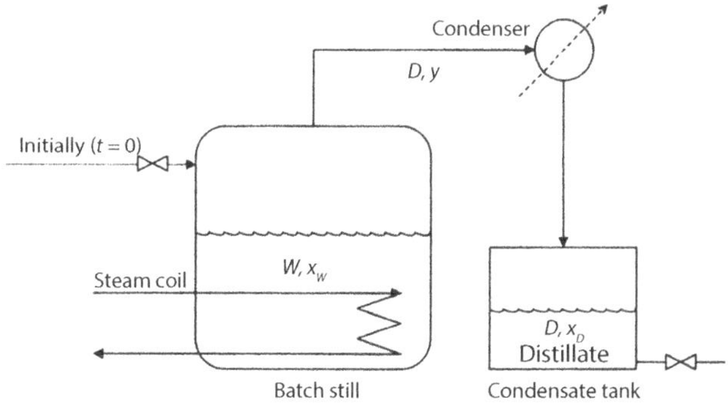

Figure 10.6 Batch distillation diagram.

vapor generated by boiling the liquid is withdrawn and enters a condenser. The condensed product is collected as distillate, D, with composition x_D, and the liquid remaining in the pot, W, has composition x_W. Total and componential material balances around a batch distillation unit are shown below.

$$F = W + D \tag{10.20}$$

$$Fx_F = Wx_W + Dx_D \tag{10.21}$$

Note that W, x_W, D, and x_D all vary throughout the distillation process.

A convenient method for mathematically representing a binary batch distillation process is known as the Rayleigh equation [6, 7]. This equation relates the composition and amount of material remaining in the batch to an initial feed charge, F, and composition, x_F

$$\ln\left(\frac{W}{F}\right) = -\int_{x_W}^{x_F} \frac{dx}{y^* - x} \tag{10.22}$$

where y^* = mole fraction of vapor in equilibrium with liquid of composition x.

At the desired distillate composition, the distillation is stopped. At this time, the moles remaining in the still is denoted W_{final}, with composition $x_{W,final}$. As such, Equation (10.22) may be re-written as shown below

$$W_{final} = F\ exp\left[-\int_{x_{W,final}}^{x_F} \frac{dx}{y^* - x}\right] \tag{10.23}$$

The above equation may be solved numerically by plotting $1/(y^* - x)$ vs. x and integrating between the limits $x_{W,final}$ and x_F to determine the area (A) under the curve. The above can therefore be written as

$$W_{final} = F\ exp[-A] \tag{10.24}$$

where A = area under the curve [6, 7]. This problem will be revisited in Part III – Numerical Methods.

References

1. *Webster's New World Dictionary*, 2nd ed., Prentice-Hall, Upper Saddle River, NJ, 1971.
2. L. Theodore, *Air Pollution Control Equipment Calculations*, John Wiley & Sons, Hoboken, NJ, 2008.
3. H. Mickley, T. Sherwood, and C. Reed, *Applied Mathematics in Chemical Engineering*, 2nd ed., McGraw-Hill, New York City, NY, 1957.

4. R. Ketter and S. Prawel, *Modern Methods of Engineering Computation*, McGraw-Hill, New York City, NY, 1969.
5. L. Theodore, personal notes, East Williston, NY, 1970.
6. L. Theodore and J. Barden, *Mass Transfer Operations*, A Theodore Tutorial, Theodore Tutorials, East Williston, NY, originally published by the USEPA/APTI, RTP, NC, 1996.
7. L. Theodore and F. Ricci, *Mass Transfer Operations for the Practicing Engineer*, John Wiley & Sons, Hoboken, NJ, 2010.
8. L. Theodore, *Chemical Engineering: The Essential Reference*, McGraw-Hill, New York City, NY, 2014.
9. L. Theodore, R. Dupont, K. Ganesan, *Unit Operations in Environmental Engineering*, Scrivener-Wiley, Salem, MA, 2017.

11

Differential Calculus

Once again, let the function $f(x)$ be defined for $a<x<b$. For each x in this range, the derivative $y' = f'(x)$ is defined by the equation

$$f'(x) = \lim_{\Delta x \to 0} \frac{f(x+\Delta x) - f(x)}{\Delta x} \tag{11.1}$$

provided this limit exists. It should be noted that the limit cannot exist if $f(x)$ is discontinuous at the value of x considered; however, continuity alone does not imply existence of the derivative.

The differential dy of the function $y = f(x)$ may be approximated as follows:

$$\Delta y = f'(x)\Delta x \tag{11.2}$$

In this equation, Δx can be replaced by dx so that one has

$$dy = f'(x)dx$$

or

$$\frac{dy}{dx} = f'(x) \tag{11.3}$$

This derivative can be interpreted graphically as the slope of the tangent to the curve $y = f(x)$ at point (x, y). The equation of the tangent at the point (x_1, y_1) is then

$$y - y_1 = f'(x_1)(x - x_1) \tag{11.4}$$

Thus dx and dy can be interpreted as changes in x and y at (x, y).

The second derivative $f''(x)$ is defined as the derivative of the first derivative, and higher derivatives are correspondingly defined. The notations

$$\frac{d^2y}{dx^2}, \frac{d^3y}{dx^3}, \frac{d^4y}{dx^4}, \ldots, \frac{d^ny}{dx^n} \tag{11.5}$$

or

$$y'', y''', y^{iv}, \ldots, y^{(n)} \tag{11.6}$$

are also used to represent higher derivatives, respectively.

The basic properties of the derivative are summarized in the following theorems where $u = f(x)$ and $v = g(x)$, are defined for $a < x < b$.

$$(u+v)' = u' + v' \tag{11.7}$$

$$(u \cdot v)' = uv' + vu' \tag{11.8}$$

$$\left(\frac{u}{v}\right)' = \frac{vu' - uv'}{v^2} \tag{11.9}$$

For example, repeated application of the rule for the derivative of the product xy gives the rule:

$$y' = nx^{n+1} \tag{11.10}$$

$$y = x^n \tag{11.11}$$

for any positive integer n. In addition, if $y' = 0$, then $y = c$, (c = constant)

Four sections complement the presentation of this chapter. Section numbers and subject titles follow:

11.1: Differential Operations
11.2: Ordinary Differential Equations
11.3: Partial Differential Equations
11.4: Maxima/Minima

11.1 Differential Operations

The following differential operations in Table 11.1 are valid for all differentiable functions of *x* where *c* is a constant and *ln* is the base of the natural logarithm.

Table 11.1 Differential operations.

	Operation
1.	$\frac{d(c)}{dx} = 0$
2.	$\frac{d(x)}{dx} = 1$
3.	$\frac{d(cx)}{dx} = c$
4.	$\frac{d(cx^n)}{dx} = ncx^{n-1}$
5.	$\frac{d(e^x)}{dx} = e^x$
6.	$\frac{d(\ln x)}{dx} = \frac{1}{x}$
7.	$\frac{d(\log x)}{dx} = \log \frac{e}{x}$
8.	$\frac{d(\sin x)}{dx} = \cos x$
9.	$\frac{d(\cos x)}{dx} = -\sin x$
10.	$\frac{d(\tan x)}{dx} = \sec^2 x$
11.	$\frac{d(\cot x)}{dx} = -\csc^2 x$
12.	$\frac{d(\sec x)}{dx} = \tan x \sec x$
13.	$\frac{d(\csc x)}{dx} = -\cot x \csc x$

(Continued)

Table 11.1 Cont.

	Operation
14.	$\dfrac{d(\sin^{-1} x)}{dx} = (1-x^2)^{-\frac{1}{2}}$
15.	$\dfrac{d(\cos^{-1} x)}{dx} = -(1-x^2)^{-\frac{1}{2}}$
16.	$\dfrac{d(\tan^{-1} x)}{dx} = (1+x^2)^{-1}$
17.	$\dfrac{d(cu)}{dx} = c\left(\dfrac{du}{dx}\right);\ u = f(x)$
18.	$\dfrac{d(uv)}{dx} = u\left(\dfrac{dv}{dx}\right) + v\left(\dfrac{du}{dx}\right);\ v = f(x)$
19.	$\dfrac{d\left(\dfrac{u}{v}\right)}{dx} = \left(\dfrac{1}{v}\right)\left(\dfrac{du}{dx}\right) - \left(\dfrac{u}{v^2}\right)\left(\dfrac{dv}{dx}\right)$
20.	$\dfrac{d(u^n)}{dx} = nu^{n-1}\left(\dfrac{du}{dx}\right)$
21.	$\dfrac{d(u^{1/2})}{dx} = \left[\dfrac{1}{2}u^{(1/2)}\right]\left(\dfrac{du}{dx}\right)$
22.	$\dfrac{d\left(\dfrac{1}{u}\right)}{dx} = -\left(\dfrac{1}{u^2}\right)\left(\dfrac{du}{dx}\right)$
23.	$\dfrac{d\left(\dfrac{1}{u^n}\right)}{dx} = -\left(\dfrac{n}{u^{n+1}}\right)\left(\dfrac{du}{dx}\right)$
24.	$\dfrac{d(\ln u)}{dx} = \left(\dfrac{1}{u}\right)\left(\dfrac{du}{dx}\right)$
25.	$\dfrac{d(c^u)}{dx} = (c^u)(\ln c)\left(\dfrac{du}{dx}\right)$
26.	$\dfrac{d(e^u)}{dx} = e^u\left(\dfrac{du}{dx}\right)$

11.2 Ordinary Differential Equations

The natural laws in any scientific or technological field are not regarded as precise and definitive until they have been expressed in mathematical form. Such a form, often an equation, is a relation between the quantity of interest, say pollutant emissions, and independent variables such as concentration or flow rates on which the emission(s) depend. When it happens that this equation involves one or more of its derivatives it is called a differential equation.

When a function involved in the equation depends only on one variable, its derivatives are ordinary derivatives and the differential equation is called an *ordinary differential equation*. When the function depends on several independent variables, then the equation is called a *partial differential equation*. The theories of ordinary and partial differential equations are quite different. In almost every respect the latter is more difficult to treat analytically. Ordinary differential equations arise from those systems in which all the quantities may be taken as a function of a *single* independent variable. It is usually easy to test for the presence of a single independent variable from knowledge of the physical situation at hand.

For example, in a single-pass countercurrent heat exchanger [1] at a water treatment facility operating under fixed terminal conditions, it is necessary only to specify the position in the exchanger by means of one variable, such as the distance x from the cold end, and all the other variables at this point are fixed by the physical laws controlling the system. In a continuously operated absorption tower [2], liquid and gas compositions vary throughout the tower, but if the input and outlet conditions are fixed, it suffices (if channeling effects are neglected and concentrations are assumed equal at all points across any cross section) to specify position in order to calculate the properties of both the liquid and gas streams.

As noted above, ordinary differential equations generally arise from a description of situations involving one independent variable. The most important cases in practice involve, in addition to the independent variable, one dependent variable, and the most general differential equation for two variables may be written thusly

$$f\left(x,\ y,\ \frac{dy}{dx},\ \frac{d^2y}{dx^2},\ \frac{d^3y}{dx^3},\ \ldots,\ \frac{d^ny}{dx^n}\right)=0 \tag{11.12}$$

Any value of y which, when substituted into Equation (11.12), reduces the left-hand side to zero is called a "solution" of the differential equation.

The *order* of a differential equation is the order of the highest derivative occurring within the equation. The *degree* of a differential equation is the power to which the highest derivative is raised when the equation has been rationalized and cleared of fractions. The differential equation

$$\frac{d^2y}{dx^2}=f(x)\left[1+\left(\frac{dy}{dx}\right)^2\right]^{\frac{1}{2}} \tag{11.13}$$

is of the second order and the second degree, as may be seen by applying the preceding rules after squaring both sides.

The solution of a differential equation involving the most general possible relationship among the variables is known as the "general solution" or the "primitive." When the constants in the general solution are allowed to take on values corresponding to the conditions in a specific problem, the resulting equation is known as a "particular solution" of the differential equation because it is now applicable to but one particular case. The limitations introduced in obtaining a particular solution from a general solution are known collectively as the "boundary conditions" of the problem.

Whenever a relationship exists between two variables x and y such that for every value of x there exists at least one value of y, y is said to be a function of x, and the functional relation may be expressed symbolically as

$$f(x, y) = 0 \tag{11.14}$$

Equation (11.14) can be solved for either *x in terms of y or y in terms of x.*

Solving differential equations analytically is one of the major problems faced by environmental engineers and scientists as such a wide variety of applications lead to differential equations, and some cannot be solved analytically. The classical initial value problem is to find a function $y(x)$ that satisfies the first order differential equation $y' = f(x, y)$ and takes on the initial value form of $y(x_0) = y_0$. A broad variety of methods have been devised for the approximate solution to this classical problem, most of which have then been generalized for treating solutions to these problems [3–6].

11.3 Partial Differential Equations

Suppose now that the end conditions in the two preceding examples, i.e., the heat exchanger and absorber, are not steady or constant with time. The rate of flow of the cold stream to the heat exchanger may vary with time, or the rate of flow of the solvent liquid to the absorption tower may vary with time, per some known relation. In order to specify conditions completely at any point in either unit, the values of *two* independent variables – position and either time or rate of flow of the variable stream – must be specified. In this case, the instantaneous behavior of the system will be expressed by an equation involving the variables and their partial derivatives. The analytical solution of partial differential equations is available in the literature [3–6].

It is fair to say that the analysis of situations involving two or more independent variables frequently results in a partial differential equation. The mathematical techniques employed to formulate the partial differential equation are similar to those used to set up an ordinary differential equation. The details of the two processes and, in particular, the manipulations involved are, however, surprisingly different. The general function of n independent variables

$$y = f(x_1, x_2, \ldots, x_n) \tag{11.15}$$

may be reduced to a function of x_1 alone by holding the remaining variables x_2, x_3, …, x_n constant and allowing x_1 to vary. The function of x_1 thus created may have a derivative defined and computed by the ordinary methods applicable to functions of a single variable. This derivative is called the first partial derivative of f or y with respect to x_1, and its notation is the symbol

$$f'_{x_1}, \text{ or } \left(\frac{df}{dx_1}\right)_{x_2 \ldots x_n}, \text{ or } \frac{\partial f}{\partial x_1}$$

This partial derivative is defined as the limit

$$\frac{\partial f(x_1, x_2, \ldots, x_n)}{\partial x_1} = \lim_{\Delta x_1 \to 0} \frac{f(x_1 + \Delta x_1, x_2, \ldots, x_n) - f(x_1, x_2, \ldots, x_n)}{\Delta x_1} \tag{11.16}$$

Partial derivatives are defined with respect to each of the variables $x_1, x_2, x_3, \ldots, x_n$ in an analogous manner.

Particular attention must be devoted to the notational aspects of partial differentiation, inasmuch as the customary symbols, carelessly employed, may become so nondescript with reference to the actual operations as to result in considerable ambiguity and confusion. The chief difficulty in this respect is with the aforementioned common partial derivative symbol

$$\left(\frac{\partial f}{\partial x_1}\right)_{x_2 \ldots x_n}$$

Thus, ∂f indicates that some function f of several independent variables is to be partially differentiated with respect to the one of these independent variables which is indicated in ∂x_1, the remaining variables in the function, as indicated in the subscript are, being held constant. Clearly ∂f standing alone has no definite meaning, for f may be partially differentiated with respect to any one of its arguments $x_1, x_2, \ldots, x_n$. Until it is indicated by ∂x_1 which of the n partial derivatives is to be taken, ∂f has only an indefinite operational significance. Thus, the above term

$$\left(\frac{\partial y}{\partial x_1}\right)_{x_2 \ldots x_n}$$

indicates three things:

1. The function that is to be differentiated is y.
2. The variable x_1 with respect to which differentiation is to be performed.
3. The variables that are to be held constant during the differentiation (usually indicated as subscripts).

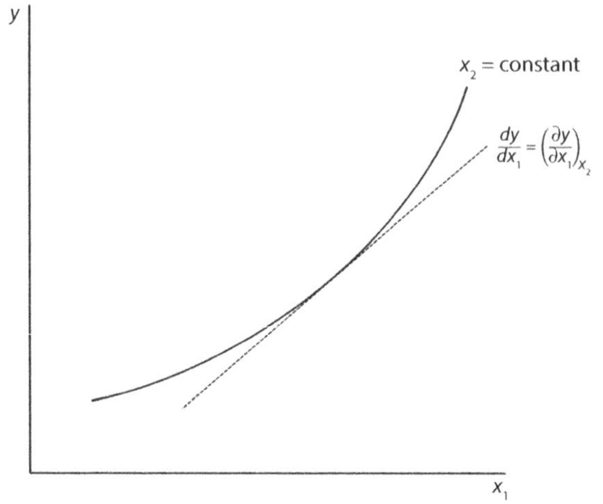

Figure 11.1 A plot of x_1 vs. y.

The partial derivative

$$\left(\frac{\partial y}{\partial x_1}\right)_{x_2} \tag{11.17}$$

or

$$\left(\frac{\partial y}{\partial x_2}\right)_{x_1} \tag{11.18}$$

may be demonstrated graphically (see previous Chapter). Refer to Figure 11.1 which provides a plot of y vs x_1. The slope of any point along the curve represents

$$\left(\frac{\partial y}{\partial x_1}\right)_{x_2}$$

since x_2 is constant. The derivative can be generated for various values of x_2 as pictured in Figure 11.2. If the four slopes calculated keeping x_2 constant at A, B, C, and D, i.e., the partial of y with respect to x_1 evaluated at x_1^*, are plotted vs the complementary four values of x_2, i.e., a, b, c, and d – as shown in Figure 11.3 – the slope at any point along the curve would represent

$$\left(\frac{\partial y/\partial x_1}{\partial x_2}\right)_{x_1=x_1^*}$$

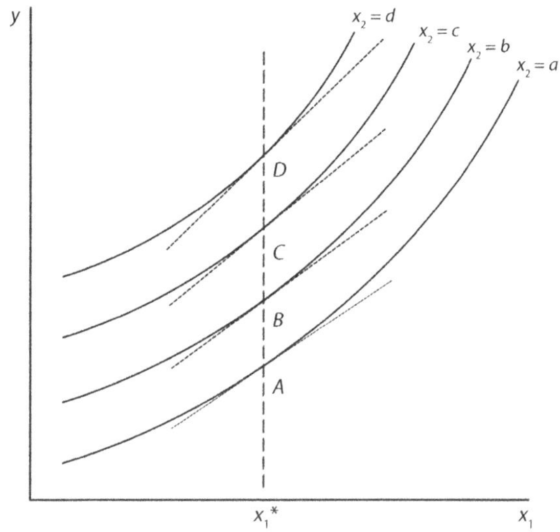

Figure 11.2 Partial derivative of y with respect to x_1 keeping x_2 constant at a, b, c, and d.

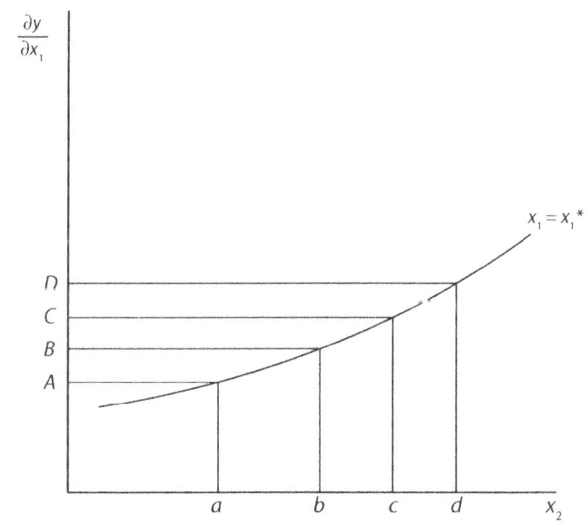

Figure 11.3 Partial derivative of y with respect to x_1 vs. x_2.

Not surprisingly, an equation involving partial derivatives is defined as a partial differential equation. The equation

$$\frac{\partial^2 T}{\partial x \partial y} = 0 \tag{11.19}$$

is a partial differential equation. Such equations are of importance in environmental engineering and science because the significant independent variables are frequently functions of more than one independent variable, and the basic differential expressions for the natural laws are, therefore, partial differential equations. The usual problem is to determine a particular relation between y, x_1, and x_2 expressed as $y = f(x_1, x_2)$, that satisfies the basic differential equation and also satisfies some particular condition specified by the practical problem at hand.

The *order* of a partial differential equation is the order of the highest derivative appearing in the equation. For example, the equation

$$\frac{\partial^4 T}{\partial x^2 \partial y^2} + \frac{\partial^2 T}{\partial x^2} + T^3 = 0 \tag{11.20}$$

is fourth order. A partial differential equation is said to be *linear* if no powers or products of the dependent variable or its partial derivatives are present. If one of the coefficients in the equation is a function of the dependent variable y, then the equation would be classified as *nonlinear*. Homogenous partial differential equations are ones in which all terms contain derivatives of the same order. The solution of a partial differential equation must also satisfy both the original differential equation and its boundary conditions. A discussion of the number and type of boundary conditions necessary and sufficient to ensure a solution of a partial differential equation is beyond the scope of this chapter [1, 3, 4]. However, in most environmental engineering applications, the physical system enables the necessary boundary conditions to be established relatively easily.

No general formalized analytical procedure for the solution of an arbitrary partial differential equation is known. Details are articulated in the literature [3–6] but one should not rule out guessing at a solution [7].

Illustrative Example 11.1

If

$$y = f(x, u, v) = xuv + u - 2v$$

evaluate

$$\frac{\partial y}{\partial x}, \frac{\partial y}{\partial u}, \text{ and } \frac{\partial y}{\partial v}$$

if both u and v are functions of x.

Solution

$$\frac{\partial y}{\partial x} = uv$$

$$\frac{\partial y}{\partial u} = xv + 1$$

$$\frac{\partial y}{\partial v} = xu - 2$$

11.4 Maxima and Minima

Since the derivative dy/dx, or $f'(x)$, represents the rate of change of y with change in x, it is evident that if the function passes through a maximum or minimum the derivative will be *zero*. If the occurrence of such a maximum or minimum is to be determined and located, the derivative is equated to zero, thus giving the condition for which the maximum or minimum exists. This procedure is of considerable value in environmental engineering calculations, particularly those involving optimization, since the location of a maximum or minimum is frequently of significant importance. This is particularly true in plant design, since the optimum operating and design variables are to be determined in order that the maximum profit and/or minimum cost will result.

In Figure 11.4, the curve *ABCDE* represents the function $y = f(x)$. If values of y in the neighborhood of points A and C are all less than the values of y at A and C, the function is said to go through a *maximum* at these points. Similarly, because the values of y in the neighborhood of B are all greater than the value of B, the function is said to go through a *minimum* at B. This curve emphasizes that the terms "maximum" and "minimum" do not necessarily denote the greatest and least possible values a function may assume.

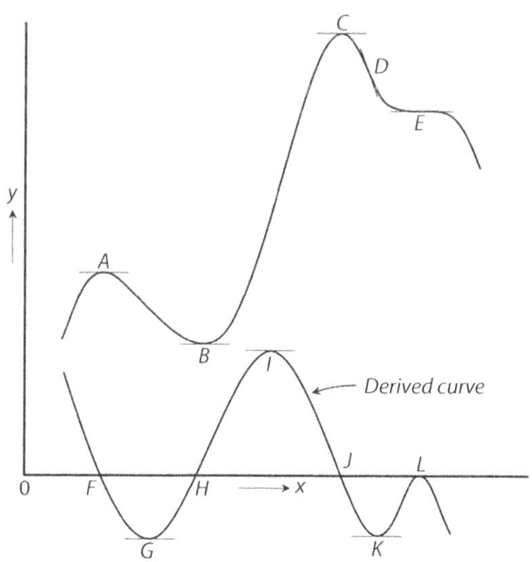

Figure 11.4 Maximum and minimum of a function with derived curve.

As noted above, it is evident geometrically that the slope of the curve is zero at the maximum and minimum points, and since the slope is given by the derivative dy/dx, or $f'(x)$, these points may be determined and located through solution of the equation $f'(x) = 0$. The roots of this equation merely locate the maximum and minimum points and cannot distinguish between them. Furthermore, the condition locates such points as E. One method of distinguishing between points A, C, and E is to calculate values of $f(x)$ in their immediate vicinity. These are known as points of *inflection*. A convenient rule of these relations is as follows:

$$\text{Maximum:} \quad \frac{dy}{dx} = 0, \quad \frac{d^2y}{dx^2} < 0 \tag{11.21}$$

$$\text{Minimum:} \quad \frac{dy}{dx} = 0, \quad \frac{d^2y}{dx^2} > 0 \tag{11.22}$$

Calculations to determine the optimum conditions from the point of view of costs and monetary return are termed *economic balances*. The basic criterion of the true optimum is maximum return on the investment, but the problem can frequently simplify to one of determining the minimum cost, the maximum production from a piece of equipment, the minimum power, etc. The quantity considered is expressed as a function of the design or operating variables. On differentiating and equating to zero, one obtains the condition which determines the optimum value of the variable under consideration.

Illustrative Example 11.2 [8]

Consider the two-stage reversible adiabatic compression [8] of a gas from an initial pressure P_1 to a final pressure P_3. If the fixed charges on the compressors are assumed to be essentially independent of the interstage pressure employed, then the optimum operation involves the determination of the interstage pressure for which the total power requirement is a *minimum*. For a fixed gas flow, this corresponds to the minimum work for the two stages. If the gas enters at T_1 and is cooled to T_1 between stages, the total work is given by

$$W = NRT\left(\frac{k}{k-1}\right)\left[\left(\frac{P_2}{P_1}\right)^{\left(\frac{k-1}{k}\right)} + \left(\frac{P_3}{P_2}\right)^{\left(\frac{k-1}{k}\right)} - 2\right]; \; T_1 = T \tag{11.23}$$

where (in consistent units)

N = pound moles of gas compressed

R = molal gas constant

k = ratio of heat capacity at constant pressure to heat capacity at constant volume for the gas compressed

T_1 = inlet gas temperature
W = total work, ft·lb_f

Outline how to determine P_2 for minimum power required.

Solution

If the above quantity is to be a minimum, then the derivative must be zero:

$$\frac{dW}{dP_2} = NRT\left(\frac{k}{k-1}\right)\left[\left(\frac{k-1}{k}\right)P_1^{(1-k)/k}P_2^{-(1/k)} - \left(\frac{k-1}{k}\right)P_3^{(k-1)/k}P_2^{(1-2k)/k}\right] = 0$$

The above equation can now be solved for P_2. This example will be revisited several times in Part V – Optimization.

References

1. L. Theodore, *Heat Transfer Applications for the Practicing Engineer*, John Wiley & Sons, Hoboken, NJ, 2012.
2. L. Theodore and F. Ricci, *Mass Transfer Operations for the Practicing Engineer*, John Wiley & Sons, Hoboken, NJ, 2010.
3. H. Mickley, T. Sherwood, and C. Reed, *Applied Mathematics in Chemical Engineering*, 2nd ed., McGraw-Hill, New York City, NY, 1957.
4. L. Theodore, *Chemical Engineering: The Essential Reference*, McGraw-Hill, New York City, NY, 2014.
5. R. Ketter and S. Prawel, *Modern Methods of Engineering Computation*, McGraw-Hill, New York City, NY, 1969.
6. L. Lapidus, *Digital Computation for Chemical Engineers*, McGraw-Hill, New York City, NY, 1962.
7. L. Theodore, personal notes, East Williston, NY, 1970.
8. J. Happel, *Chemical Process Economics*, John Wiley & Sons, Hoboken, NJ, 1958.

12

Integral Calculus

The reader is reminded that Chapter 10 primarily keyed on graphical integration. That material is now extended to include the analytical approach to integration. As noted earlier, numerical integration receives treatment in Part III.

Four sections complement the presentation of this chapter. Section numbers and subject titles follow:

12.1: Analytical Integration
12.2: Indefinite Integrals
12.3: Definite Integrals
12.4: Integration Applications

12.1 Analytical integration

The definite integral $\int_a^b f(x)dx$ of a function $f(x)$ for $a \le x \le b$, is defined by the limiting process:

$$I = \int_a^b f(x)dx = \lim_{\substack{n \to \infty \\ \max \Delta x_i \to 0}} \sum_{i=1}^{n} f(x_i^*)\Delta x_i \tag{12.1}$$

Here one denotes $x_i (i = 0, 1, \ldots, n)$ as a sequence of values of x in the interval, such that $a = x_0 < x_1 < x_2 < \ldots < x_n = b$ and by Δ_{xi} the difference $x_i - x_{i-1}$. The value x_{i^*} is any value of x such that

$$x_{i-1} \leq x_i^* \leq x_i \tag{12.2}$$

As illustrated in Figure 12.1 the limiting process is understood as follows: there exists a value such that, for n sufficiently large and the largest Δx_i sufficiently small, the sum

$$\sum_{i=1}^{n} f(x_i^*)\Delta x_i = f(x_1^*)\Delta x_1 + \ldots f(x_n^*)\Delta x_n \tag{12.3}$$

differs from I by as little as desired, no matter how the values x_i^* are chosen. The number I is then the value of the integral on the left of Equation (12.4). The existence of this limit can be established if $f(x)$ is continuous for $a \leq x \leq b$.

If $f(x)$ is continuous and positive throughout the interval, the integral can be interpreted as the area of the part of the $\overline{xy}$ plane bounded by the x axis, the graph of $y = f(x)$, and the lines $x = a$, $x = b$. If I denotes this area, one then has

$$I = \int_a^b f(x)dx;\ f(x) \geq 0 \tag{12.4}$$

The definite integral satisfies certain basic laws:

$$\int_a^b [f(x) + g(x)]dx = \int_a^b f(x)dx + \int_a^b g(x)dx \tag{12.5}$$

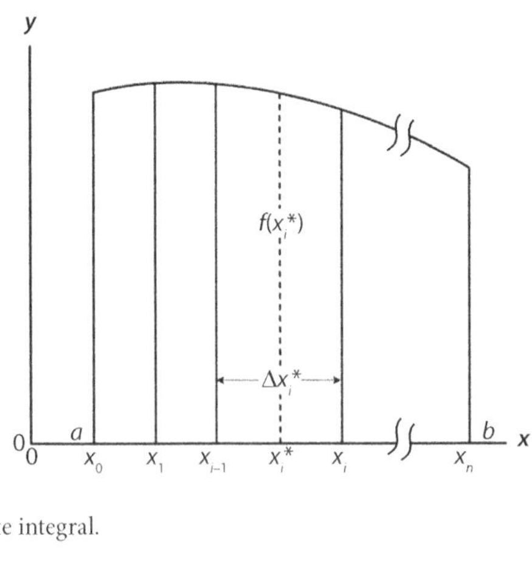

Figure 12.1 The definite integral.

$$\int_a^b cf(x)dx = c\int_a^b f(x)dx;\ c = constant \tag{12.6}$$

$$\int_a^b f(x)dx + \int_b^c f(x)dx = \int_a^c f(x)dx \tag{12.7}$$

12.2 Indefinite Integrals

If $f'(x)$ is the derivative of $f(x)$, the antiderivative of $f'(x)$ is $f(x)$. Symbolically, the indefinite integral of $f'(x)$ is

$$\int f'(x)dx = f(x) + c \tag{12.8}$$

where c is an arbitrary constant to be determined. The following relationships in Table 12.1 hold (a and b are constants) by virtue of the known formulas for differentiation.

Table 12.1 Indefinite integrals.

	Operation
1.	$\int a\ dx = ax$
2.	$\int af(x)dx = a\int f(x)dx$
3.	$\int (u \pm v \pm w \pm \ldots)dx = \int u\ dx \pm \int v\ dx \pm \int w\ dx \pm \ldots$
4.	$\int (du \pm dv \pm dw) = \int du \pm \int dv \pm \int dw$
5.	$\int u\ dv = uv - \int v\ du$ [integration by parts]
6.	$\int u^n\ du = \frac{u^{n+1}}{n+1},\ n \neq -1$
7.	$\int \frac{du}{u} = \ln u\ if\ u > 0$ or $\ln(-u) if\ u < 0$ $= \ln\lvert u \rvert$
8.	$\int e^u du = e^u$

(*Continued*)

Table 12.1 Cont.

	Operation
9.	$\int a^u\, du = \int e^{u\ln a} du = \frac{e^{u\ln a}}{\ln a} = \frac{a^u}{\ln a},\ a>0,\ a\neq 1$
10.	$\int \sin x\, dx = -\cos x$
11.	$\int \cos x\, dx = \sin x$
12.	$\int \tan x\, dx = \ln\sec x = -\ln\cos x$
13.	$\int \cot x\, dx = \ln\sin x$
14.	$\int \sec x\, dx = \ln(\sec x + \tan x) = \ln\tan\left(\frac{x}{2}+\frac{\pi}{4}\right)$
15.	$\int \csc x\, dx = \ln(\csc x - \cot x) = \ln\tan\frac{x}{2}$
16.	$\int \sec^2 x dx = \tan x$
17.	$\int \csc^2 x dx = -\cot x$
18.	$\int \tan^2 x\, dx = \tan x - x$
19.	$\int \cot^2 x\, dx = -\cot x - x$
20.	$\int \frac{dx}{ax+b} = \frac{1}{a}\ln(ax+b)$
21.	$\int \frac{x\, dx}{ax+b} = \frac{x}{a} - \frac{b}{a^2}\ln(ax+b)$
22.	$\int \frac{x^2\, dx}{ax+b} = \frac{(ax+b)^2}{2a^3} - \frac{2b(ax+b)}{a^3} + \frac{b^2}{a^3}\ln(ax+b)$
23.	$\int \frac{x^3\, dx}{ax+b} = \frac{(ax+b)^3}{3a^4} - \frac{3b(ax+b)^2}{2a^4} + \frac{3b^2(ax+b)}{a^4} - \frac{b^3}{a^4}\ln(ax+b)$
24.	$\int \frac{dx}{x(ax+b)} = \frac{1}{b}\ln\left(\frac{x}{ax+b}\right)$

(Continued)

Table 12.1 Cont.

	Operation
25.	$\int \frac{dx}{x^2(ax+b)} = -\frac{1}{bx} + \frac{a}{b^2}\ln\left(\frac{ax+b}{x}\right)$
26.	$\int \frac{dx}{x^3(ax+b)} = \frac{2ax-b}{2b^2x^2} + \frac{a^2}{b^3}\ln\left(\frac{x}{ax+b}\right)$
27.	$\int \frac{dx}{(ax+b)^2} = \frac{-1}{a(ax+b)}$
28.	$\int \frac{x\,dx}{(ax+b)^2} = \frac{b}{a^2(ax+b)} + \frac{1}{a^2}\ln(ax+b)$

12.3 Definite Integrals

The concept and derivation of the definite integral are completely different from those for the indefinite integral. These are by definition different types of operations. However, the formal operation $\int$ as it turns out treats the integrand in the same way for both.

Consider the function $f(x) = 10 - 10e^{-2x}$. Define $x_1 = a$ and $x_n = b$, and suppose it is desirable to compute the area between the curve and coordinate axis $y = 0$ and bounded by $x_1 = a$, $x_n = b$. Obviously, this area could be approximated as closely as desired by a sufficiently large number of rectangles. Thus, the value of the definite integral depends on the limits a, b and any selected variable coefficients in the function - but not on x. There are certain restrictions on the integration definition: "The function $f(x)$ must be continuous on the finite interval (a, b) with at most a finite number of discontinuities," which must be observed before integration formulas can be generally applied. Two techniques for determining when integration is valid under these conditions are available in the literature [1–4].

Some useful *analytical* integrals in environmental engineering and science are provided below in Table 12.2; note that the equations are definite integrals, i.e., the upper and lower limits are specified [5].

Note, once again, that there are limits on the integrals.

Illustrative Example 12.1

Find

$$\int_0^{\frac{\pi}{2}} 2\sin x\,dx$$

Table 12.2 Definite integrals.

	Operation
1.	$\int_0^x a\,dx = ax$
2.	$\int_0^x \frac{1}{1-x}dx = \ln\left(\frac{1}{1-x}\right)$
3.	$\int_0^x \frac{1}{(1-x)^2}dx = \frac{1}{1-x}$
4.	$\int_0^x \frac{1}{(1-x)^2}dx = \frac{x}{1-x}$
5.	$\int_0^x \frac{1}{1+\varepsilon x}dx = \left(\frac{1}{\varepsilon}\right)\ln(1+\varepsilon x)$
6.	$\int_0^x \frac{1+\varepsilon x}{1-x}dx = (1+\varepsilon)\ln\left(\frac{1}{1-x}\right) - \varepsilon x$
7.	$\int_0^x \frac{1+\varepsilon x}{(1-x)^2}dx = \frac{(1+\varepsilon)(x)}{1-x} - \varepsilon\ln\left(\frac{1}{1-x}\right)$
8.	$\int_0^x \frac{(1+\varepsilon x)^2}{(1-x)^2}dx = 2\varepsilon(1+\varepsilon)\ln(1-x) + \varepsilon^2 x + \frac{(1+\varepsilon)^2(x)}{(1-x)}$
9.	$\int_0^x \frac{1}{(1-x)(\theta_B - x)}dx = \frac{1}{(\theta_B - 1)}\ln\left[\frac{(\theta_B - x)}{(\theta_B)(1-x)}\right]$ for $\theta_B \neq 1$
10.	$\int_0^x \frac{1}{ax^2+bx+c}dx = -\frac{2}{(2ax+b)} + \frac{2}{b}$ for $b^2 = 4ac$
11.	$\int_0^x \frac{1}{ax^2+bx+c}dx = \frac{1}{(a)(p-q)}\ln\left[\frac{(q)(x-p)}{(p)(x-q)}\right]$ for $b^2 > 4ac$ where p and q are roots of the equation $ax^2+bx+c=0$, i.e., $p,\ q = \frac{-b \pm (b^2-4ac)^{\frac{1}{2}}}{2a}$
12.	$\int_0^x \frac{a+bx}{c+gx}dx = \frac{bx}{g} + \frac{ag-bc}{g^2}\ln\lvert c+gx\rvert$

(*Continued*)

Table 12.2 Cont.

	Operation
13.	$\int_0^x \frac{x}{ax+b}dx = \frac{x}{a} + \frac{b}{a^2}\ln(ax+b)$
14.	$\int_0^x \frac{x^2}{ax+b}dx = \frac{(ax+b)^2}{2a^3} - \left(\frac{2b}{a^2}\right)(ax+b) + \frac{b^2}{a^3}\ln(ax+b)$
15.	$\int_0^x \frac{1}{(x)(ax+b)}dx = \frac{1}{b}\ln\left(\frac{x}{ax+b}\right)$
16.	$\int_0^\infty \frac{1}{x^2+a^2}dx = \frac{\pi}{2a}$
17.	$\int_0^a \frac{1}{\sqrt{a^2-x^2}} = \frac{\pi}{2}$
18.	$\int_0^a \sqrt{a^2-x^2}\,dx = \frac{\pi a^2}{4}$

Solution

Apply (10) in Table 12.1.

$$2\int_0^{\frac{\pi}{2}} \sin x\ dx = 2[-\cos x]_0^{\frac{\pi}{2}} = -2\left(\cos\frac{\pi}{2} - \cos 0\right) = 2$$

Illustrative Example 12.2

Find

$$\int_0^4 \frac{1}{(x-2)^2}dx$$

Solution

Direct application of the general integration formula would yield the *incorrect* value.

$$\int_0^4 \frac{1}{(x-2)^2}dx = \left[-\frac{1}{x-2}\right]_0^4 = -1$$

It should be noted that

$$f(x) = \frac{1}{(x-2)^2}$$

becomes unbounded as $x = 2$, i.e., the integral diverges and hence is said not to exist.

12.4 Integration Applications

Illustrative Example 12.3

Calculate the surface area and volume of the *unit* cube described in Section 8.4.

Solution

Refer to Chapter 8. There are six surfaces on the cube. The surface in the yz-plane is given by

$$f_x = \int_{z=0}^{1} \int_{y=0}^{1} dy\, dz$$

$$f(x) = 1$$

There are two such surfaces – one at $x = 0$ and another at $x = 1$. The remaining four surfaces can be determined in a similar manner.

$$f_y = 1$$

$$f_z = 1$$

The total surface area is $6 \times 1 = 6$ units of area.

The volume of the cube is given by

$$\int_{z=0}^{1} \int_{y=0}^{1} \int_{x=0}^{1} dx\, dy\, dz = 1$$

unit of volume.

Illustrative Example 12.4

Calculate the volume V of a cylinder of radius a and height h.

Solution

Refer to Chapter 8.

$$d\tau = r\ dr\ d\phi\ dz$$
$$V = \iiint_{\tau} d\tau$$

Substituting

$$V = \int_{z=0}^{h}\int_{\phi=0}^{2\pi}\int_{r=0}^{a} r\ dr\ d\phi\ dz$$
$$= h\int_{\phi=0}^{2\pi}\int_{r=0}^{a} r\ dr\ d\phi = 2\pi h\int_{r=0}^{a} r\ dr = 2\pi h\left[\frac{r^2}{2}\right]_0^a = \pi a^2 h$$

Illustrative Example 12.5

Evaluate the surface area A of a sphere of radius a.

Solution

Refer to Chapter 8.

$$df = r^2 \sin\theta\ d\theta\ d\phi$$
$$A = \int_{\phi=0}^{2\pi}\int_{\theta=0}^{\pi} df = \int_0^{2\pi}\left[\int_0^{\pi} r^2 \sin\theta\ d\theta\right] d\phi$$
$$= 2\pi\int_0^{\pi} r^2 \sin\theta\ d\theta = 2\pi r^2\int_0^{\pi} \sin\theta\ d\theta = 2\pi r^2[-\cos\theta]\Big|_0^{\pi} = 2\pi r^2[1+1] = 4\pi r^2$$

Therefore, the surface of A a sphere where $r = a$ is

$$A = 4\pi a^2$$

Illustrative Example 12.6

The velocity of a fluid in a horizontal cylinder of radius R is represented in cylindrical coordinates by the equation

$$v = 100\left[1-\left(\frac{r}{R}\right)^2\right];\ \text{ft/s}$$

Calculate the average velocity of the fluid flowing in the cylinder [6].

Solution

Based on the problem statement

$$v_z = 100\left[1 - \left(\frac{r}{R}\right)^2\right]$$

The velocity v_z varies over the cross-sectional area of the cylinder f. A differential area element in this cross section is given by (see Chapter 8)

$$df = r\ dr\ d\phi$$

By definition,

$$\bar{v}_z = \frac{\iint v_z df}{\iint df}$$

Substituting

$$\bar{v}_z = \frac{\int_0^{2\pi}\int_0^R \left[100 - 100\left(\frac{r}{R}\right)^2\right] r\ dr\ d\phi}{\int_0^{2\pi}\int_0^R r\ dr\ d\phi} = 2\pi \frac{\int_0^R \left[100r - 100\left(\frac{r^3}{R^2}\right)\right] dr}{\pi R^2}$$

$$= 2\pi \frac{\left[50r^2 - \frac{25r^4}{R^2}\right]_0^R}{\pi R^2} = 2\pi \frac{(50R^2 - 25R^2)}{\pi R^2} = 50 \text{ ft/s}$$

Note that the maximum velocity is located at $r = 0$ and $v_z = 100$ ft/s at that point. Also note that the ratio of the average to the maximum velocity is 1/2 [7].

Illustrative Example 12.7

The temperature T in a solid sphere of radius a is given by $T = 100r$. Compute the average temperature of the sphere.

Solution

Refer to Chapter 8. By definition

$$\bar{T} = \frac{\iiint_\tau T d\tau}{\iiint_\tau d\tau}$$

where

$$d\tau = r^2 \sin\theta dr\, d\theta\, d\phi$$

and

$$T = 100r^2$$

The integration limits are carefully chosen to insure $d\tau$ encompasses the entire volume element.

$$\bar{T} = \frac{\int_{\phi=0}^{2\pi}\int_{\theta=0}^{\pi}\int_{r=0}^{a}(100r^2)r^2 \sin\theta dr\, d\theta\, d\phi}{\int_{\phi=0}^{2\pi}\int_{\theta=0}^{\pi}\int_{r=0}^{a} r^2 \sin\theta dr\, d\theta\, d\phi}$$

$$= \frac{\int_0^{2\pi}\int_0^{\pi} 20a^5 \sin\theta\, d\theta\, d\phi}{\frac{4\pi a^3}{3}} = \frac{\int_0^{2\pi} 40a^5\, d\phi}{\frac{4\pi a^3}{3}}$$

$$= \frac{80\pi a^5}{\frac{4\pi a^3}{3}} = 60a^2$$

Illustrative Example 12.8

The local velocity of fluid flow through a tube of radius a is given by

$$v_z = \frac{g_c \Delta p}{4\mu L}(a^2 - r^2)$$

Calculate the volumetric flow rate, Q, the local velocity in terms of Q, and the ratio of the average to maximum velocity.

Solution

The volumetric flow rate is given by [6]

$$Q = \iint_f v_z\, df$$

where (see also Chapter 8)

$$df = r\, dr\, d\phi$$

Thus,

$$Q = \int_{\phi=0}^{2\pi}\int_{r=0}^{a} \frac{g_c \Delta p}{4\mu L}(a^2 - r^2) r \, dr \, d\phi = \frac{\pi g_c \Delta p}{2\mu L}\int_0^a (a^2 r - r^3) dr$$

$$= \frac{\pi g_c \Delta p}{2\mu L}\left[\frac{a^2 r^2}{2} - \frac{r^4}{4}\right]_0^4 = \frac{\pi g_c \Delta p a^4}{8\mu L}$$

and

$$\frac{\Delta p}{L} = \frac{8\mu Q}{g_c \pi a^4}$$

Substituting the above equation into the equation for the local velocity of the fluid flow leads to

$$v_z = \frac{2Q}{\pi a^2}\left[1 - \left(\frac{r}{a}\right)^2\right]$$

The maximum velocity can easily be shown to be located at $r = 0$. Therefore,

$$v_{z_{max}} = \frac{2Q}{\pi a^2}$$

The average velocity is given by

$$\bar{v}_z = \frac{\int_0^{2\pi}\int_0^a \frac{2Q}{\pi a^2}\left[1 - \left(\frac{r}{a}\right)^2\right] r \, dr \, d\phi}{\int_0^{2\pi}\int_0^a r dr d\phi}$$

$$= \frac{\frac{2Q}{\pi a^2}(2\pi)\int_0^a \left[1 - \left(\frac{r}{a}\right)^2\right] r \, dr}{\pi a^2}$$

$$= \frac{4Q}{\pi a^4}\left[\frac{r^2}{2} - \frac{r^4}{4a^2}\right]_0^a = \frac{Q}{\pi a^2}$$

$$\bar{v}_z = \frac{Q}{\pi a^2} = \frac{Q}{A}$$

so that

$$v_z = 2\bar{v}_z\left[1-\left(\frac{r}{a}\right)^2\right] = v_{z_{max}}\left[1-\left(\frac{r}{a}\right)^2\right]$$

Finally

$$\frac{\bar{v}_z}{v_{z_{max}}} = \frac{1}{2}$$

This result (as expected) is in agreement with the result of Illustrative Example 12.6.

References

1. H. Mickley, T. Sherwood, and C. Reed, *Applied Mathematics in Chemical Engineering*, 2nd ed., McGraw-Hill, New York City, NY, 1957.
2. L. Theodore, *Chemical Engineering: The Essential Reference*, McGraw-Hill, New York City, NY, 2014.
3. R. Ketter and S. Prawel, *Modern Methods of Engineering Computation*, McGraw-Hill, New York City, NY, 1969.
4. L. Lapidus, *Digital Computation for Chemical Engineers*, McGraw-Hill, New York City, NY, 1962.
5. L. Theodore, personal notes, East Williston, NY, 1970.
6. L. Theodore, "Transport Phenomena for Engineers," Theodore Tutorials, East Williston, NY, originally published by the International Textbook Company, Scranton, PA, 1971.
7. P. Abulencia and L. Theodore, *Fluid Flow for the Practicing Chemical Engineer*, John Wiley & Sons, Hoboken, NJ, 2009.

13

Matrix Algebra [1]

Contributing Authors: Megan Menzel and Christopher Testa

The simulation of complex environmental systems, whether they are described by linear or non-linear systems, will often require the use of computers. Analog computers in the past were quite useful in handling systems that can be modelled in terms of ordinary differential equations. If the mathematical model contains partial differential equations, a numerical solution will usually be required, and these can best be handled on a digital computer. Matrix methods efficiently manipulate the many process variables and large sets of algebraic equations often found in complex systems.

This chapter covers the basic definitions and theorems concerned with matrix analysis with simple examples to illustrate their applications. Following a section on definitions, an introductory review of determinants precedes the discussion of matrices.

Four sections complement the presentation of this chapter. Section numbers and subject titles follow:

13.1: Definitions
13.2: Rules for Determinants and Matrices
13.3: Rank and Solution of Linear Equations
13.4: Linear Equations

13.1 Definitions

A matrix is defined as an array of elements arranged in rows and columns. The elements may be real or complex, constants or functions. Several examples are shown below:

$$\begin{bmatrix} 1 & 5 & 3 \\ 2 & 1 & 0 \end{bmatrix} \tag{13.1}$$

$$\begin{bmatrix} 1+i & 0 \\ 0 & 2+3i \\ 2+i & 1-i \end{bmatrix} \tag{13.2}$$

$$\begin{bmatrix} \cos t + i \sin t & \cos t + i \sin 2t \\ \cos 2t + i \sin t & \cos 2t + i \sin 2t \\ \cos 3t + i \sin t & \cos 3t + i \sin 2t \end{bmatrix} \tag{13.3}$$

The order of a matrix is denoted by $m \; x \; n$ where m is the number of rows and n the number of columns. The matrices in equations (13.2) and (13.3) are of the order 3×2 while the matrix in equation (13.1) is of the order 2×3.

Elements of a matrix are identified by a double subscript, the first denoting *row* position and the second denoting *column* position. Thus, if A is a matrix containing elements a_{ij}, then

$$A = [a_{ij}] = \begin{matrix} a_{11} + a_{12} + a_{13} + \ldots + a_{1n} \\ \vdots \\ a_{m1} + a_{m2} + a_{m3} + \ldots + a_{mn} \end{matrix} \tag{13.4}$$

and a_{23} is the elements of A in the second row, third column. When a matrix X contains a single column and is thus of order $m \; x \; 0$, it is called a *vector*.

$$X = \begin{bmatrix} x_1 \\ x_2 \\ \vdots \\ x_m \end{bmatrix} \tag{13.5}$$

The n columns of an $m \; x \; n$ order matrix may also be considered vectors. For example, the third column of the $m \; x \; n$ order matrix A may be considered the vector A_3, in which case

$$A_3 = \begin{bmatrix} x_1 \\ x_2 \\ \vdots \\ x_m \end{bmatrix} \tag{13.6}$$

Two matrices are equal if they are of the same order and the corresponding elements are equal. If a matrix is multiplied by a scalar constant k, then each element within the matrix is multiplied by k. Thus, if

$$A = [a_{ij}] \tag{13.7}$$

then

$$kA = [ka_{ij}]$$

Matrices may be added provided they are of the same order. If

$$A = [a_{ij}] \text{ and } B = [b_{ij}] \tag{13.8}$$

then

$$A + B = [a_{ij} + b_{ij}]$$

If all elements of a matrix are zero, the matrix is considered a *null matrix*.

A square matrix of (order $n \times n$) with zeros for all elements but the diagonal is called a *diagonal matrix*.

$$A = \begin{bmatrix} a_{11} & 0 & 0 \\ 0 & a_{22} & 0 \\ 0 & 0 & a_{33} \end{bmatrix} \tag{13.9}$$

If the elements of a diagonal matrix consist of ones, it is called a *unit matrix*, *identity matrix*, or *idem matrix*.

$$A = \begin{bmatrix} 1 & 0 & 0 \\ 0 & 1 & 0 \\ 0 & 0 & 1 \end{bmatrix} \tag{13.10}$$

The unit matrix can be represented concisely in terms of the kronecher delta function, defined as

$$\delta_{ij} = \begin{cases} 0 & i \neq j \\ 1 & i = j \end{cases} \tag{13.11}$$

In terms of δ_{ij},

$$I = [\delta_{ij}] \tag{13.12}$$

A general diagonal matrix may be written as

$$A = [\delta_{ij} a_{ij}] \tag{13.13}$$

The product of two matrices A and B in the order $A\ B$ can be found provided A has as many columns as B has rows. The first factor, A, is labelled the prefactor, while the second factor, B, is called the postfactor. If the product of B premultiplied by A is designated by C,

$$C = AB \tag{13.14}$$

The rule for multiplication is that the element c_{ij} is formed by multiplying corresponding elements of the ith row of A with the jth column of B and summing. Consider the example below:

$$\begin{bmatrix} 1 & 1 \\ 0 & 2 \\ 1 & 0 \end{bmatrix} \begin{bmatrix} 1 & 1 \\ 2 & 1 \end{bmatrix} = \begin{bmatrix} (1x1)+(1x2) & (1x1)+(1x1) \\ (0x1)+(2x2) & (0x1)+(2x1) \\ (1x1)+(0x2) & (1x1)+(0x1) \end{bmatrix} \tag{13.15}$$

In effect,

$$\begin{matrix} A & x & B & = & C \\ 3x2 & x & 2x2 & = & 3x2 \\ (m)x(p) & x & (p)x(n) & = & (m)x(n) \end{matrix}$$

The elements of C are formed in accordance with the rule

$$c_{ij} = \sum_{k=1}^{p} a_{ij} b_{ij} \tag{13.16}$$

Matrix multiplication obeys the distributive law,

$$A(B+C) = AB + AC \tag{13.17}$$

and the associative law

$$A(BC) = (AB)C \tag{13.18}$$

However, the reverse rule cannot generally be applied, i.e.,

$$AB \neq BA \tag{13.19}$$

In fact, the product may exist in one order, but not the other.

The determinant of a square matrix of order n is denoted

$$\det A = |a_{ij}| = |A| \tag{13.20}$$

The determinant of a matrix A is a scalar function of the elements of A. It is the sum of all the products formed by taking one element from each row and column (see Equations (13.21) and (13.22) below).

13.2 Rules for Determinants and Matricies

A determinant consists of a certain expression associated with a square array of n^2 quantities, where the number n is the *order* of the determinant [2]. A second order determinant is equal to the difference between the product of the elements on the principle diagonal and the product of the elements on the other diagonal:

$$\begin{vmatrix} a & b \\ c & d \end{vmatrix} = ad - bc \tag{13.21}$$

Similarly, a third order determinant has an expansion containing six terms:

$$\begin{vmatrix} a_1 & b_1 & c_1 \\ a_2 & b_2 & c_2 \\ a_3 & b_3 & c_3 \end{vmatrix} = a_1 \begin{vmatrix} b_2 & c_2 \\ b_3 & c_3 \end{vmatrix} - b_1 \begin{vmatrix} a_2 & c_2 \\ a_3 & c_3 \end{vmatrix} + c_1 \begin{vmatrix} a_2 & b_2 \\ a_3 & b_3 \end{vmatrix} \tag{13.22}$$
$$= a_1b_2c_3 - a_1c_2b_3 - b_1a_2c_3 + b_1c_2a_3 + c_1a_2b_3 - c_1b_2a_3$$

A number of rules follow from the definition of a determinant. Some of these are listed below:

1. Multiplication of a determinant by a scalar k is equivalent to multiplying every element of a single row or column by k.
2. An interchange of k rows (or columns) results in a multiplication of the determinant by $(-1)^k$.
3. If two rows (or columns) are proportional, the determinant must vanish.

4. Two determinants may be added if they are identical, except for one row (or one column). The sum is obtained as shown below

$$\begin{vmatrix} a_{11} & a_{12} & a_{13} \\ a_{21} & a_{22} & a_{23} \\ a_{31} & a_{32} & a_{33} \end{vmatrix} + \begin{vmatrix} a_{11} & x_1 & a_{13} \\ a_{21} & x_2 & a_{23} \\ a_{31} & x_3 & a_{33} \end{vmatrix} = \begin{vmatrix} a_{11} & a_{12}+x_1 & a_{13} \\ a_{21} & a_{22}+x_2 & a_{23} \\ a_{31} & a_{32}+x_3 & a_{33} \end{vmatrix} \tag{13.23}$$

5. The *minor* of an element a_{ij} is defined as the determinant of the matrix remaining after striking out the ith row and j^{th} column of A. The *cofactor* of a_{ij}, designated A_{ij}, is defined as

$$A_{ij} = (-1)^{i+j}(\text{minor of } a_{ij}) \tag{13.24}$$

It can be shown from the above rules that

$$\sum_{j=1}^{n} a_{ij} A_{kj} = \delta_{ik} \, |a_{ij}| \tag{13.25}$$

$$\sum_{i=1}^{n} a_{ij} A_{ik} = \delta_{jk} \, |a_{ij}| \tag{13.26}$$

Equation (13.25) is the Laplace expansion of a determinant in terms of cofactors of elements of the ith row or j^{th} column.

The transpose of a matrix A^T is the matrix formed by interchanging rows and columns. If $A = [a_{ij}]$, then

$$A^T = [a_{ji}] \text{ and } a_{ij}^T = a_{ji} \tag{13.27}$$

The transpose of a vector (a column matrix) is a *row* matrix.

$$X^T = [x_1, x_2, \ldots, x_n] \tag{13.28}$$

An *adjoint matrix* can be defined for square matrices in the following manner: Given the square matrix $A = [a_{ji}]$, the adjoint matrix of A is

$$adj\ A = \begin{bmatrix} A_{11} & A_{12} & . & . & . & A_{1n} \\ A_{21} & A_{22} & . & . & . & A_{21} \\ . & . & & & & . \\ . & . & & & & . \\ . & . & & & & . \\ A_{m1} & A_{m2} & . & . & . & A_{mn} \end{bmatrix} \tag{13.29}$$

where A_{ij} is the cofactor of a_{ij}.

It should be noted that the adj A is formed from A by replacing each element by its cofactor and transposing the resulting matrix. An example of an adjoint matrix is

$$A = \begin{bmatrix} 1 & 0 & 1 \\ 1 & 2 & 1 \\ 0 & 1 & 0 \end{bmatrix} \tag{13.30}$$

where

$$\begin{aligned} A_{11} &= (-1)^{1+1}((2x0)-(1x1)) = -1 \\ A_{12} &= (-1)^{1+2}((1x0)-(0x1)) = 0 \\ A_{13} &= (-1)^{1+3}((1x1)-(0x2)) = 1 \end{aligned} \tag{13.31}$$

Similarly,

$$\begin{aligned} &A_{21} = 1, \quad A_{22} = 0, \; A_{23} = -1 \\ &A_{31} = -2, \; A_{32} = 0, \; A_{33} = 2 \end{aligned}$$

Thus,

$$adj\ A = \begin{bmatrix} -1 & 1 & -2 \\ 0 & 0 & 0 \\ 1 & -1 & 2 \end{bmatrix} \tag{13.32}$$

It should also be noted that $(adj\ A)_{kj} = A_{jk}$. Therefore, if $C = A\ adj\ A$,

$$c_{ij} = \sum_{k=1}^{n} a_{ij} (adj\ A)_{kj} \tag{13.33}$$

and

$$c_{ij} = \sum_{k=1}^{n} a_{ij} A_{jk} \tag{13.34}$$

However, from Laplace's expansion, which is provided in equation (13.25), one may write

$$c_{ij} = \delta_{ij} \left|a_{ij}\right| = \delta_{ij} \det A \tag{13.35}$$

In the preceding example, det $A = 0$ and A *adj* A is the null matrix of order 3. Since det A is a scalar it may be factored out of the matrix so that

$$\begin{aligned} A\ adj\ A &= \det A[\delta_{ij}] \\ &= (\det A)I \end{aligned} \tag{13.36}$$

where I is once again the identity matrix.

The operation of division is not defined for matrices. However a reciprocal or inverse matrix is defined, which is analogous to the reciprocal of a scalar. For example, in scalar arithmetic the reciprocal of x has the property that

$$(x)(\text{reciprocal of } x) = 1 \tag{13.37}$$

Likewise, in matrix algebra, the inverse of A has the property that

$$A(\text{inverse } A) = I \tag{13.38}$$

The inverse A is denoted symbolically by A^{-1}. Since division is not defined, it is improper to say that $A^{-1} = 1/A$. If both sides of equation (13.35) are multiplied by A

$$(AA^{-1})A = IA$$

$$A(A^{-1}A) = A \tag{13.39}$$

Therefore,

$$A^{-1}A = I \tag{13.40}$$

If Equation (13.39) is divided by det A, one obtains

$$A\left(\frac{adj\ A}{\det A}\right) = I \tag{13.41}$$

and it is clear from equation (13.40) that

$$A^{-1} = \left(\frac{adj\ A}{\det A}\right) \tag{13.42}$$

If det $A = 0$, as in the previous example, the matrix is said to be *singular* and it cannot possess an inverse. However, it may possess an adjoint as demonstrated.

The transpose and inverse matrices follow an interesting *reversal rule* given in the equation below. If $C = AB$, one may write

$$C^T = A^T B^T \tag{13.43}$$

and

$$C^{-1} = A^{-1} B^{-1} \tag{13.44}$$

13.3 Rank and Solution of Linear Equations

The *rank* of matrix A of order $m \times n$ is equal to the order of the largest square matrix contained in A whose determinant is non-vanishing, i.e., the maximum number of independent rows or columns. This square matrix is obtained by striking out rows and columns of A. As an example of the determination of rank, consider the following matrix.

$$A = \begin{bmatrix} 2 & 1 & 2 & 3 \\ 1 & 1 & 2 & 3 \\ 0 & 2 & 4 & 6 \end{bmatrix} \tag{13.45}$$

All third order square matrices contained in A have at least two proportional columns, hence by the above rule, all third order determinants must vanish. The matrix A has many non-vanishing second order determinants, the elements of one being a_{11}, a_{21}, a_{14}, and a_{24}; hence, it is of rank 2.

According to Sylvester's law of nullity, given two square matrices A and B of order n and of rank r_a and r_b, respectively, then the rank of the product $A\ B$, designated r_{ab}, must satisfy the following two inequalities.

$$r_{ab} \geq r_a + r_b - n \tag{13.46}$$

$$r_{ab} \leq \text{smaller of the ranks of } r_a \text{ and } r_b$$

In order to illustrate Sylvester's law of nullity consider the following example

$$A = \begin{bmatrix} 1 & 0 & 0 \\ 0 & 2 & 0 \\ 0 & 0 & 1 \end{bmatrix} \text{ of } r_A = 3 \tag{13.47}$$

$$B = \begin{bmatrix} 1 & 0 & 0 \\ 0 & 1 & 0 \\ 0 & 0 & 1 \end{bmatrix} \text{ of } r_B = 2 \qquad (13.48)$$

$$AB = \begin{bmatrix} 1 & 0 & 0 \\ 0 & 2 & 0 \\ 0 & 0 & 0 \end{bmatrix} \text{of } r_{AB} = 2 \qquad (13.49)$$

$$r_{AB} = 2 \geq 3 + 2 - 3; \; r_{AB} = 2 \leq 2 \qquad (13.50)$$

The above is also valid when A and B are not square; that is, for any two matrices A and B. Thus,

$$r_{AB} \leq \text{smaller of rank } r_A \text{ and } r_B \qquad (13.51)$$

It can also be shown that

$$r_{AB} = r_B \qquad (13.52)$$

if A is a non-singular square matrix.

13.4 Linear Equations

Consider the following set of linear equations

$$\begin{aligned} a_{11}x_1 + a_{12}x_2 + \ldots + a_{1n}x_n &= b_1c_1 \\ a_{21}x_1 + a_{22}x_2 + \ldots + a_{2n}x_n &= b_2c_2 \\ &\ldots \\ a_{m1}x_1 + a_{m2}x_2 + \ldots + a_{mn}x_n &= b_mc_m \end{aligned} \qquad (13.53)$$

Equation (13.53) may be written in matrix form as

$$\begin{bmatrix} a_{11} & a_{12} & \ldots & a_{1n} \\ a_{21} & a_{22} & \ldots & a_{2n} \\ \ldots & & & \\ a_{m1} & a_{m2} & \ldots & a_{mn} \end{bmatrix} \begin{bmatrix} x_1 \\ x_2 \\ \ldots \\ x_n \end{bmatrix} = \begin{bmatrix} b_1c_1 \\ b_2c_2 \\ \ldots \\ b_mc_m \end{bmatrix} \qquad (13.54)$$

Or more concisely as

$$Ax = c \qquad (13.55)$$

One notes that, in the form of equation (13.55), the matrix A operates on the unknown vector x to produce vector c. This operation is called a *linear transformation*, and A is the *coefficient matrix* for the transformation.

Another matrix of theoretical importance in determining the solvability of a linear transformation is the augmented coefficient matrix, denoted *aug A*.

$$aug\ A = \begin{bmatrix} a_{11} & a_{12} & \cdots & a_{1n} & b_1 \\ a_{21} & a_{22} & \cdots & a_{2n} & b_2 \\ \cdots & \cdots & \cdots & \cdots & \cdots \\ a_{m1} & a_{m2} & \cdots & a_{mn} & b_m \end{bmatrix} \quad (13.56)$$

References

1. J. Famularo, personal communication to L. Theodore (with permission), Engelwood Cliffs, NJ, 1973.
2. C. Wiley, *Advanced Engineering Mathematics*, McGraw-Hill, New York City, NY, 1951.

14

Laplace Transforms

Contributing Author: Megan Menzel

The Laplace transform of a function $f(t)$ is denoted by $F(s)$ and is represented by the following equation:

$$L\{f(t)\} = F(s) = \int_0^\infty f(t)e^{-st}\,dt \tag{14.1}$$

The Laplace transform of a function $f(t)$ is defined for positive values of t as a function of the new variable s. The transform exists if $f(t)$ satisfies the following three conditions:

1. $f(t)$ is continuous or piecewise continuous in any interval $t_1 \leq t \leq t_2$, where $t_1 > 0$.
2. $t^n|f(t)|$ is bounded near $t = 0$ when approached from positive values of t for some number n where $n < 1$.
3. e^{-s^*t} is bounded for large values of t for some number s^*.

The function $f(t)$ is *piecewise continuous* in the range $t_1 \leq t \leq t_2$. if it is possible to divide the range into a finite number of intervals in such a way that $f(t)$ is continuous within each interval and approaches finite values as either end of the interval is approached from within the interval. Thus, a piecewise continuous function may have a number of finite discontinuities.

The Laplace transform method can reduce the solution of an ordinary or partial differential equation to essentially an algebraic procedure. In effect, the two advantages of the Laplace transformation in solving differential equations are the way in which it reduces the problem to one in algebra and by the automatic way in which it takes care of initial conditions without the necessity of constructing a general solution and then specifying the arbitrary constants which it contains. Material is presented in this chapter that allows one to employ tables of transforms which can be used, like tables of logarithms, to manipulate functions, and by means of which one can recover the proper function of t from its Laplace transform.

Five sections complement the presentation of this chapter. Section numbers and subject titles follow:

14.1: Laplace Transform Theorems
14.2: Laplace Transforms of Special Functions
14.3: Splitting Proper Rational Fractions into Partial Fractions
14.4: Converting an Ordinary Differential Equation into an Algebraic Equation
14.5: Converting a Partial Ordinary Differential Equation into an Ordinary Differential Equation

14.1 Laplace Transform Theorems

The utility of the Laplace transform is based primarily upon the following four theorems.

1. The Laplace transform of a sum is the sum of the transforms of the individual terms, i.e.,

$$\begin{aligned} L\{f(t)+g(t)\} &= \int_0^\infty (f+g)e^{-st}\,dt \\ &= \int_0^\infty fe^{-st}\,dt + \int_0^\infty ge^{-st}\,d = \mathcal{L}(f)+\mathcal{L}(g) \end{aligned} \tag{14.2}$$

2. The Laplace transform of a constant times a function is the constant times the transform of the function, i.e.,

$$\begin{aligned} L(cf) &= \int_0^\infty (cf)e^{-st}\,dt \\ &= c\int_0^\infty fe^{-st}\,dt = c\mathcal{L}(f) \end{aligned} \tag{14.3}$$

3. If $f(t)$ is a function of exponential order which is continuous and whose derivative is continuous, the Laplace transform of the derivative of $f(t)$, i.e. $f'(t)$, is given by

$$L(f') = \int_0^\infty f' e^{-st} dt \tag{14.4}$$

4. If $f(t)$ is of exponential order and continuous, then the Laplace transform of $\int_a f(t)dt$ is given by the formula

$$L\left[\int_a^t f(t)dt\right] = \frac{1}{s}L(f) + \frac{1}{s}\int_a^0 f(t)dt \tag{14.5}$$

Theorem 4, like Theorem 3, is also useful in finding the *inverses* of transforms. Thus, if

$$L(F) = \frac{1}{s}\phi(s) \tag{14.6}$$

and if $f(t)$ is a function such that

$$L(f) = \phi(s) \tag{14.7}$$

then by Theorem 4

$$\begin{aligned} L\left[\int_0^t f(t)dt\right] &= \frac{1}{s}L(f) \\ &= \frac{1}{s}\phi(s) = L(F) \end{aligned} \tag{14.8}$$

Therefore,

$$F(t) = \int_0^t f(t)dt \tag{14.9}$$

In effect, if one chooses, one may neglect the factor $1/s$ in a transform, determine the inverse $f(t)$ of what remains, and then integrate $f(t)$ from 0 to t to find the required function $F(t)$. In the same way, the factor $1/s^k$ can be neglected in a transform, provided that the inverse of the remaining portion is integrated k times from 0 to t.

14.2 Laplace Transforms of Specific Functions

The most difficult problem in practical environmental engineering and science applications involving Laplace transforms is the determination of the function which corresponds to a known transform. A short list of transforms is provided below in Table 14.1. This material, together with additional techniques to be described shortly, will handle most analytical problems. When all other methods of finding the inverse transform fail, numerical integration [1,2] techniques may be employed. These procedures are described in Part III of this book. Fortunately, the inversion process is unique; i.e., if one function $f(t)$ corresponding to the known transform can be found, it is the correct one.

Table 14.1 Special Laplace Transforms.

	$f(s)$	$F(t)$
1.	$\frac{1}{s}$	1
2.	$\frac{1}{s^2}$	t
3.	$\frac{1}{s^n}; n=1, 2, 3, \ldots$	$\frac{t^{n-1}}{(n-1)!}; \; 0!=1$
4.	$\frac{1}{s-a}$	e^{at}
5.	$\frac{1}{(s-a)^n}; n=1, 2, 3, \ldots$	$\frac{t^{n-1}e^{at}}{(n-1)!}; \; 0!=1$
6.	$\frac{1}{s^2+a^2}$	$\frac{\sin at}{a}$
7.	$\frac{s}{s^2+a^2}$	$\cos at$
8.	$\frac{1}{(s-b)^2+a^2}$	$\frac{e^{bt}\sin at}{a}$
9.	$\frac{s-b}{(s-b)^2+a^2}$	$e^{bt}\cos at$
10.	$\frac{1}{s^2-a^2}$	$\frac{\sinh at}{a}$

(*Continued*)

Table 14.1 Cont.

	$f(s)$	$F(t)$
11.	$\frac{s}{s^2-a^2}$	$\cos h\, at$
12.	$\frac{1}{(s-b)^2-a^2}$	$\frac{e^{bt}\sinh at}{a}$
13.	$a f_1(s) + b f_2(s)$	$a f_1(t) + b f_2(t)$
14.	$f(s/a)$	$a f(at)$
15.	$f(s-a)$	$e^{at} f(t)$
16.	$e^{-as} f(s)$	$u(t-a) = \begin{cases} f(t-a) & t>a \\ 0 & t<a \end{cases}$
17.	$s f(s) - F(0)$	$f'(t)$
18.	$s^2 f(s) - sF(0) - F'(0)$	$f''(t)$
19.	$f'(s)$	$-t f(t)$
20.	$f'(s)$	$t^2 f(t)$
21.	$f^{(n)}(s)$	$(-1)^n t^n f(t)$
22.	$\frac{f(s)}{s}$	$\int_0^t f(u)du$
23.	$\frac{f(s)}{s^n}$	$\int_0^t \cdots \int_0^t f(u)du^n = \int_0^t \frac{(t-u)^{n-1}}{(n-1)!} f(u)du$
24.	$f(s)g(s)$	$\int_0^t f(u)g(t-u)du$

14.3 Splitting Proper Rational Fractions into Partial Fractions

This section presents useful rules for inverting Laplace transforms back to the time domain. An expression of the form P/Q, where P and Q are polynomials in s, is called a rational fraction. If P is of a lower degree than Q, P/Q is called a *proper* fraction. If P is not of a lower degree than Q, then P/Q is called an *improper* fraction. In general, if P and Q are two rational, integral algebraic functions of s, and the fraction P/Q is expressed as the algebraic sum of a simpler fraction with denominators containing the factors of Q, the fraction P/Q can be resolved into partial fractions. The following rules are applicable for splitting a proper rational fraction into partial fractions.

1. Corresponding to every non-repeated linear factor $as + b$ of the denominator, there exists a partial fraction

$$\frac{A}{as+b} \tag{14.1}$$

where A is a constant.

2. Corresponding to every repeated linear factor $(as + b)^p$ of the denominator, there exists fractions of the form

$$\frac{A_1}{as+b}+\frac{A_2}{(as+b)^2}+\ldots+\frac{A_p}{(as+b)^p} \tag{14.2}$$

where $A_1, A_2, \ldots, A_p$ are constants.

3. Corresponding to every non-repeated irreducible quadratic factor $as^2 + bs + c$ of the denominator, there exists a fraction of the form

$$\frac{As+B}{as^2+bs+c} \tag{14.3}$$

where A and B are constants.

4. Corresponding to every repeated quadratic factor $(as^2 + bs + c)^p$ when $as^2 + bs + c$ has no linear factors, there exist fractions of the form

$$\frac{A_1s+B_1}{as^2+bs+c}+\frac{A_2s+B_2}{(as^2+bs+c)^2}+\ldots+\frac{A_ps+B_p}{(As^2+bs+c)^p} \tag{14.4}$$

Finally, one must express an improper fraction as the sum of an integral function and a proper fraction, and then split the proper fraction into partial fractions.

Illustrative Example 14.1

Convert

$$\frac{5s+3}{(s+1)^2(2s+1)}$$

into partial fractions.

Solution

As noted, it is important to resolve partial fractions prior to taking the inverse of the Laplace transform. Let

$$\frac{5s+3}{(s+1)^2(2s+1)} = \frac{A}{s+1} + \frac{B}{(s+1)^2} + \frac{C}{2s+1}$$

Setting $s = -1$ and $s = -0.5$ in succession, one obtains $B = C = 2$. Equating the known absolute terms on both sides of the above equation leads to

$$3 = A + B + C \text{ or } A = -1$$

Therefore,

$$\frac{5s+3}{(s+1)^2(2s+1)} = \frac{-1}{s+1} + \frac{2}{(s+1)^2} + \frac{2}{2s+1}$$

14.4 Converting An Ordinary Differential Equation (ODE) Into An Algebraic Equation

As noted earlier, the Laplace transform method of analysis can be employed to convert an ordinary differential equation into an algebraic equation. The illustrative example that follows, adopted from Mickley *et al.* [2], illustrates this method. The next section outlines how a partial differential equation can be "reduced" to an ordinary differential equation.

Illustrative Example 14.2

Consider the ordinary differential equation

$$\frac{dT}{dt} - T = e^{at}$$

with the boundary condition

$$T = -1 \text{ at } t = 0$$

A solution is desired which is valid for positive values of t.

Solution

The Laplace transform method proceeds by multiplying both sides of the initial equation by $e^{-pt}\,dt$ (provided that $p > a$) and integrating the result from zero to infinity:

$$\int_0^{\infty} e^{-pt}\frac{dT}{dt}dt - \int_0^{\infty} e^{-pt}T\,dt = \int_0^{\infty} e^{-pt}e^{at}\,dt$$

$$\int_0^\infty e^{-pt}e^{at}\,dt = -\left[\frac{e^{-(p-a)t}}{p-a}\right]_0^\infty = \frac{1}{p-a}$$

The first integral above may be integrated by parts:

$$\int_0^\infty e^{-pt}\frac{dT}{dt}dt = [e^{-pt}T]_0^\infty + p\int_0^\infty e^{-pt}T\,dt$$

$$= 1 + p\int_0^\infty e^{-pt}T\,dt$$

This applies provided that $e^{-pt}T \to 0$ as $t \to \infty$, following the introduction of the boundary condition. When the above equations are combined one obtains

$$(p-1)\int_0^\infty e^{-pt}T\,dt = \frac{1}{p-a} - 1$$

or

$$\int_0^\infty e^{-pt}T\,dt = \frac{a+1-p}{(p-1)(p-a)}$$

The problem that remains is to determine the function of T whose Laplace transform is given by the right hand side (RHS) of the above equation. First, expand the right-hand side by the method of partial fractions to give

$$\int_0^\infty e^{-pt}T\,dt = \frac{a+1-p}{(p-1)(p-a)}$$

$$= \left(\frac{1}{a-1}\right)\left(\frac{1}{p-a}\right) - \left(\frac{a}{a-1}\right)\left(\frac{1}{p-1}\right)$$

It then follows that

$$T = \left(\frac{1}{a-1}\right)e^{at} - \left(\frac{a}{a-1}\right)e^{t}$$

This is the solution to the above ordinary differential equation. Direct substitution and comparison with the initial conditions demonstrates that this is in fact the solution of this ODE.

14.5 Converting a Partial Differential Equation (PDE) into an Ordinary Differential Equation (ODE)

The following partial differential equation describes the temperature profile in an infinite solid medium [3].

$$\frac{\partial T}{\partial t} = a\frac{\partial^2 T}{\partial x^2} \tag{14.5}$$

The solution? The separation of variables method [2] is usually applicable to finite-media systems. However, this technique often produces meaningless results when used for semi-infinite or infinite media. Equation (14.5) can be solved by using Laplace transforms (or Fourier integrals) [3]. An outline of the Laplace transform method of solution is presented below.

Begin by multiplying both sides of Equation (14.5) by e^{-pt} and integrating from 0 to ∞.

$$\int_0^\infty e^{-pt}\frac{\partial T}{\partial t}dt = \int_0^\infty a\frac{\partial^2 T}{\partial x^2}e^{-pt}dt \tag{14.6}$$

The left-hand side (LHS) of (14.6) is integrated by parts to give

$$\begin{aligned} LHS &= e^{-pt}\,T\Big|_0^\infty + p\int_0^\infty e^{-pt}T\,dt \\ &= -T(0) + p\overline{T}; T(0) = \text{initial temperature} \end{aligned} \tag{14.7}$$

where the integral

$$\int_0^\infty e^{-pt}\,T\,dt \tag{14.8}$$

now is represented by $\overline{T}$ and defined as the Laplace transform of T. One notes that $\overline{T}$ is a function of p and T. The RHS of equation (14.6) becomes

$$\begin{aligned} RHS &= a\frac{\partial^2}{\partial x^2}\int_0^\infty Te^{-pt}dt \\ &= a\frac{\partial^2 \overline{T}}{\partial x^2} = a\frac{d^2\overline{T}}{dx^2} \end{aligned} \tag{14.9}$$

The resulting equation is

$$-T(0)+p\bar{T}=\propto \frac{d^2\bar{T}}{dx^2} \tag{14.10}$$

Note that this operation has converted the PDE in Equation (14.5) to an ODE and eliminated time as a variable. It is an order of magnitude easier to integrate. This ODE can now be solved subject to revised boundary conditions. The remainder of this solution is not presented, but is available in the literature [3]

References

1. L. Theodore, *Chemical Engineering: The Essential Reference*, McGraw-Hill, New York City, NY, 2014.
2. H. Mickley, T. Sherwood, and C. Reed, *Applied Mathematics in Chemical Engineering*, 2nd ed., McGraw-Hill, New York City, NY, 1957.
3. L. Theodore, *Transport Phenomena for Engineers*, Theodore Tutorials, East Williston, NY, originally published by the International Textbook Company, Scranton, PA, 1971.

Part III
NUMERICAL ANALYSIS

Early in one's career, the environmental engineer/scientist learns how to use equations and mathematical methods to obtain exact analytical answers to a large range of relatively simple problems. Unfortunately, these techniques are often inadequate for solving real-world problems, although the reader should note that one rarely needs exact answers in technical practice. Most real-world applications involving calculations are usually inexact because they have been generated from data or parameters that are measured, and hence represent only approximations. What one is likely to require in a realistic situation is not an exact answer but rather one having reasonable accuracy from an engineering point of view.

Numerical methods provide a step-by-step procedure that ultimately leads to an answer and a solution to a particular problem. The method usually requires a large number of calculations and is therefore ideally suited for digital computation. The subject of numerical methods was taught in the past as a means of providing environmental engineers and scientists with ways to solve complicated mathematical expressions that they could not solve otherwise. However, with the advent of computers, these solutions have become readily obtainable. Because of the breadth of the subject matter, the reader should note that only the seven numerical methods of primary interest to the environmentalist receive treatment in this part. Details on the remaining methods are available in the literature.

There are seven chapters in Part III. The chapter numbers and accompanying titles are listed below:

Chapter 15: Trial and Error Solutions
Chapter 16: Non-linear Algebraic Equations
Chapter 17: Simultaneous Linear Algebraic Equations
Chapter 18: Differentiation
Chapter 19: Integration
Chapter 20: Ordinary Differential Equations
Chapter 21: Partial Differential Equations

15

Trial-and-Error Solutions

This introductory (and abbreviated) chapter to Part III is primarily concerned with trial-and-error solutions. Although most of the material presented was developed in earlier times, it does serve as an excellent introduction to the more sophisticated numerical methods employed today.

Five sections complement the presentation of this chapter. Section numbers and subject titles follow:

15.1: Square Root Calculations
15.2: Quadratic and Cubic Equations
15.3: Two or More Simultaneous Non-Linear Equations
15.4: Higher Order Algebraic Equations
15.5: Other Approaches

15.1 Square Root Calculations

The following iterative procedure can be employed to find the square root of a positive real number, n. Let x_0 be the first guess. First compute n/x_0. If $n/x_0 = x_0$, the calculation is completed since $x_0^2 = n$. However, the probability that this will occur is small. If $n/x_0 < x_0$ so that $n < x_0^2$, one can conclude that x_0 is too large. If x_0 is too large, $n/x_0 < \sqrt{n}$. Thus,

$$\frac{n}{x_0} < \sqrt{n} < x_0 \tag{15.1}$$

Similarly, one can establish that if $n/x_0 > x_0$ then $x_0 < \sqrt{n}$ and $x_0 < \sqrt{n} < n/x_0$, i.e.,

$$\frac{n}{x_0} \leq n \leq x_0 \tag{15.2}$$

In either solution ($x_0 > \sqrt{n}$ or $x_0 < \sqrt{n}$) the square root of n is between n/x_0 and x_0. Now guess again:

$$x_1 = \frac{1}{2}\left[x_0 + \left(\frac{n}{x_0}\right)\right] \tag{15.3}$$

This average at least appears reasonable. This procedure will converge to $\sqrt{n}$ as it is used repetitively employing Equation (15.4).

$$x_{i+1} = \frac{1}{2}\left[x_i + \left(\frac{n}{x_i}\right)\right] \tag{15.4}$$

Illustrative Example 15.1

Find $\sqrt{90}$ using $x_0 = 1$ and Equation (15.4).

Solution

Calculations subsequent to $x_0 = 1$ are provided below.

$$x_1 = \frac{1}{2}\left(1 + \frac{90}{1}\right) = 45.5$$

$$x_2 = \frac{1}{2}\left(45.5 + \frac{90}{45.5}\right) = 23.739$$

$$x_3 = \frac{1}{2}\left(23.739 + \frac{90}{23.739}\right) = 13.7651$$

$$x_4 = \frac{1}{2}\left(13.7651 + \frac{90}{13.7651}\right) = 10.15169$$

$$x_5 = \frac{1}{2}\left(10.15169 + \frac{90}{10.15169}\right) = 9.508604$$

$$x_6 = \frac{1}{2}\left(9.508604 + \frac{90}{9.508604}\right) = 9.4868579$$

$$x_7 = \frac{1}{2}\left(9.4868579 + \frac{90}{9.4868579}\right) = 9.4868330$$

$$x_8 = \frac{1}{2}\left(9.4868330 + \frac{90}{9.4868330}\right) = 9.4868329$$

The last approximation, x_8, is accurate to six significant figures.

Another procedure may also be employed. It is based on a starting value for $\sqrt{n}$ of $x_0 = n/2$. For $\sqrt{90}$, $n/2 = 45 = x_0$. This approach produces the following results.

$$x_1 = \frac{1}{2}\left(45 + \frac{90}{45}\right) = 23.5$$

$$x_2 = \frac{1}{2}\left(23.5 + \frac{90}{23.5}\right) = 13.66$$

$$x_3 = \frac{1}{2}\left(13.66 + \frac{90}{13.66}\right) = 10.124$$

$$x_4 = \frac{1}{2}\left(10.124 + \frac{90}{10.124}\right) = 9.5069$$

$$x_5 = \frac{1}{2}\left(9.5069 + \frac{90}{9.5069}\right) = 9.48685$$

$$x_6 = \frac{1}{2}\left(9.48685 + \frac{90}{9.48685}\right) = 9.486833$$

$$x_7 = \frac{1}{2}\left(9.486833 + \frac{90}{9.486833}\right) = 9.4868329$$

This produced a result accurate to six figures in seven iterations. This method was programmed in the past as a subroutine on digital computers.

Useful iterative methods should have the following requirements:

1. A means of making a satisfactory first guess; in many applications, physical or other considerations may provide this guess
2. A means of systematically improving on the previous approximation
3. A criterion (or choice of several criteria) for stopping the iteration when sufficient accuracy has been obtained

Brute force trial-and-error procedures are employed and occasionally used in not only other algebraic equations but also in the solution of ordinary and partial differential equations, simultaneous linear equations, and simultaneous nonlinear equations. This chapter will primarily be limited to a discussion of iterative methods used in the solution of algebraic and transcendental equations.

15.2 Quadratic and Cubic Equations

A problem commonly encountered in environmental engineering and scientific practice is that of determining the root of a non-linear equation of the form $f(x) = 0$. In this section, two methods of finding the *real* root of such equations are considered.

1. The trial-and-error method
2. The linear-interpolation method

The Newton-Raphson method is addressed in the next chapter.

If $f(x)$ is a polynomial with real coefficients, there are several ways to find both the *real* and *complex* roots of the polynomial. One such method for solving $f(x) = 0$ involves a trial-and-error calculation. In the simplest of the trial-and-error approaches, one calculates values of $f(x)$ for successive values of x until a sign change occurs for $f(x)$, i.e., a positive (+) value to a negative (−) value, or vice versa. The sign change indicates that the root solution has been passed. A closer approximation of the root may then be obtained by reverting to the last x value preceding the sign change and, beginning with this x value, again determining values of $f(x)$ for successive values of x using a smaller increment than was used initially until the sign of $f(x)$ changes again. This procedure is repeated with progressively smaller increments of x until a sufficiently accurate value of the root is obtained. If additional roots are desired, the incrementation of x can be continued until the next root is approximately located by another sign change of $f(x)$, etc.

As with many numerical methods, care must be exercised in selecting the initial value by which x is to be incremented so that a root is not passed by in a situation when two roots are in close proximity. However, this is usually not troublesome if fairly small increments are used in the initially selected value.

Consider first the quadratic equation

$$ax^2 + bx + c = 0 \tag{15.5}$$

The above procedures may be employed to find the solution(s), though the equation can also be solved algebraically to obtain an exact solution.

$$x = \frac{-b \pm \sqrt{b^2 - 4ac}}{2a} \tag{15.6}$$

If a, b, c are real, and if D is defined as

$$D = b^2 - 4ac \tag{15.7}$$

then the solutions are

1. Real and unequal if $D > 0$
2. Real and equal if $D = 0$
3. Complex conjugate if $D < 0$

In addition, if x_1, x_2 are the roots, then

$$x_1 + x_2 = -\frac{b}{a} \tag{15.8}$$

$$x_1 x_2 - \frac{c}{a} \tag{15.9}$$

Note also that if $b >>> c$, one root is approximately b/a.

Consider the second case of a cubic equation

$$x^3 + a_1 x^2 + a_2 x + a_3 = 0 \tag{15.10}$$

The trial-and-error method suggested above may again be employed. However, an algebraic (exact) solution exists. The following four terms are first defined:

$$Q = \frac{3a_2 - a_1^2}{9} \tag{15.11}$$

$$R = \frac{9a_1 a_2 - 27a_3 - 2a_1^3}{54} \tag{15.12}$$

$$S = \sqrt{R + \sqrt{Q^3 + R^2}} \tag{15.13}$$

$$T = \sqrt{R - \sqrt{Q^3 + R^2}} \tag{15.14}$$

The solution takes the form

$$x_1 = S + T - \frac{1}{3}a_1 \tag{15.15}$$

$$x_2 = -\frac{1}{2}(S+T) - \frac{1}{3}a_1 + \frac{1}{2}i\sqrt{3}(S-T) \tag{15.16}$$

$$x_3 = -\frac{1}{2}(S+T) - \frac{1}{3}a_1 - \frac{1}{2}i\sqrt{3}(S-T) \tag{15.17}$$

If a_1, a_2, a_3 in Equation (15.7) are real and if

$$D = Q^3 + R^2 \tag{15.18}$$

then:

1. One root is real and two are complex conjugate if $D > 0$
2. All roots are real and at least two are equal if $D = 0$
3. All roots are real and unequal if $D < 0$

If $D < 0$, one obtains

$$x_1 = 2\sqrt{-Q}\cos\left(\frac{1}{3}\theta\right) \tag{15.19}$$

$$x_2 = 2\sqrt{-Q}\cos\left(\frac{1}{3}\theta + 120^\circ\right) \tag{15.20}$$

$$x_3 = 2\sqrt{-Q}\cos\left(\frac{1}{3}\theta + 240^\circ\right) \tag{15.21}$$

where

$$\cos\theta = -\frac{R}{\sqrt{-Q^3}} \tag{15.22}$$

In addition

$$x_1 + x_2 + x_3 = -a_1 \tag{15.23}$$

$$x_1x_2 + x_2x_3 + x_3x_1 = a_2 \tag{15.24}$$

$$x_1x_2x_3 = -a_3 \tag{15.25}$$

where x_1, x_2, x_3 are the three roots.

Illustrative Example 15.2

Solve the cubic equation

$$3x^2 + 4x^2 + 6x - 3 = 0$$

Solution

Here is one approach to solving this equation. Substituting $x = 0$, the left-hand-side (LHS) of the equation equals −3, while if $x = 1$, it equals +2. Because of the continuity of polynomial functions, one of the roots lies between 0 and 1. The graphical interpolation shown in Figure 15.1 can be used to *approximate* the correct answer.

The trial-and-error calculation may, of course, be continued until an answer of the desired accuracy is obtained. Note that only one root of the equation has been located. More can be found by repeating the procedure. It is often easy to pick out the proper root in problems involving actual physical phenomena.

15.3 Two or More Simultaneous Non-Linear Equations

Two or more simultaneous equations of the second degree or higher may also be solved conventionally by trial-and-error techniques. The procedure to be used varies with each problem, although for second-order equations having two unknowns, one can always assume values for one unknown and then calculate the

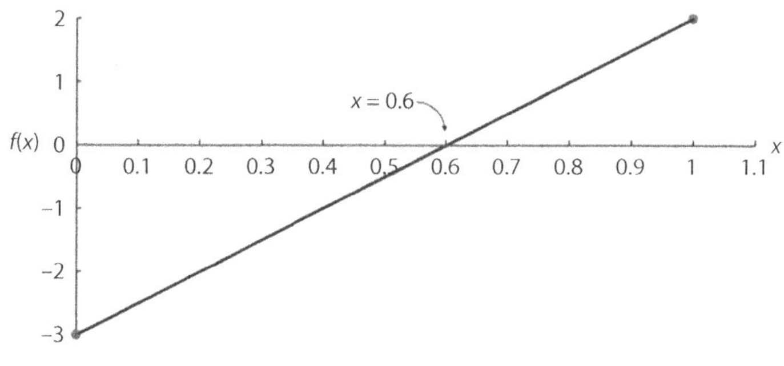

Figure 15.1 Graphical interpolation.

values of the other one. In a graphical solution, the intersection of the curves of the two equations marks their simultaneous solution.

Illustrative Example 15.3

Equations (1) and (2) below describe the equilibrium conversion (x and y) of two chemical reactions [1] provided below.

$$2 = \frac{(x-3)(xy)}{(x-y)(4-x)} \tag{1}$$

$$1.33 = \frac{y^2 + yx^2}{(x-y-4)(x-6)} \tag{2}$$

Solve the equations.

Solution

Although it is possible to solve Equations (1) and (2) analytically, a graphical trial-and-error solution may be expedient. The trial-and-error approach can be conducted in a number of ways. One method is to assume values for x and calculate the corresponding y as shown in Table 15.1. These values can also be plotted and the intersection of the two lines provides the solution of the equations. Referring to Table 15.1, can the reader figure out the solution?

15.4 Higher Order Algebraic Equations

Consider the following n^{th} order algebraic equation

$$f(x) = a_n x^n + a_{n-1}x^{n-1} + \ldots + a_1 x + a_0 = 0 \tag{15.26}$$

Table 15.1 Calculated results for Illustrative Example 15.3.

x (assumed)	*y* (calculated)	
	Equation (1)	Equation (2)
0	0	10.8
1	1.5	7.5
1.5	2.8	5.7
2	4	4
2.5	4.3	2.4
3	3	0.7
4	0	

The solution to this equation is demonstrated by solving the following 3rd order equation

$$f(x) = x^3 + 6.6x^2 - 29.05x + 22.64 = 0 \quad (15.27)$$

Divide through by x, drop $22.64/x$, and solve the resulting quadratic equation. This provides a starting value for the trial-and-error solution.

$$x^2 + 6.6x - 29.05 = 0 \quad (15.28)$$

The solution is

$$x = -9.6 \quad (15.29)$$

One now notes that

$$f(-9.6) = 22.04 \quad (15.30)$$

This provides an excellent starting value in an attempt to obtain a solution to Equation (15.27)

15.5 Other Approaches

Environmental engineers and scientists occasionally solve equations which have no simple analytical solution. For example, consider the equation [2]

$$e^{x^2} - x = 14.2 \quad (15.31)$$

Once again, the most convenient way to solve this equation is to assume values of x until the left side of the equation equals the right. However, the equation should be inspected carefully to determine the general effect of a variation in x. Perhaps one term is substantially smaller than the other so that it may be neglected for a first estimation of x. In Equation (15.31), x is probably small compared to e^{x^2}. As a first approximation for x, assume $x = 0$ in Equation (15.31) so that $e^{x^2} = 14.2$ and $x = 1.63$. But

$$e^{x^2} - x = 14.2 - 16.3 = 12.6 \neq 14.2 \quad (15.32)$$

Inspection of Equation (15.31) indicates that a somewhat larger value of x should be used. For the second trial, set $x = 1.70$. Then

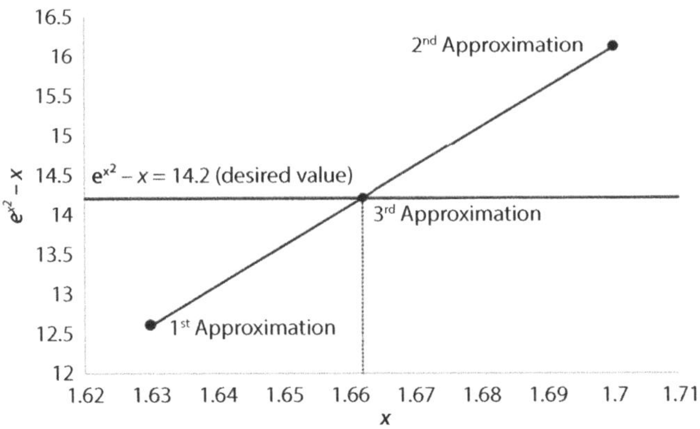

Figure 15.2 Graphical estimation in a trial-and-error procedure.

$$e^{1.7^2} - 1.7 = 17.8 - 1.7 = 16.1 \neq 14.2 \tag{15.33}$$

The second trial assumed an x which is too large. The two points can be plotted on Figure 15.2. Although the function is highly non-linear, a straight line is drawn between the fist and second approximations. The third approximation is taken at the intersection of this line and the line $(e^{x^2} - x = 14.2)$. Applying a third approximation with $x = 1.665$ gives

$$e^{(1.665)^2} - 1.665 = 15.9 - 1.665 = 14.2 \tag{15.34}$$

The third trial has effectively done the job. In many environmental engineering calculations it is not necessary to plot the trials for a new estimate. The experienced engineer can often make mental estimates and interpolations.

References

1. L. Theodore, *Chemical Reactor Analysis and Applications for the Practicing Engineer*, John Wiley & Sons, Hoboken, NJ, 2011.
2. J. Famularo, personal communication to L. Theodore (with revision), Englewood Cliffs, NJ, 1970.

16

Nonlinear Algebraic Equations

Many areas of environmental engineering and scientific analysis require the use of efficient techniques for determining the roots of algebraic (and/or transcendental) equations. It is for this reason that the subject of the solution to a nonlinear algebraic equation is considered in this chapter. Although several algorithms are available in the literature, this presentation will focus on the *Newton-Raphson* (NR) method of evaluating the root(s) of a nonlinear algebraic equation.

Three sections complement the presentation of this chapter. Section numbers and subject titles follow:

16.1: The Reguli-Falsi (False Position) Method
16.2: The Newton-Raphson Method
16.3: Newton's Second Order Method

16.1 The Reguli-Falsi (False Position) Method

The Reguli-Falsi method (referred to by some as the False Position linear-interpolation method) involves obtaining values for two roots, i.e., values (x_1, y_1) and (x_2, y_2) to the equation $f(x) = 0$ by trial-and-error. From a plot of x vs. y and applying similar "triangles" to a third root (x_3, y_3),

$$\frac{y_2 - y_1}{x_2 - x_1} = \frac{y_3 - y_1}{x_3 - x_1} \tag{16.1}$$

For $y_3 = 0$, one obtains

$$x_3 = \frac{x_1 y_2 - x_2 y_1}{y_2 - y_1} = x_1 - \frac{y_1(x_2 - x_1)}{y_2 - y_1} \tag{16.2}$$

If y is expressed as $y = f(x)$, the above equation becomes

$$x_3 = \frac{x_1 f(x_2) - x_2 f(x_1)}{f(x_2) - f(x_1)} = x_1 - \frac{f(x_1)(x_2 - x_1)}{f(x_2) - f(x_1)} \tag{16.3}$$

Theodore [1] also provides the following "algorithm".

1. Assume a value of x_1, (hopefully near to the actual value of the root).
2. Evaluate $f(x_1)$.
3. Find an x_2 such that the sign of $f(x_2)$ is different from that of $f(x_1)$.
4. Plot $f(x)$ vs x.
5. Draw a straight line between the two points and use the point where the chord crosses the axis at $f(x) = 0$ as x_3.
6. Evaluate $f(x_3)$.
7. Use a new chord, $f(x_3)$ to $f(x_2)$ to evaluate x_4.
8. The procedure is repeated employing Equation (16.4).

$$x_{n+1} = \frac{x_n f(x_n) - x_n f(x_{n-1})}{f(x_n) - f(x_{n-1})} \tag{16.4}$$

9. Continue until

$$|x^{n+1} - x^n| < \varepsilon_1 \tag{16.5}$$

or

$$|f(x^{n+1}) - f(x^n)| < \varepsilon_2 \tag{16.6}$$

Note: ε (error tolerance) will be defined later in Equation (16.20)

Thus, if values of $f(x)$ for two relatively close trial values of x are known and that one of the functional values is positive while the other is negative, at least one root will be contained within this range of x. This is illustrated in Figure 16.1.

There is also a modification to the False Position method [1]. If

$$x_1 = 1^{st} \text{ value of } x$$

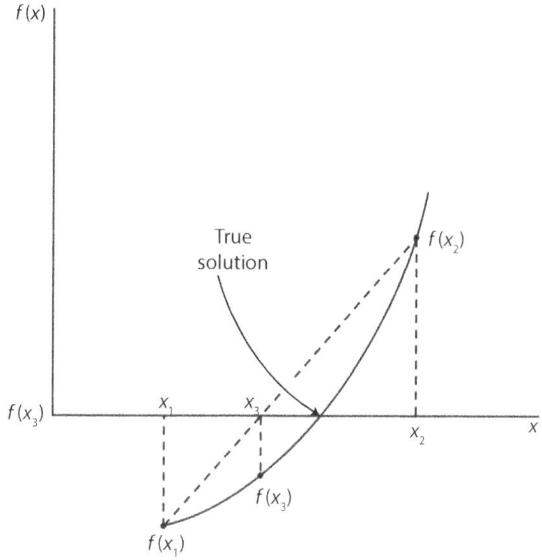

Figure 16.1 Graphical illustration of reguli-falsi method.

$$x_2 = 2^{\text{nd}} \text{ value of } x$$

$$x_3 = \text{new value}$$

locate $f(x_1)$ and $f(x_2)$ as before and compute (see Equation (16.4))

$$x_3 = x_2 - \left[\frac{f(x_2)}{f(x_2) - f(x_1)} \right] (x_2 - x_1) \tag{16.7}$$

Theodore [1] also provides a second modification. If

$$[f(a) f(b)] < 0 \tag{16.8}$$

set

$$b = x \tag{16.9}$$

$$f(b) = f(x) \tag{16.10}$$

$$f(a) = \frac{f(x)}{2} \tag{16.11}$$

and test for convergence. If satisfactory, terminate the calculation; if not satisfactory, return to Equation (16.7). If Equation (16.8) ≥ 0, set

$$a = x \tag{16.12}$$

$$f(a) = f(x) \tag{16.13}$$

$$f(b) = \frac{f(x)}{2} \tag{16.14}$$

and check as above for convergence.

Illustrative Example 16.1

Solve the following equation

$$f(x) = x^2 - 4x + 1 = 0$$

using the interpolation method. You have been informed that $3 \leq x \leq 4$.

Solution

The appropriate equation below is a modified form of Equation (16.4).

$$x_{n+1} = x_n - \frac{(x_n - x_{n-1})f(x_n)}{f(x_n) - f(x_{n-1})}$$

Apply the above equation to solve for x_3., i.e., n = 2.

$$x_3 = x_2 - \frac{(x_2 - x_1)f(x_2)}{f(x_2) - f(x_1)}$$

Set $x_1 = 3$ and $x_2 = 4$.

$$f(x_1) = 3^2 - (4)(3) + 1 = -2$$

$$f(x_2) = 4^2 - (4)(4) + 1 = 1$$

Therefore

$$x_3 = 4 - \frac{(1)(1)}{(1) + (2)} = 4 - \frac{1}{3} = 3.667$$

Once again employ Equation (16.4) for $x_{n+1} = x_4$

$$x_4 = x_3 - \frac{(x_3 - x_2) f(x_3)}{f(x_3) - f(x_2)}$$

For this trial, $x_3 = 3.667$, $x_2 = 4$, $f(x_3) = -0.222$, $f(x_2) = 1$. Therefore

$$\begin{aligned} x_4 &= 3.667 - \frac{(3.667 - 4)(-0.222)}{-0.222 - 1} \\ &= 3.667 - \frac{(-0.333)(-0.222)}{(-1.222)} = 3.667 + 0.061 \\ &= 3.728 \end{aligned}$$

Proceeding further is left as an option for the reader. For reference the true solution for this root is $\sqrt{3} + 2$ or 3.732.

16.2 Newton-Raphson Method

Although several algorithms are available for the solution to a nonlinear algebraic equation, the presentation in this section will key on the Newton-Raphson method (or the Newton Method of Tangents) of evaluating the root(s) of a nonlinear algebraic equation. The solution to the equation

$$f(x) = 0 \tag{16.15}$$

is obtained by guessing a value for x (x_{old}) that will satisfy the above equation. This value is continuously updating (x_{new}) using the equation

$$x_{new} = x_{old} - \frac{f(x_{old})}{f'(x_{old})}; \; f' = \text{derivative} \tag{16.16}$$

until either little or no change in ($x_{new} - x_{old}$) is obtained. One can express this operation graphically (see Figure 16.2). Noting that

$$f'(x_{old}) = \frac{df(x)}{dx} \approx \frac{\Delta f(x)}{\Delta x} = \frac{f(x_{old}) - 0}{x_{old} - x_{new}} \tag{16.17}$$

one may rearrange Equation (16.17) to yield Equation (16.18) below. The x_{new} then becomes the x_{old} in the next calculation.

This method as mentioned above is also referred to as Newton's Method of Tangents and is a widely used method for improving a first approximation to a

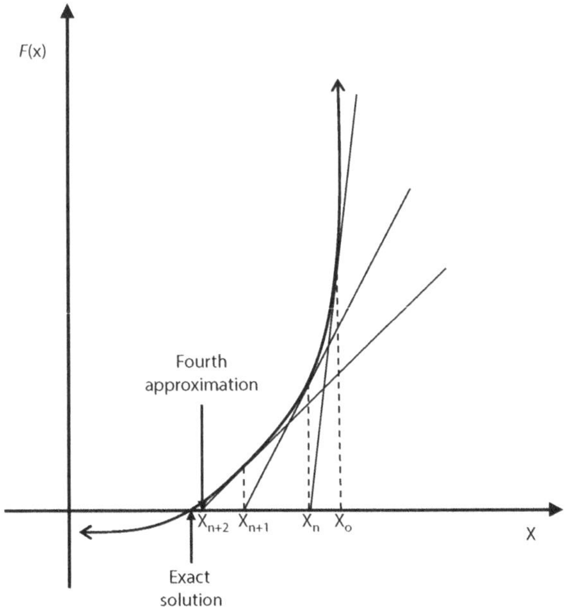

Figure 16.2 Graphical expression of the Newton-Raphson method.

root to the aforementioned equation of the form $f(x) = 0$. The above development can be rewritten in subscripted form to (perhaps) better accommodate a computer calculation. Thus,

$$f'(x_n) = \frac{f(x_n)}{x_n - x_{n+1}} \tag{16.18}$$

from which

$$x_{n+1} = x_n - \frac{f(x_n)}{f'(x_n)}$$

or

$$x_2 = x_1 - \frac{f(x_1)}{f'(x_1)} \tag{16.19}$$

where x_{n+1} is again the improved estimate of x_n, the solution to the equation $f(x) = 0$. The value of the function and the value of the derivative of the function are determined at $x = x_n$ for this procedure and the new approximation to the root, x_{n+1} is obtained. The same procedure is repeated with the new approximation to obtain a better approximation of the root. This continues until successive values for the approximate root differ by less than a prescribed small value, ε, which

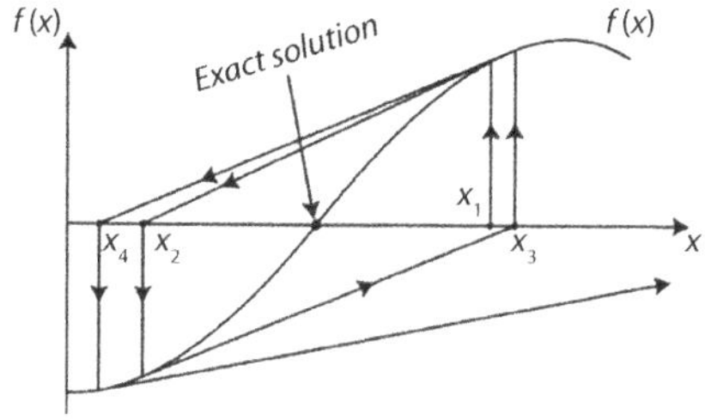

Figure 16.3 Failure of the Newton-Raphson method.

controls the allowable error (or tolerance) in the root. Relative to the previous estimate, ε is given by

$$\varepsilon = \frac{x_{n+1} - x_n}{x_n} \tag{16.20}$$

Despite its popularity, the method suffers for two reasons. First, an analytical expression for the derivative, i.e., $f'(x_n)$ is required. In addition to the problem of having to compute an analytical derivative value at each iteration, one would expect Newton's method to converge fairly rapidly to a root in the majority of cases. However, as is common with some numerical methods, it may fail occasionally in certain instances. A possible initial oscillation followed by a displacement away from a root is illustrated in Figure 16.3. Note, however, that the method would have converged (in this case) if the initial guess had been somewhat closer to the exact root. Thus, it can be seen that the initial guess may be critical to the success of the calculation.

In addition to the aforementioned False-Position and Second Order Newton methods, one may also employ the Wegstein method and the Half-interval method.[1, 2]

Illustrative Example 16.2

The vapor pressure, p', for a new synthetic chemical at a given temperature has been determined to take the form [2]:

$$p' = T^3 - 2T^2 + 2T^{-1};\ p' = atm$$

where the actual temperature, t (in K), is given by

$$t = 10^3 T \text{ or } T = 10^{-3}\text{t}$$

Solve the above equation for the actual temperature in K for $p' = 1$. Earlier studies indicate that t is in the 1000–1200K range.

Solution

If $p' = 1$, one may then write

$$f(T) = T^3 - 2T^2 + 2T - 1 = 0$$

Assume an initial temperature t_1. Set $t_1 = 1100$, so that

$$T_1 = (1100)(10^{-3}) = 1.1$$

Obtain the analytical derivative, $f'(T)$.

$$f'(T) = 3T^2 - 4T + 2$$

Calculate $f(T_1)$ and $f'(T_1)$.

$$f(1.1) = T^3 - 2T^2 + 2T - 1 = (1.1)^3 - 2(1.1)^2 + 2(1.1) - 1 = 0.111$$

$$f'(1.1) = 3T^2 - 4T + 2 = 3(1.1)^2 - 4(1.1) + 2 = 1.23$$

Use the Newton-Raphson method to estimate T_2. Employ Equation (16.19).

$$T_2 = T_1 - \frac{f(T_1)}{f'(T_1)}$$

Substituting

$$T_2 = 1.1 - \frac{0.111}{1.23} = 1.0098$$

Calculate T_3.

$$f(T_2) = 0.0099; \quad f'(T_2) = 1.0198; \quad T_3 = 1.0001$$

Finally, calculate the best estimate (based on two iterations) of t.

$$t = 10^3;\ TK = 1000.1\text{K}$$

Illustrative Example 16.3

It can be shown that the average velocity in a system involving the flow of water at 60°F is given by [3]

$$v = \sqrt{\frac{2180}{213.5\, Re^{-\left(\frac{1}{4}\right)} + 10}}$$

For water at 60°F, $Re = 12{,}168v$.

Calculate the average velocity, v (ft/s), using the Newton-Raphson method of solution [4].

Solution

Substitute the expression of Reynolds number as a function of velocity into the velocity equation.

$$v = \sqrt{\frac{2180}{213.5\, Re^{-\left(\frac{1}{4}\right)} + 10}} = \sqrt{\frac{2180}{213.5(12{,}168v)^{-\left(\frac{1}{4}\right)} + 10}}$$
$$= \sqrt{\frac{2180}{\left[213.5/(12{,}168v)^{\frac{1}{4}}\right] + 10}}$$

Manipulate the above equation into one that is easier to differentiate. Squaring both sides gives

$$v^2 = \frac{2180}{[213.5/(12{,}168v)^{\frac{1}{4}}] + 10} = \frac{2180(12{,}168v)^{\frac{1}{4}}}{213.5 + 10(12{,}168v)^{\frac{1}{4}}}$$

Cross multiplying leads to

$$213.5v^2 + 10(12{,}168v)^{\frac{1}{4}} v^2 - 2180(12{,}168v)^{\frac{1}{4}} = 0$$

or

$$f(v) = 213.5v^2 + 105.03v^{2.25} - 22{,}896.08v^{0.25} = 0$$

The analytical derivative of $f(v)$ is

$$f'(v) = 427v + 236.313v^{1.25} - 5724.02v^{-0.75}$$

Make an initial guess of $v = 5$ ft/s and substitute into the above equations:

$$f_1(v) = f_1(5) = -24{,}973.8$$

$$f_1'(v) = f_1'(5) = 2189.97$$

Calculate the next guess using Equation (16.19).

$$v_2 = v_1 - \frac{f_1(v)}{f_1'(v)} = 5 - \frac{-24,973.8}{2189.97} = 16.40 \text{ ft/s}$$

Solve for v_3, etc., until the result converges.

$$v_3 = 11.56 \text{ ft/s}$$

$$v_4 = 10.22 \text{ ft/s}$$

$$v_5 = 10.09 \text{ ft/s}$$

$$v_6 = 10.09 \text{ ft/s}$$

The average velocity is therefore approximately 10.09 ft/s.

Illustrative Example 16.4

Solve the following equation using the N–R method.

$$f(x) = x^3 - 3x^2 - 10x + 24$$

Solution

One may calculate $f'(x)$ as

$$f'(x) = 3x^2 - 6x - 10$$

Initially assume $x_1 = 5$. Thus,

$$f(5) = f(x_1) = 125 - 75 - 50 + 24 = 24$$

$$f'(5) = f'(x_1) = 75 - 30 - 10 = 35$$

and

$$x_2 = x_1 - \frac{f(x_1)}{f'(x_1)} = 5 - \frac{24}{35} = 4.313$$

For $x_2 = 4.313$

$$f(x_2) = f(4.313) = 80 - 55.8 - 43.1 + 24 = 5.1$$

$$f'(x_2) = f'(4.313) = 55.8 - 25.9 - 10 = 19.9$$

so that

$$x_3 = x_2 - \frac{f(x_2)}{f'(x_2)} = 4.313 - \frac{5.1}{19.9} = 4.057$$

One may choose to continue the iteration.

Illustrative Example 16.5

Solve the equation provided in Illustrative Example 16.1 employing the Newton-Raphson method.

Solution

Apply the N–R method of solution to the equation $x^2 - 4x + 1 = 0$. Employ

$$x_{n+1} = x_n - \frac{f(x_n)}{f'(x_n)}$$

For this example start with $x_1 = 3$. Then

$$f(x_n) = f(x_1) = f(3) = -2$$

$$f'(x_n) = f'(x_1) = f'(3) = 2$$

so that

$$x_2 = 3 - \frac{(-2)}{2} = 3 + 1 = 4$$

For the second iteration,

$$f(x_2) = 1$$

$$f'(x_2) = 4$$

so that

$$x_3 = 4 - \frac{1}{4} = 3.75$$

Proceeding further is left as an exercise for the reader. Again, for reference, the true solution rounded to three decimal places is 3.732.

16.3 Newton's Second Order Method

Newton's Second Order method requires the determination of the first derivative of $f'(x_1)$, for the first assumed value of x_1, *and* the second derivative, $f''(x_1)$. The algorithm (derivation not presented) is with $f' = y'$ and $f'' = y''$.

$$x_2 = x_1 - \left[\frac{y_1}{y_1' - \left(\frac{y_1'' y_1}{2y_1'} \right)} \right] \tag{16.21}$$

If y is replaced by $f(x)$, the above algorithm becomes

$$x_2 = x_1 - \left[\frac{f(x_1)}{f'(x_1) - \frac{f''(x_1)f(x_1)}{2f'(x_1)}} \right] \tag{16.22}$$

which may be rewritten as

$$x_2 = x_1 - \left[\frac{2f'(x_1)f(x_1)}{2f'(x_1)^2 - f''(x_1)f(x_1)} \right] \tag{16.23}$$

As one might suppose, there are a host of other procedures/methods for solving nonlinear algebraic equations. It should also be mentioned that brute-force and trial-and-error approaches have merit, particularly if one has a reasonable idea of the correct root. However, this method was discussed in the previous chapter.

Illustrative Example 16.6

Solve the equation

$$f(x) = x^3 - 3x^2 - 10x + 24$$

using Newton's Second Order method.

Solution

Calculate the first and second derivative of $f(x)$

$$f'(x) = 3x^2 - 6x - 10$$

$$f''(x) = 6x - 6$$

Assume an educated value for x_1, say $x_1 = 5$. Then

$$f(5) = 24$$

$$f'(5) = 35$$

$$f''(5) = 24$$

Substitute into Equation (16.23).

$$x_2 = 5 - \left[\frac{2(24)(35)}{2(35)^2 - (24)(24)}\right] = 5 - \frac{1680}{1874}$$
$$= 4.104$$

Similarly

$$f(x_2) = f(4.104) = 1.55$$

$$f'(x_2) = f'(4.104) = 15.9$$

$$f''(x_2) = f''(4.104) = 18.6$$

so that

$$x_3 = 4.104 - \left[\frac{2(1.55)(15.9)}{2(15.9)^2 - (1.4)(18.6)}\right] = 4.104 - \frac{49.29}{479.58}$$
$$= 4.001$$

Proceed or terminate, as desired.

References

1. L. Theodore, personal notes, East Williston, NY, 1970.
2. L. Theodore, F. Ricci, and T. VanVliet, *Thermodynamics for the Practicing Engineer*, John Wiley & Sons, Hoboken, NJ, 2009.
3. P. Abulencia and L. Theodore, *Fluid Flow for the Practicing Chemical Engineer*, John Wiley & Sons, Hoboken, NJ, 2009.

17

Simultaneous Linear Algebraic Equations

Environmentalists frequently encounter problems involving the solution of sets of simultaneous linear algebraic equations. Such problems arise in a host of areas that include electrical-circuit analysis, heat transfer, vibrations, chemical reactions, etc. The most common methods of solving simultaneous equations involve the elimination of unknowns by combining equations. When three or more simultaneous equations are being solved, one must resort to other methods. For example, the method of determinants with expansion by minors is completely impractical when large numbers of equations must be solved simultaneously.

Note that in the presentation to follow there are an equal number of unknowns and equations. If there are more unknowns than equations, the system has an infinite number of solutions. If there are fewer equations than unknowns, then the system of equations does not have a solution. Because of the breadth of the subject matter, the reader should also note that only three numerical methods receive treatment in the chapter. Information on the remaining methods can be found in the literature [1,2].

Four sections complement the presentation of this chapter. Section numbers and subject titles follow:

17.1: Notation for Solving Simultaneous Linear Algebraic Equations
17.2: Gauss Elimination Method

17.1 Notation For Solving Simultaneous Linear Algebraic Equations

The engineer often encounters problems that not only contain more than two or three simultaneous algebraic equations, but also those that are sometimes nonlinear as well. Therefore, there is an obvious need for providing systematic methods of solving simultaneous linear and simultaneous nonlinear equations [1]. This section will address linear sets of equations. Information on nonlinear sets is available in the literature [2].

Consider the following set of n equations:

$$\begin{aligned} a_{11}x_1 + a_{12}x_2 + \ldots + a_{1n}x_n &= c_1 \\ a_{21}x_1 + a_{22}x_2 + \ldots + a_{2n}x_n &= c_2 \\ &\ldots \\ a_{m1}x_1 + a_{m2}x_2 + \ldots + a_{mn}x_n &= c_n \end{aligned} \tag{17.1}$$

where a is the coefficient of the variable x and c is a constant. The above set is considered to be linear as long as none of the x terms are nonlinear, for example, x_2^2 or $\ln x_1$. Thus, a linear system requires that all terms in x be linear.

The above system of linear equations may be set in matrix form (see Part II):

$$\begin{bmatrix} a_{11} & a_{12} & \ldots & a_{1n} \\ a_{21} & a_{22} & \ldots & a_{2n} \\ \ldots & \ldots & \ldots & \ldots \\ a_{m1} & a_{m2} & \ldots & a_{mn} \end{bmatrix} \begin{bmatrix} x_1 \\ x_2 \\ \ldots \\ x_n \end{bmatrix} = \begin{bmatrix} c_1 \\ c_2 \\ \ldots \\ c_n \end{bmatrix} \tag{17.2}$$

However, it is often more convenient to represent Equation (17.2) in the *augmented matrix* provided in Equation (17.3).

$$\begin{bmatrix} a_{11} & a_{12} & \ldots & a_{1n} & c_1 \\ a_{21} & a_{22} & \ldots & a_{2n} & c_2 \\ \ldots & \ldots & \ldots & \ldots & \ldots \\ a_{m1} & a_{m2} & \ldots & a_{mn} & c_n \end{bmatrix} \tag{17.3}$$

As noted above, only the first three methods are discussed in this chapter.

The *coefficient matrix* of the set of equations in Equation (17.1) is

$$\begin{bmatrix} a_{11} & a_{12} & \dots & a_{1n} \\ a_{21} & \dots & \dots & a_{2n} \\ \dots & \dots & \dots & \dots \\ a_{31} & \dots & \dots & a_{3n} \end{bmatrix} \tag{17.4}$$

If the constants of Equation (17.1) are added to the coefficient matrix as a column of elements in the position shown in Equation (17.3), the *augmented matrix* is formed. A comparison of Equations (17.1) and (17.4) shows that a coefficient matrix is a matrix whose elements are the coefficients of the unknowns in the set of the equations.

In many instances in computer programming, it may be found convenient to express the column of constants $c_1, c_2, c_3, \ldots, c_n$, as simply an additional column of a_{ij}'s. In such an instance the matrix of Equation (17.3) might be expressed as

$$B = \begin{bmatrix} a_{11} & a_{12} & \dots & a_{1n} & a_{1,n+1} \\ a_{21} & a_{22} & \dots & a_{2n} & a_{2,n+2} \\ \dots & \dots & \dots & \dots & \dots \\ a_{m1} & \dots & \dots & a_{mn} & a_{m,n+1} \end{bmatrix} \tag{17.5}$$

If $c_1, c_2, c_3, \ldots, c_n$ are not all zero, the set of equations is *nonhomogeneous*, and one will find that all the equations must be independent to obtain unique solutions. If the constants $c_1, c_2, c_3, \ldots, c_n$ are all zero, the set of equations is *homogenous*, where nontrivial solutions exist only if all the equations are not independent.

In this chapter, the solution of both homogenous and nonhomogeneous sets of linear algebraic equations shall be considered since both types appear in environmental applications. The sets of equations which are to be addressed will involve n unknowns in n equations, which have the general form provided in Equation (17.1).

Methods of solution available for solving these linear sets of equations include:

1. Gauss elimination
2. Gauss-Jordan reduction
3. Gauss-Seidel
4. Cramer's rule
5. Cholesky's method

17.2 Gauss Elimination Method

Using Gauss' elimination method, a set of n equations and n unknowns is reduced to an *equivalent* triangular set (an equivalent set is a set having identical solution values), which is then easily solved by "back substitution," a simple procedure which will be illustrated in an illustrative example.

Gauss' scheme begins by reducing a set of simultaneous equations, such as those given in Equation (17.1), to an equivalent triangular set such as the following:

$$\begin{aligned} a_{11}x_1 + a_{12}x_2 + a_{13}x_3 + a_{14}x_4 + \ldots + a_{1n}x_n &= c_1 \\ a'_{22}x_2 + a'_{23}x_3 + a'_{24}x_4 + \ldots + a'_{2n}x_n &= c_2 \\ a''_{33}x_3 + a''_{34}x_4 + \ldots + a''_{3n}x_n &= c_3 \\ \ldots\ldots\ldots\ldots\ldots\ldots\ldots\ldots\ldots\ldots & \\ a^{m-2}_{m-1,n-1}x_{n-1} + a^{m-2}_{m-1,n}x_n &= c^{n-2}_{n-1} \\ a^{m-1}_{mn}x_n &= c^{n-1}_{n} \end{aligned} \tag{17.6}$$

where the prime subscripts indicate new coefficients which are formed in the reduction process. The following steps accomplish the actual reduction. First, the lead equation in Equation (17.1) is divided by the coefficient of x_1 to obtain

$$x_1 + \frac{a_{12}}{a_{11}}x_2 + \frac{a_{13}}{a_{11}}x_3 + \ldots + \frac{a_{1n}}{a_{11}}x_n = \frac{c_1}{a_{11}} \tag{17.7}$$

Next, Equation (17.7) is multiplied by the coefficient of x_1 in the second equation found in Equation (17.1), and the resulting equation is subtracted from the original second equation in Equation (17.1), which eliminates x_1. Equation (17.7) is then multiplied by the coefficient of x_1 in the third equation of Equation (17.1), and the resulting equation is subtracted from the third equation to eliminate x_2. This process is repeated until one can solve for x_n directly. The algorithm for solving an n x m set of equations is provided in Figure 17.1.

17.3 Gauss-Jordan Reduction Method

The second method presented for the solution of simultaneous linear algebraic equations – a variation of the gauss elimination – is one in which the unknowns are reduced by combining equations. Such a method is known as a *reduction* method. It is referred to by some as Gauss's method or the Gauss-Jordan method if a particular systematic scheme is used in the reduction process. This method involves the "reduction" of the simultaneous equations to the solution of one linear equation with one unknown.

Carnahan and Wilkes [1] solved the following two simultaneous equations using the Gauss-Jordan reduction method:

$$3x_1 + 4x_2 = 29 \tag{17.8}$$

$$6x_1 + 10x_2 = 68 \tag{17.9}$$

Application of the aforementioned procedure follows.

1. Divide Equation (17.8) through by the coefficient of x_1:

$$x_1 + \frac{4}{3}x_2 = \frac{29}{3} \tag{17.10}$$

2. Subtract a suitable multiple, in this case 6, of equation (17.10) from Equation (17.9), so that x_1 is eliminated. Equation (17.10) remains untouched, leaving:

$$x_1 + \frac{4}{3}x_2 = \frac{29}{3} \tag{17.11}$$

$$2x_2 = 10 \tag{17.12}$$

3. Divide Equation (17.12) by the coefficient of x_2, that is, solve Equation (17.12)

$$x_2 = 5 \tag{17.13}$$

4. Subtract a suitable factor of Equation (17.13) from Equation (17.11) so that x_2 is eliminated. When $(4/3)x_2 = 20/3$ is subtracted from Equation (17.11), one obtains

$$x_1 = 3 \tag{17.14}$$

The augmented matrix from Equation (17.3) is

$$\left[\begin{array}{cc|c} 3 & 4 & 29 \\ 6 & 10 & 68 \end{array}\right]$$

$$\left[\begin{array}{cc|c} 1 & 4/3 & 29/3 \\ 6 & 10 & 68 \end{array}\right]$$

$$\left[\begin{array}{cc|c} 1 & 4/3 & 29/3 \\ 0 & 2 & 10 \end{array}\right]$$

$$2x_2 = 10$$

$$x_2 = 5$$

$$\begin{bmatrix} 1 & 0 & \vline & 3 \\ 0 & 2 & \vline & 10 \end{bmatrix}$$

$$x_1 = 3$$

Illustrative Example 17.1

Solve the following set of linear algebraic equations using Gauss elimination.

$$3x_1 - 2x_2 + x_3 = 7$$

$$x_1 - 4x_2 - 2x_3 = 21$$

$$2x_1 - 3x_2 - 4x_3 = 9$$

Solution

First, set up the augmented matrix.

$$\begin{bmatrix} 3 & -2 & 1 & 7 \\ 1 & 4 & -2 & 21 \\ 2 & -3 & -4 & 9 \end{bmatrix}$$

In order to convert the first column of elements below the diagonal to zero, start by setting the first row as the pivot row. For the remaining rows, multiply each element by a_{11}/a_{n1} and subtract each row from the pivot row. Thus:

Row 2 coefficients:

$$3 - (3)(1) = 0$$

$$-2 - (3)(4) = -14$$

$$1 - (3)(-2) = 7$$

$$7 - (3)(21) = -56$$

Row 3 coefficients:

$$3 - \frac{(3)(2)}{2} = 0$$

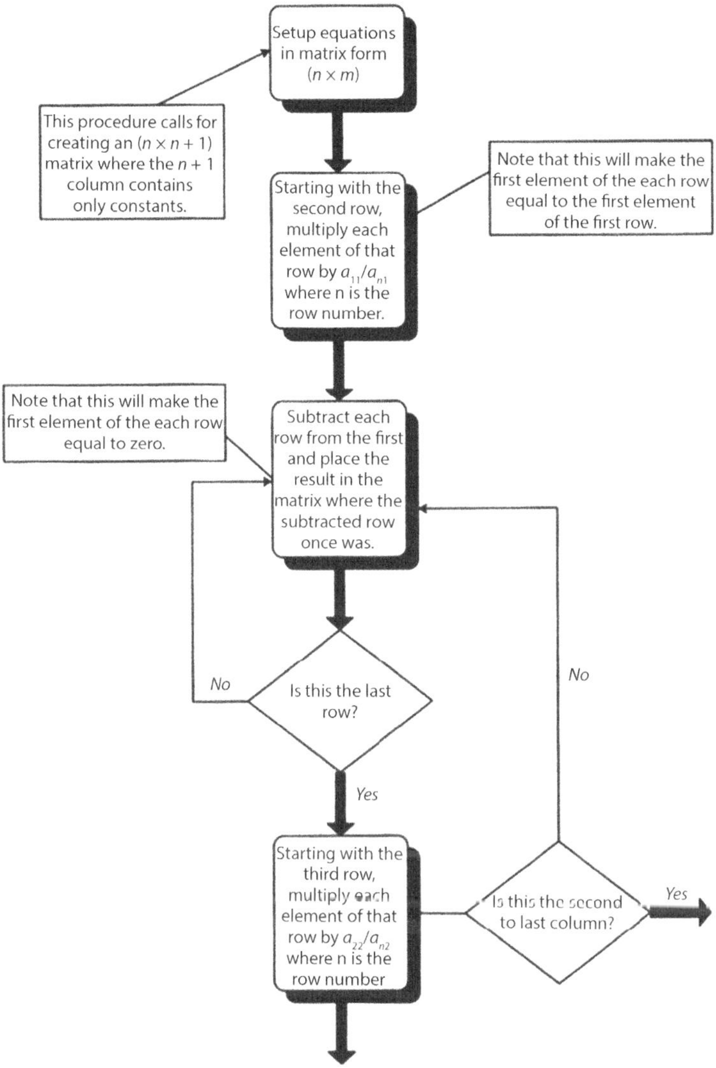

Figure 17.1 Algorithm for Gauss Elimination with back substitution.

$$-2-\frac{(3)(-3)}{2}=2.5$$

$$1-\frac{(3)(-4)}{2}=7$$

$$7-\frac{(3)(9)}{2}=-6.5$$

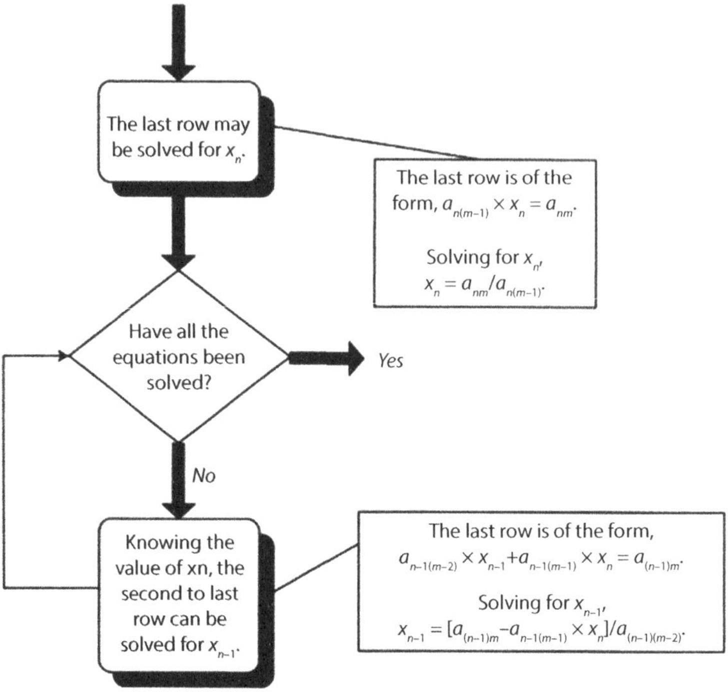

Figure 17.1 Algorithm for Gauss Elimination with back substitution (continued).

The resulting matrix is:

$$\begin{bmatrix} 3 & -2 & 1 & 7 \\ 0 & -14 & 7 & -56 \\ 0 & 2.5 & 7 & -6.5 \end{bmatrix}$$

Now starting with the second column and the second row, perform the same procedure again, with the second row as the pivot row.

Row 3 coefficients:

$$14 - \frac{(-14)(2.5)}{2.5} = 0$$

$$7 - \frac{(-14)(7)}{2.5} = 46.2$$

$$-56 - \frac{(-14)(-6.5)}{2.5} = -92.4$$

This produces the following matrix:

$$\begin{bmatrix} 3 & -2 & 1 & 7 \\ 0 & -14 & 7 & -56 \\ 0 & 0 & 46.2 & -92.45 \end{bmatrix}$$

At this point, back substitution may be employed. The following results are obtained:

$$46.2x_3 = -92.4;\ x_3 = -2$$

$$-14x_2 + 2(-7) = -56;\ x_2 = 3$$

$$3x_1 + (-2)(3) + 1(-2) = 7;\ x_1 = 5$$

Gauss elimination is useful for systems that contain fewer than 30 equations. Larger systems become subject to roundoff error where numbers are truncated by computers performing the calculations.

17.4 Gauss-Seidel Method

Another approach to solving an equation or series/sets of equations is to make an educated guess. If the first assumed value(s) does not work, the value is updated. By carefully noting the influence of these guesses on each variable, the answers or correct set of values for a system of equations can be approached. The reader should note that when this type of iterative procedure is employed, a poor guess does not prevent the correct solution from ultimately being obtained.

Ketter and Prawel provide the following example [1]. Consider the equations below:

$$4\boldsymbol{x}_1 + 2x_2 + 0x_3 = 2$$

$$2x_1 + 10\boldsymbol{x}_2 + 4x_3 = 6$$

$$0x_1 + 4x_2 + 5\boldsymbol{x}_3 = 5$$

The reader may choose to assume as a starting point $x_1 = x_2 = x_3 = 0$, for the *non-bold* terms. Solving each equation for the bold terms (found on the diagonal), one obtains

$$x_1 = 0.50;\ x_2 = 0.60;\ x_3 = 1.00$$

These computed values can now be used as new guesses to update the initially assumed x values.

$$x_1 = \boldsymbol{x}_1 - \frac{2}{4}x_2 + \frac{0}{4}x_3 = 0.50 - \frac{2}{4}(0.60) = 0.20$$

$$x_2 = \boldsymbol{x}_2 - \frac{2}{10}x_1 + \frac{4}{10}x_3 = 0.60 - \frac{2}{10}(0.50) + \frac{4}{10}(1.00) = 0.90$$

$$x_3 = \boldsymbol{x}_3 - \frac{0}{5}x_1 + \frac{4}{5}x_2 = 1.00 - \frac{0}{5}(0.50) + \frac{4}{5}(0.60) = 1.48$$

The right-hand side of these equations may be viewed as residuals. The procedure is repeated until convergence is achieved with the residuals approaching zero. More rapid convergence techniques are available in the literature [2].

References

1. R. Ketter and S. Prawler, *Modern Methods of Engineering Computations,* McGraw-Hill, New York City, NY, 1969.
2. B. Carnahan and J. Wilkes, *Digital Computing and Numerical Methods,* John Wiley & Sons, Hoboken, NJ, 1973.

18

Differentiation

Numerical differentiation should be avoided whenever possible, particularly when data are empirical and subject to appreciable observation errors. Errors in data can affect numerical derivatives, i.e., differentiation is a significantly "roughening" process. When such a calculation must be made, it is usually desirable to first smooth the data to a certain extent (if possible).

Derivatives of a function $f(x)$ may be evaluated numerically by any one of several methods involving two data points. Three simple formulas that may be employed to evaluated the derivative, $f(x) = d[f(x)]/dx$, are given below.

$$f'(x) = \frac{1}{h}[f(x+h) - f(x)]; \text{ forward difference} \tag{18.1}$$

$$f'(x) = \frac{1}{2h}[f(x+h) - f(x-h)]; \text{ central difference} \tag{18.2}$$

$$f'(x) = \frac{1}{h}[f(x) - f(x-h)]; \text{ backward difference} \tag{18.3}$$

The term h represents an incremental change in x.

Three sections complement the presentation of this chapter. Section numbers and subject titles follow:

18.1: Employing Two and Three Point Formulas
18.2: Employing Five Point Formulas
18.3: Method of Least Squares

18.1 Employing Two and Three Point Formulas

Several differentiation methods [1] are available for generating expressions for a derivative using three data points. Consider the problem of determining the benzene concentration time gradient dC/dt at $t = 4.0$ s; refer to Table 18.1.

Method 1. This method involves the selection of any three data points and calculating the slope m of the two extreme points. This slope is approximately equal to the slope at the point lying in the middle. The value obtained is the equivalent of the derivative at that point 4. Using data points from 3.0s to 5.0s, one obtains

$$\begin{aligned} Slope = m &= \frac{C_5 - C_3}{t_5 - t_3} \\ &= \frac{1.63 - 2.70}{5.0 - 3.0} = -0.535 \end{aligned} \tag{18.4}$$

Method 2. This method involves determining the average of two slopes. Using the same points chosen above, two slopes can be calculated, one for points 3 and 4 and the other for points 4 and 5. Adding the two results and dividing them by 2 will provide an approximation of the derivative at point 4. For the points used in this method, the results are as follows.

Table 18.1 Concentration-Time data.

Time, s	Data point	Concentration of benzene, mg/L
0.0	0	7.46
1.0	1	5.41
2.0	2	3.80
3.0	3	2.70
4.0	4	2.01
5.0	5	1.63
6.0	6	1.34
7.0	7	1.17

$$m_1 = slope_1 = \frac{C_4 - C_3}{t_4 - t_3}$$
$$= \frac{2.01 - 2.70}{4.0 - 3.0} = -0.69 \tag{18.5}$$

$$m_2 = slope_2 = \frac{C_5 - C_4}{t_5 - t_4}$$
$$= \frac{1.63 - 2.01}{5.0 - 4.0} = -0.38 \tag{18.6}$$

$$m_{avg} = slope_{avg} = \frac{-0.69 + (-0.38)}{2} = -0.535 \tag{18.7}$$

Method 3. Method 3 consists of using any three data points (in this case the same points chosen before) and fitting a curve to it. The equation for the curve is obtained by employing a second-order equation and solving it with the three data points. The equation can be quickly provided by Microsoft Excel's "add trend line" function. The result is

$$C = 0.155t^2 - 1.775t + 6.63 \tag{18.8}$$

The derivative of the equation is then calculated and evaluated at any point. (Here, point 4 is used).

$$\frac{dC}{dt} = 0.31t - 1.775 \tag{18.9}$$

Evaluating the derivative at $t = 4.0$ yields

$$Slope = \frac{dC}{dt} = 0.31(4) - 1.775 = -0.535 \tag{18.10}$$

18.2 Employing Five Point Formulas

There are two methods to employ here and they are very similar to each other. They are based on five data points used to generate coefficients. For this development, represent C and t by f and x (as it appears in the literature [2]), respectively.

The first method uses five data points to generate a five coefficient (fourth order) model using an equation in the form

$$f(x) = A + Bx + Cx^2 + Dx^3 + Ex^4 \tag{18.11}$$

This method is known as interpolating. A set of equations is used to evaluate numerical derivatives from the interpolating polynomial. The equations are listed below.

$$f'(x_0) = \frac{(-25f_0 + 48f_1 - 36f_2 + 16f_3 - 3f_4)}{12h} \tag{18.12}$$

$$f'(x_1) = \frac{(-3f_0 - 10f_1 + 18f_2 - 6f_3 + f_4)}{12h} \tag{18.13}$$

$$f'(x_i) = \frac{(f_{i-2} - 8f_{i-1} + 8f_{i+1} - f_{i+2})}{12h} \tag{18.14}$$

$$f'(x_{n-1}) = \frac{(-f_{n-4} + 6f_{n-3} - 18f_{n-2} + 10f_{n-1} + 3f_n)}{12h} \tag{18.15}$$

$$f'(x_n) = \frac{(3f_{n-4} - 16f_{n-3} + 36f_{n-2} - 48f_{n-1} + 25f_n)}{12h} \tag{18.16}$$

where $h = x_{i+1} - x_i$, f_i = function evaluated at i.

For example, the equation obtained for "the five data set" from 1.0 to 5.0 s, i.e., $t = 1.0, 2.0, 3.0, 4.0$, and 5.0 s, using the author's [2] approach gives

$$f(x) = -0.0012x^4 + 0.002x^3 + 0.02616x^2 - 2.34x + 7.467$$

All these equations can be evaluated for each value of x and $f(x)$. The value for the derivative for point 4.0 is –0.5448.

The second and last method also uses five data points but only three coefficients are generated for a second-order polynomial equation of the form

$$f = A + Bx + Cx^2 \tag{18.17}$$

Another set of equations are used to evaluate the derivative at each point using this method. The equations are provided below.

$$f'(x_0) = \frac{(-54f_0 + 13f_1 + 40f_2 + 27f_3 - 26f_4)}{70h} \tag{18.18}$$

$$f'(x_1) = \frac{(-34f_0 + 3f_1 + 20f_2 + 17f_3 - 6f_4)}{70h} \tag{18.19}$$

$$f'(x_i) = \frac{(-2f_{i-2} - f_{i-1} + f_{i+1} + 2f_{i+2})}{10h} \tag{18.20}$$

$$f'(x_{n-1}) = \frac{(6f_{n-4} - 17f_{n-3} - 20f_{n-2} - 3f_{n-1} + 34f_n)}{70h} \tag{18.21}$$

$$f'(x_n) = \frac{(26f_{n-4} - 27f_{n-3} - 40f_{n-2} - 13f_{n-1} + 54f_n)}{70h} \tag{18.22}$$

At point 4.0, the solution for the derivative using this method is –0.6897.

18.3 Method of Least Squares

The following method uses the method of least squares (see Part IV). In this case, all data points are used to generate a second-order polynomial equation [3]. This equation is then differentiated and evaluated at the point where the value of the derivative is required. For example, Microsoft excel can be employed to generate the regression equation

$$C = 0.1626t^2 - 1.9905t + 7.3108 \tag{18.23}$$

Once all the coefficients are known, the equation has only to be analytically differentiated.

$$\frac{dC}{dt} = 0.3252t - 1.9905 \tag{18.24}$$

Set

$$t = 4.0 \tag{18.25}$$

and substitute

$$\begin{aligned} \frac{dC}{dt} &= (0.3252)(4.0) - 1.9905 \\ &= -0.6897 \end{aligned} \tag{18.26}$$

* * * * *

Comparing all the values obtained for the derivative at $t = 4.0$ s, it can be observed that the answers are very close to each other. It is important to remember that these are approximate values and that they vary depending on the approach and the number of data points used to generate the equations.

References

1. L. Perez, term project submitted to L. Theodore, Manhattan College, Bronx, NY, 2003.
2. F. Lavery, *The Perils of Differentiating Engineering Data Numerically*, Chem. Eng., New York City, NY, 1979.
3. S. Shaefer and L. Theodore, *Probability and Statistics for Environmental Science*, CRC Press/Taylor & Francis Group, Boca Raton, FL, 2007.

19

Integration

Numerous environmental engineering and science problems require the solution of integral equations. In a general sense, the problem is to evaluate the function on the right-hand side (RHS) of Equation (19.1)

$$I = \int_a^b f(x)dx \tag{19.1}$$

where I is the value of the integral. As discussed earlier, there are two key methods employed in their solution: analytical and numerical. If $f(x)$ is a simple function, it may be integrated analytically. For example, if $f(x) = x^2$, then

$$I = \int_a^b x^2 dx = \frac{1}{3}(b^3 - a^3) \tag{19.2}$$

If, however, $f(x)$ is a function too complex to integrate analytically {e.g., $\log[\tan h(e^{x^3-2})]$}, one may resort to any of the numerical methods available. Two simple numerical integration methods that are commonly employed in practice are the trapezoidal rule and Simpson's rule. These are described below. However, there are other numerical integration methods, including:

1. Romberg's method
2. Composite formulas
3. Gregory's formulas
4. Taylor's theorem
5. Method of undetermined coefficients
6. Richardson's extrapolation

Details are available in the literature [1–3].

Finally, the reader should once again realize that analytical approaches yield closed form and/or exact solutions. Numerical methods provide discrete and/or inexact answers. Thus, the analytical approach should always be attempted first even though numerical methods have become the preferred choice.

Three sections complement the presentation of this chapter. Section numbers and subject titles follow:

19.1:Trapezoidal Rule
19.2: Simpson's Rule
19.3: Comparing the Trapezoidal Rule with Simpson's Rule

19.1 Trapezoidal Rule

In order to use the trapezoidal rule to evaluate the integral given by Equation (19.1), as

$$I = \int_a^b f(x)dx \tag{19.1}$$

use the equation

$$I = \frac{h}{2}[y_0 + 2y_1 + 2y_2 + \ldots + 2y_{n-1} + y_n] \tag{19.3}$$

where h is the incremental change in x. Thus,

$$h = \Delta x \tag{19.4}$$

and

$$y_0 = f(x_0) = f(x = a) \tag{19.5}$$

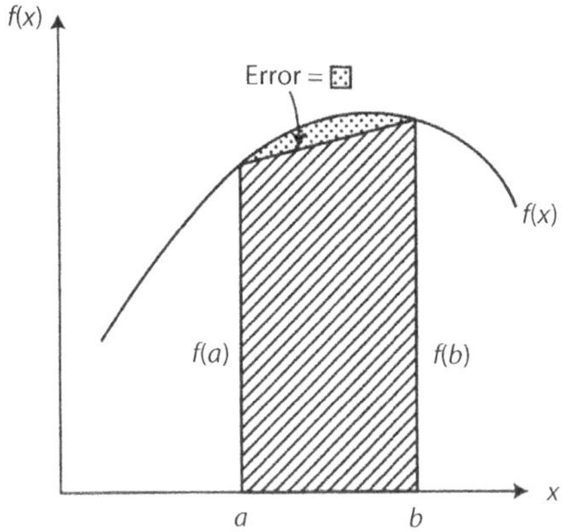

Figure 19.1 Trapezoidal rule error.

$$y_n = f(x_n) = f(x = b) \tag{19.6}$$

$$h = \frac{b - a}{n} \tag{19.7}$$

This method is known as the trapezoidal rule because it approximates the area under the function $f(x)$ – which is generally curved – with a two-point trapezoidal rule calculation. The error associated with this rule is illustrated in Figure 19.1.

There is an alternative available for improving the accuracy of this calculation. The interval $(a - b)$ can be subdivided into smaller intervals. The trapezoidal rule can be applied repeatedly in turn over each subdivision.

Illustrative Example 19.1

The volume requirement of a reactor [4, 5] undergoing conversion X for the reactant (the principal component from the bottom of a distillation column) [6] is described by the following integral:

$$V = 6.0 \times 10^{-3} \int_0^{0.45} \frac{(1 - 0.125X)^3 \, dX}{10^{-4}(1 - X)(1 - 0.5X)^2}; \text{ liters}$$

Calculate the volume using the trapezoidal rule method of integration. Discuss the effect of varying the increment in conversion ΔX (e.g., if $\Delta X = 0.45$, 0.09, 0.05, 0.01, 0.005, 0.001).

Solution

Regarding integration, an algorithm for applying the trapezoidal rule is given in Figure 19.2.

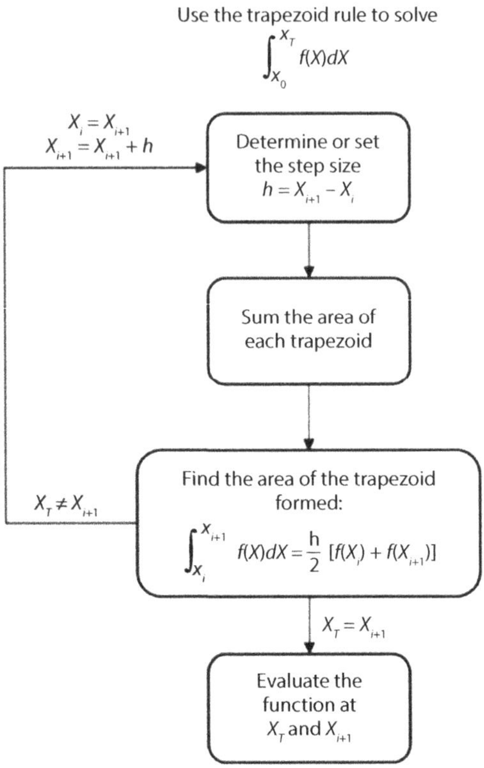

Figure 19.2 Algorithm for trapezoid rule.

For an increment size of conversion $X = 0.45$, the step size for the reactor volume is:

$$h = X_1 - X_0 = 0.45 - 0.00 = 0.45$$

Evaluate the function above at X_0 and X_1:

$$V_0 = f(X_0) = f(0) = 6.0\times10^{-3}\left[\frac{(1-0.125X)^3 dX}{10^{-4}(1-X)(1-0.5X)^2}\right]_{X=0.0}$$

$$= 6.0\times10^{-3}\left[\frac{(1)^3}{10^{-4}(1)(1)^2}\right] = 60$$

$$V_1 = f(X_1) = f(0.45) = 6.0\times10^{-3}\left[\frac{(1-0.125X)^3 dX}{10^{-4}(1-X)(1-0.5X)^2}\right]_{X=0.45}$$

$$= 6.0\times10^{-3}\left[\frac{(1-0.125(0.45))^3}{10^{-4}(1-0.45)(1-0.5(0.45))^2}\right] = 152.67$$

Table 19.1 Trapezoid rule for various step sizes.

Step size (ΔX)	Volume
0.45	47.85094
0.09	43.13532
0.05	42.98842
0.01	42.92521
0.005	42.92324
0.001	42.92260

The two point trapezoid rule is given by

$$\int_{X_1}^{X_2} f(X)dX = \frac{h}{2}[f(X_0)+f(X_1)]$$

Therefore,

$$\begin{aligned} V &= \frac{h}{2}[f(X_0)+f(X_1)] \\ &= \frac{h}{2}[V_0+V_1] \\ &= \frac{0.45}{2}[60+152.67] \\ &= 47.85L \end{aligned}$$

The trapezoid rule is often the quickest but least accurate way to perform a numerical integration by hand. However, if the step size is decreased the answer should converge to the analytical solution. Note that for smaller step sizes, the results of each numerical integration must be added together to obtain the final answer. The results (to seven significant figures) for various step sizes are listed in Table 19.1. Can the reader comment on whether convergence has been achieved?

19.2 Simpson's Rule

A higher-degree interpolating polynomial scheme can be employed for more accurate results. One of the more popular integration approaches is Simpson's rule. For Simpson's 3-point (or one-third) rule, one may use the equation

$$I = \frac{h}{3}[y_a + 4y_{(b+a)/2} + y_b] \tag{19.8}$$

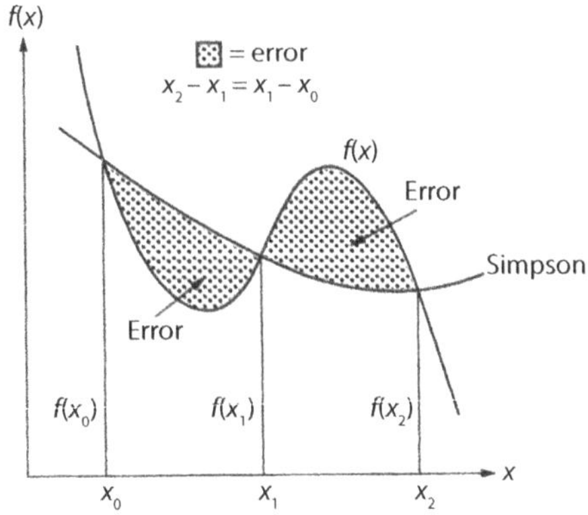

Figure 19.3 Simpson's rule analysis and error.

For the general form of Simpson's rule (where n is an even integer), the equation is

$$I = \frac{h}{3}(y_0 + 4y_1 + 2y_2 + 4y_3 + \ldots + 4y_{n-1} + y_n) \tag{19.9}$$

This method also generates an error, although it is usually smaller than that associated with the trapezoidal rule. A diagrammatic representation of the error for a 3-point calculation is provided in Figure 19.3.

Illustrative Example 19.2

Evaluate the integral below using Simpson's 3-point rule. The term I in this application represents the volume requirement for a tubular flow reactor.[4, 5]

$$I = \int_0^{0.468} \left[\frac{(1-0.4X)^2}{(1-X)(1-0.4X) - 1.19X^2} \right] dX$$

Solution

Write the 3-point rule. See Equation (19.9).

$$\begin{aligned} I &= \frac{h}{3}[y_a + 4y_{(b+a)/2} + y_b] \\ &= \frac{h}{3}[f(X=a) + 4f(X=(b+a)/2) + f(X=b)] \end{aligned}$$

Evaluate h.

$$h = \frac{0.468}{2} = 0.234$$

Calculate y_a, $y_{(b+a)/2}$, and y_b. For $X = a = 0$,

$$y(x=0) = y(0) = \frac{(1-0.4(0))^2}{(1-(0))(1-0.4(0))-1.19(0)^2} = 1$$

Similarly,

$$y(X = 0.234) = 1.306$$

$$y(X = 0.468) = 3.846$$

Finally, calculate the integral I.

$$\begin{aligned} I &= \frac{h}{3}[y_a + 4y_{(b+a)/2} + y_b] \\ &= \frac{0.234}{3}[1 + 4(1.306) + 3.846] = 0.785 \end{aligned}$$

19.3 Comparing the Trapezoidal and Simpson's Rules

Comparing the integrations in the two previous sections is accomplished via a two part illustrative example.

Illustrative Example 19.3

Given the following integral:

$$I = \int_0^4 (25 - x^2)^{-\frac{1}{2}} dx$$

1. Calculate I by the trapezoidal rule for 2 and 4 intervals to three decimal places.
2. Calculate I by Simpson's one-third rule for 2 and 4 intervals to three decimal places.

Compare the results.

Table 19.2 Values of I vs x; Illustrative Example 19.3.

x	I
0	0.200
1	0.204
2	0.218
3	0.250
4	0.333

Solution

Referring to Table 19.2, tabulate I at 5 points (values of x).

1. By the trapezoidal rule with 2 intervals, i.e., $h = 2$.

$$I = \frac{2}{2}(0.200 + 2(0.218) + 0.333)$$

$$I = 0.969$$

For 4 intervals, $h = 1$ so that

$$I = \frac{1}{2}[0.200 + 2(0.204) + 2(0.218) + 2(0.250) + 0.333]$$

$$I = 0.939$$

2. By Simpson's rule with 2 intervals, i.e., $h = 2$.

$$I = \frac{2}{3}(0.200 + 4(0.218) + 0.333)$$

$$I = 0.937$$

For 4 intervals, $h = 1$ so that

$$I = \frac{1}{3}[0.200 + 4(.204) + 2(0.218) + 4(0.250) + 0.333]$$

$$I = 0.928$$

Which is more accurate? One needs the analytical solution for comparison. One is left the exercise of determining that (2) is a better approximation.

In general, comparing the trapezoid rule and Simpson's one-third rule, one notes that the trapezoid rule requires $n + 1$ data points for obtaining the area in n subintervals of width Δx, whereas Simpson's one-third rule requires $n + 1$ data points for obtaining the area in $2n$ subintervals of width $(\Delta x)/2$.

References

1. R. L. Ketter and S. P. Prawler, *Modern Methods of Engineering Computations*, McGraw-Hill, New York City, NY, 1969.
2. J. Reynolds, class notes (with permission), Manhattan College, Bronx, NY, 2001.
3. J. Famularo, class notes (with permission), Manhattan College, Bronx, NY, 1981.
4. L. Theodore, *Chemical Reaction Analysis and Applications for the Practicing Engineer*, John Wiley & Sons, Hoboken, NJ, 2011.
5. L. Theodore, *Chemical Reaction Kinetics*, A Theodore Tutorial, Theodore Tutorials, East Williston, NY, originally published by USEPA/APTI, RTP, NC, 1994.
6. L. Theodore and F. Ricci, *Mass Transfer Operations for the Practicing Engineer*, John Wiley & Sons, Hoboken, NJ, 2010.

20

Ordinary Differential Equations

The natural laws in environmental engineering and science are not regarded as precise and definitive until they have been expressed in mathematical form. Such a form, often an equation, is a relation between the quantity of interest, say, pollutant concentration, and independent variables such as time and position on which the concentration depends. When it happens that this equation involves, besides the function itself, one or more of its derivatives it is called a differential equation.

When the aforementioned function involved in the equation depends only on one variable, its derivatives are ordinary derivatives and the differential equation is referred to as an ordinary differential equation. When the function depends on several independent variables, the equation is then called a partial differential equation (see next chapter). The theories of ordinary and partial differential equations are different and the latter is more "difficult" in almost every respect.

Of increasing concern to the environmental engineer and scientist is the description of dynamic or transient staged and/or lumped-parameter processes. Ordinary differential equations are usually used to represent the behavior of such processes; when combined with chemical reaction [1] or equilibrium [2] relationships, the equations are often highly nonlinear.

One often encounters differential equations which cannot be solved by any analytical methods except that of a solution in terms of a series, and this method may be difficult to apply in certain cases. Even when an analytical solution is available, it is sometimes difficult to find numerical values of corresponding pairs of the dependent and independent variables. However, a variety of methods have been devised to solve ordinary differential equations numerically. Several references contain detailed information [3–6].

Interestingly, a numerical solution of a differential equation can generally provide a table of values of the dependent variable and its derivatives over only a limited portion of the range of the independent variable. Also of interest, every differential equation of order n can be written as n first order equations. Therefore, the methods given in the early sections of this chapter will be for first-order equations, and the generalization to simultaneous systems will be subsequently discussed. The reader might also note that the analytical solution to an ordinary differential equation was not addressed in any detail in Part II, Chapter 11.

Five sections complement the presentation of this chapter. Section numbers and subject titles follow:

20.1: Finite Difference/Lumped Parameter Method
20.2: Runge-Kutta Method
20.3: Runge-Kutta-Gill Method
20.4: Several Ordinary Differential Equations
20.5: Higher Ordinary Differential Equations

20.1 Finite Difference/Lumped Parameter Method

This method is illustrated via a chemical reactor example [7]. The concentration, C, variation with length, z, in a 3 ft continuous flow tubular reactor (TF) is described by the equation:

$$\frac{dC}{C} = -\frac{k}{v}dz \tag{20.1}$$

where k is $50(h)^{-1}$, v is 50 ft/h, C is in lbmol/ft^3, z is in feet, and $C = 1.0$ lbmol/ft^3 at $z = 0$.

It has been proposed to represent the TF with 10 CSTR (continuous stirred tank reactor) staged units 0.3 ft in length. Solutions to this problem are now developed using a finite difference method of solving an ordinary differential equation *and* lumped parameter model employing a method of solution of simultaneous linear algebraic equations. The aforementioned CSTR is a perfectly mixed tank reactor whose discharge concentration is equal to the concentration within the reactor resulting in an equation containing finite differences rather than differential values [7].

A finite difference procedure is first applied to Equation (20.1). The first derivative of C with respect to z is equivalent to the finite difference in the z-direction, i.e.,

$$\frac{dC}{dz} = \frac{\Delta C}{\Delta z} = \frac{C_{n+1} - C_n}{\Delta z} \tag{20.2}$$

Inserting this into the describing equation and noting that the average concentration over the increment should be used leads to

$$\frac{C_{n+1} - C_n}{\Delta z} = -\left(\frac{k}{v_z}\right) C_{av} \tag{20.3}$$

Knowing the initial condition, each successive concentration may be found by rearranging the above equation:

$$C_{n+1} = -\left(\frac{k\Delta z}{v_z}\right) C_{av} + C_n \tag{20.4}$$

The application of this equation requires a trial-and-error procedure where the average concentration over the increment is approximated and then checked when C_{n+1} is calculated. One should note that the approximation part of this calculation can be removed by replacing C_{av} by

$$\frac{C_{n+1} + C_n}{2} \tag{20.5}$$

and solving for C_{n+1} directly in terms of C_n.

In the lumped parameter method, the reactor is divided into equal stages and each segment is considered to be a perfectly mixed vessel. This is shown schematically in Figure 20.1. The equation is now written as:

$$\frac{C_{out} - C_{in}}{\Delta z} = -\left(\frac{k}{v_z}\right) C_{out} \tag{20.6}$$

Rearranging the above equation gives

$$C_{out} = \frac{C_{in}}{1 + \dfrac{k\Delta z}{v_z}} \tag{20.7}$$

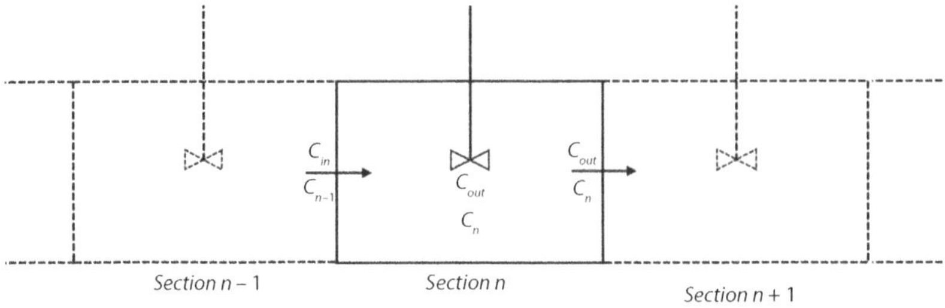

Figure 20.1 Lumped parameter method.

If the initial concentration is C_0 and the term

$$1+\frac{k\Delta z}{v_z}=R \tag{20.8}$$

then the following set of equations are generated for each stage:

$$C_1R-C_0=0 \tag{20.9}$$

$$C_2R-C_1=0 \tag{20.10}$$

$$\vdots$$

$$C_{final}R-C_{final-1}=0 \tag{20.11}$$

The above represents a series of (simultaneous) linear algebraic equations. These equations can be solved by a direct step-by-step hand calculation. Alternatively, a Gauss-Jordan or Gauss-Seidel method can also be used [3–6]. (See also Chapter 17).

The results of the analyses by the two methods are combined and presented in tabular form for comparison along with the analytical solution (see Table 20.1). The reader is left the exercise of calculating the values at $z = 3.0$ ft.

20.2 Runge-Kutta Method

The *Runge-Kutta* (RK) method is one of the most widely used techniques in engineering and science practice for solving first order differential equations. For the equation

$$\frac{dy}{dx}=f(x_n,y_n) \tag{20.12}$$

the solution takes the form

$$y_{n+1}=y_n+\frac{h}{6}(D_1+2D_2+2D_3+D_4) \tag{20.13}$$

Table 20.1 Concentration profile via three methods.

z, ft	Analytical	Finite-difference	Lumped-parameter
0.0	1.0000	1.0000	1.0000
0.3	0.8354	0.8358	0.8474
0.6	0.6980	0.7001	0.7181
0.9	0.5830	0.5863	0.6086
1.2	0.4870	0.4910	0.5157
1.5	0.4070	0.4112	0.4371
1.8	0.3400	0.3444	0.3704
2.1	0.2840	0.2884	0.3139
2.4	0.2370	0.2415	0.2660
2.7	0.1980	0.2022	0.2254

where

$$D_1 = hf(x_n, y_n) \tag{20.14}$$

$$D_2 = hf\left(x_n + \frac{h}{2}, y_n + \frac{D_1}{2}\right) \tag{20.15}$$

$$D_3 = hf\left(x_n + \frac{h}{2}, y_n + \frac{D_2}{2}\right) \tag{20.16}$$

$$D_4 = hf(x_n + h, y_n + D_3) \tag{20.17}$$

The term h represents the increment in x, the term y_n is the solution to the equation at x_n, and y_{n+1} is the solution to the equation at x_{n+1} where $x_{n+1} = x_n + h$. Thus, the RK method provides a straightforward means for developing expressions for Δy, namely, $y_{n+1} - y_n$, in terms of the function $f(x, y)$ at various "locations" along the interval in question.

For a simple equation of the form

$$\frac{dC}{dt} = a + bC \tag{20.18}$$

where at $t = 0$, $C = C_0$, the RK algorithm given above becomes (for $t = h$)

$$C_1 = C_0 + \frac{h}{6}(D_1 + 2D_2 + 2D_3 + D_4) \tag{20.19}$$

and the coefficients are then given by

$$D_1 = hf(x_n, y_n) = h(a + bC_0) \tag{20.20}$$

$$D_2 = hf\left(x_n + \frac{h}{2}, y_n + \frac{D_1}{2}\right) = h\left[a + b\left(C_0 + \frac{D_1}{2}\right)\right] \tag{20.21}$$

$$D_3 = hf\left(x_n + \frac{h}{2}, y_n + \frac{D_2}{2}\right) = h\left[a + b\left(C_0 + \frac{D_2}{2}\right)\right] \tag{20.22}$$

$$D_4 = hf(x_n + h, y_n + D_3) = h[a + b(C_0 + D_3)] \tag{20.23}$$

The same procedure is repeated to obtain values for C_2 at $t = 2h$, C_3 at $t = 3h$, and so on.

The RK method can also be used if the function in question also contains the independent variable. Consider the following equation:

$$\frac{dC}{dt} = f(t, C) \tag{20.24}$$

For this situation, one obtains

$$C_1 = C_0 + \frac{h}{6}(D_1 + 2D_2 + 2D_3 + D_4) \tag{20.25}$$

with

$$D_1 = hf(t_0, C_0) \tag{20.26}$$

$$D_2 = hf\left(t_0 + \frac{h}{2}, C_0 + \frac{D_1}{2}\right) \tag{20.27}$$

$$D_3 = hf\left(t_0 + \frac{h}{2}, C_0 + \frac{D_2}{2}\right) \tag{20.28}$$

$$D_4 = hf(t_0 + h, C_0 + D_3) \tag{20.29}$$

For example, if

$$\frac{dC}{dt} = 5C - e^{-Ct} \tag{20.30}$$

then

$$D_2 = h\left\{5\left(C_0 + \frac{D_1}{2}\right) - e^{\left[-\left(C_0+\frac{D_1}{2}\right)\left(t_0+\frac{h}{2}\right)\right]}\right\} \tag{20.31}$$

etc.

Illustrative Example 20.1 [7]

The equation describing the concentration of a reactant in a tubular flow reactor is given by

$$\frac{dC}{dt} = -0.580C + 6$$

with $C_0 = 5.0$ gmol/cm^3 at $t = 0$ min. Estimate C in gmol/cm^3 at 1 and 5 minutes. Use the Runge-Kutta method analysis [3–6].

Solution

Based on the data provided, evaluate the R–K coefficients for $t = h = 1$:

$$D_1 = 1.0(6.0 - 2.9) = 3.1$$

$$D_2 - 1.0\left[6.0 - 0.58\left(5.0 + \frac{3.1}{2}\right)\right] - 2.2$$

$$D_3 = 1.0\left[6.0 - 0.58\left(5.0 + \frac{2.2}{2}\right)\right] = 2.46$$

$$D_4 = 1.0[6.0 - 0.58(5.0 + 2.46)] = 1.67$$

Calculate C_1:

$$\begin{aligned} C_1 &= C_0 + \frac{h}{6}(D_1 + 2D_2 + 2D_3 + D_4) \\ &= 5.0 + \frac{1}{6}[3.1 + 2(2.2) + 2(2.46) + 1.67] \\ &= 7.35 \text{ gmol/cm}^3 \end{aligned}$$

To calculate C_2:

$$D_1 = 1.0[6.0 - 0.58(7.35)] = 1.74$$

$$D_2 = 1.24$$

$$D_3 = 1.38$$

$$D_4 = 0.94$$

and

$$C_2 = C_1 + \frac{h}{6}(D_1 + 2D_2 + 2D_3 + D_4)$$
$$= 7.34 + \frac{1}{6}[1.74 + 2(1.24) + 2(1.38) + 0.94]$$
$$= 8.66 \text{ gmol/cm}^3$$

Also calculate C_3, C_4, and C_5:

$$C_3 = 9.40 \text{ at } t = 3 \text{ min}$$

$$C_4 = 9.81 \text{ at } t = 4 \text{ min}$$

$$C_5 = 10.04 \text{ at } t = 5 \text{ min}$$

The reader is left the exercise of comparing the numerical solution above (including that at $t = \infty$) with that provided by the analytical solution provided below.

$$C = \frac{5}{29}\left(60 - 31e^{\frac{-29t}{50}}\right)$$

20.3 Runge-Kutta-Gill Method

The RKG method of solving an ordinary differential equation is an extension of, and is similar to, the RK method. The describing equations are given by

$$k_1 = hf(x_n, y_n) \tag{20.32}$$

$$k_2 = hf\left(x_n + \frac{h}{2}, y_n + \frac{k_1}{2}\right) \tag{20.33}$$

$$k_3 = hf\left(x_n + \frac{h}{2}, y_n + \left(-\frac{1}{2} + \frac{1}{v_2}\right)k_1 + \left(1 - \frac{1}{v_2}\right)k_2\right) \tag{20.34}$$

$$k_4 = hf\left(x_n + h,\ y_n + \left(-\frac{1}{v_2}\right)k_2 + \left(1 + \frac{1}{v_2}\right)k_3\right) \tag{20.35}$$

with

$$y_{n+1} = y_n + \frac{1}{6}(k_1 + k_4) + \frac{1}{3}\left(1 - \frac{1}{v_2}\right)k_2 + \frac{1}{3}\left(1 + \frac{1}{v_2}\right)k_3 \tag{20.36}$$

Where $v_2 = 2/\sqrt{2}$

Illustrative Example 20.2

Resolve the previous illustrative example using the RKG method.

Solution

The solution is left as an exercise for the reader , but it can be seen that both methods only diverge from the analytical solution by ~0.04%.

20.4 Several Ordinary Differential Equations

Situations may arise when there is a need to simultaneously solve *more* than one ordinary differential equation (ODE). In a more general case, one could have n dependent variables $y_1, y_2, \ldots, y_n$ with each related to a single independent variable x by the following system of n simultaneous first-order ODEs:

$$\frac{dy_1}{dx} = f_1(x,\ y_1,\ y_2,\ \ldots,\ y_n) \tag{20.37}$$

$$\frac{dy_2}{dx} = f_2(x,\ y_1,\ y_2,\ \ldots,\ y_n) \tag{20.38}$$

$$\vdots$$

$$\frac{dy_n}{dx} = f_n(x,\ y_1,\ y_2,\ \ldots,\ y_n) \tag{20.39}$$

Note that Equations (20.37–20.39) are interrelated, i.e., they are dependent on each other.

An outline to the solution is illustrated in the following two equations:

$$\frac{dC}{dt} = -Ae^{-\frac{E}{RT}}C = f(C,\ t) \tag{20.40}$$

$$\frac{dT}{dt} = -kC\frac{\Delta H}{\rho C_p} = g(C,\ t) \tag{20.41}$$

or in a more general sense

$$\frac{dy}{dx} = f(x,\ y,\ z);\ (e.g.,\ xyz) \tag{20.42}$$

$$\frac{dz}{dx} = g(x,\ y,\ z);\ (e.g.,\ x^2 y^2 e^{-z}) \tag{20.43}$$

The RK algorithm for Equations (20.42) and (20.43) is

$$y_1 = y_0 + \frac{1}{6}(RY_1 + 2RY_2 + 2RY_3 + RY_4) \tag{20.44}$$

$$z_1 = z_0 + \frac{1}{6}(RZ_1 + 2RZ_2 + 2RZ_3 + RZ_4) \tag{20.45}$$

where $y_1 - y_0 = \Delta y$, $z_1 - z_0 = \Delta z$, $h = \Delta x$ and

$$RY_1 = h\,f(x_0,\ y_0,\ z_0) \tag{20.46}$$

$$RZ_1 = h\,g(x_0,\ y_0,\ z_0) \tag{20.47}$$

$$RY_2 = h\,f\left(x_0 + \frac{h}{2},\ y_0 + \frac{RY_1}{2},\ z_0 + \frac{RZ_1}{2}\right) \tag{20.48}$$

$$RZ_2 = h\,g\left(x_0 + \frac{h}{2},\ y_0 + \frac{RY_1}{2},\ z_0 + \frac{RZ_1}{2}\right) \tag{20.49}$$

$$RY_3 = h\,f\left(x_0 + \frac{h}{2},\ y_0 + \frac{RY_2}{2},\ z_0 + \frac{RZ_2}{2}\right) \tag{20.50}$$

$$RZ_3 = h\,g\left(x_0 + \frac{h}{2},\ y_0 + \frac{RY_2}{2},\ z_0 + \frac{RZ_2}{2}\right) \tag{20.51}$$

$$RY_4 = h\,f(x_0 + h,\ y_0 + RY_3,\ z_0 + RZ_3) \tag{20.52}$$

$$RZ_4 = h\,g(x_0 + h,\ y_0 + RY_3,\ z_0 + RZ_3) \tag{20.53}$$

20.5 Higher Order Ordinary Differential Equations

Although the RK approach (and other companion methods) have traditionally been employed to solve first-order ODEs, it can also treat higher order ODEs. The procedure requires reducing an nth order ODE to n first-order ODEs. For example, if the equation is of the form [8]

$$\frac{d^2y}{dx^2} = f(y, x) \tag{20.54}$$

set

$$z = \frac{dy}{dx} \tag{20.55}$$

so that

$$\frac{dz}{dx} = \frac{d^2y}{dx^2} \tag{20.56}$$

The second order equation in Equation (20.54) has now been reduced to the two first-order ODEs in Equation (20.57):

$$\begin{aligned} \frac{d^2y}{dx^2} &= \frac{dz}{dx} \\ &= f(y, x) \end{aligned}$$

$$\frac{dy}{dx} = z \tag{20.57}$$

The procedure set forth earlier can now be applied to generate a solution to Equation (20.54). Note, however, that the first derivative (i.e., dy/dx or its estimate) is required at the start of the integration. Extending the procedure to higher order equations is left as an exercise for the reader.

The selection of increment size remains a variable to the practicing engineer. Most numerical analysis methods provided in the literature are not concerned with error analysis. In general, roundoff errors appear as demonstrated in Figure 20.2. In the limit, when the increment approaches 0, one obtains a solution approaching that analytically. Roundoff error then increases exponentially as the increment approaches 0 as the number of computations correspondingly increases. However, selecting the increment size that will minimize the error is rarely a problem; in addition, computing it is also rarely a problem.

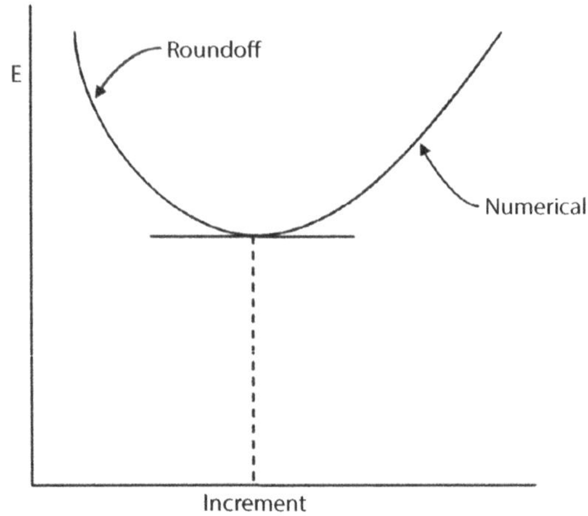

Figure 20.2 Error analysis.

References

1. L. Theodore, *Chemical Reaction Analysis and Applications for the Practicing Engineer*, John Wiley & Sons, Hoboken, NJ, 2011.
2. *L. Theodore, F. Ricci, and T. VanVliet, Thermodynamics for the Practicing Engineer*, John Wiley & Sons, Hoboken, NJ, 2009.
3. M. Moyle, *Introduction to Computers for Engineers*, John Wiley & Sons, Hoboken, NJ, 1967.
4. B. Carnahan and J. Wilkes, *Digital Computing and Numerical Methods*, John Wiley & Sons, Hoboken, NJ, 1973.
5. R. Ketter and S. Prawler, *Modern Methods of Engineering Computations*, McGraw-Hill, New York City, NY, 1969.
6. J. Reynolds, J. Jeris, and L. Theodore, *Handbook of Chemical and Environmental Engineering Calculations*, John Wiley & Sons, Hoboken, NJ, 2004.
7. L. Theodore, *Chemical Reaction Kinetics*, A Theodore Tutorial, Theodore Tutorials, East Williston, NY, originally published by USEPA/APTI, RTP, NC, 1994.
8. L. Theodore, personal notes, East Williston, NY, 1970.

21

Partial Differential Equations

Technical Consultant: Megan Menzel

Many practical problems in environmental engineering and science involve at least two independent variables i.e., the dependent variable is defined in terms of (or is a function of) more than one independent variable. The derivatives describing these independent variables are defined as partial derivatives. Differential equations containing partial derivatives are referred to as partial differential equations (PDEs).

Contrary to a widely accepted myth, an engineer's mathematical obligations do not end after formulating a problem, where it may be given to a mathematician to solve. Even if such an ideal situation should exist, it is still necessary for engineers – and this includes environmental engineers and scientists – to have a reasonable understanding of the mathematical methods and their limitations in order to interpret results.

It has been said that "the solution of a partial differential equation is essentially a guessing game." In other words, one cannot expect to be given a formal method that will yield exact solutions for all partial differential equations [1]. Fortunately, numerical methods for solving these equations were developed during the middle and latter part of the 20th century.

Some of the applications in environmental engineering and science fall naturally into partial differential equations of second order, although there are exceptions in elasticity, vibration theory, and elsewhere. The solutions of problems involving partial differential equations often revolve about an attempt to reduce the partial differential equation to one or more ordinary differential equations. The solutions of the ordinary differential equations are then combined (if possible) so that the boundary conditions as well as the original partial differential equation are simultaneously satisfied. It should be noted that most of the computational methods for partial differential equations have evolved from finite-difference approximations for the partial derivatives. Partial derivatives can be approximated by finite differences in many ways depending on accuracy requirements.

As noted, this concluding chapter in Part III is concerned with the numerical solutions of partial differential equations.

Four sections complement the presentation of this chapter. Section numbers and subject titles follow:

21.1 Partial Differential Equation (PDE) Classification

This chapter will examine partial differential equations of the general form

$$a\frac{\partial^2 T}{\partial x^2}+b\frac{\partial^2 T}{\partial x \partial y}+c\frac{\partial^2 T}{\partial y^2}=f(x, y, T) \tag{21.1}$$

where, in general, the coefficients a, b, and c are functions of x and y, and f is a function of x, y, and T. Such equations arise in environmental work involving heat transfer, boundary-layer flow, vibrations, elasticity, and so on. However, the scope of this chapter is not to present a comprehensive coverage of the various numerical methods for solving partial differential equations. Rather, the material presented will provide an introduction to the subject. The various forms of Equation (21.1) may take are classified as *elliptic*, *parabolic*, or *hyperbolic*. Some details are provided in this chapter. These three PDEs are briefly introduced below employing T (e.g., the temperature as the dependent variable) with t (time) and x, y, z (position) as the independent variables.

The parabolic equation is:

$$\frac{\partial T}{\partial t}=a\frac{\partial^2 T}{\partial z^2} \tag{21.2}$$

The elliptical equation is:

$$\frac{\partial^2 T}{\partial x^2}+\frac{\partial^2 T}{\partial y^2}=0 \tag{21.3}$$

The hyperbolic equation is:

$$\frac{\partial^2 T}{\partial t^2} = a\frac{\partial^2 T}{\partial x^2} \tag{21.4}$$

Regarding the hyperbolic equations, it should be noted that the coefficients a, b, and c in Equation (21.1) have values such that

$$b^2 - 4ac > 0 \tag{21.5}$$

The *one-dimensional wave equation*

$$\frac{\partial^2 u}{\partial t^2} = a^2\frac{\partial^2 u}{\partial x^2} \tag{21.6}$$

which is frequently encountered in the areas of applied physics, is an example of a hyperbolic partial differential equation. This equation has been used to describe the motion of various types of systems.

As noted, the preferred numerical method of solution involves finite differencing. Only the parabolic and elliptical equations are considered in the next two sections.

21.2 Parabolic Partial Differential Equations

Parabolic differential equations arise in what some refer to as *propagation* problems. In this type of system, the solution advances away from specified initial and boundary conditions. The general partial differential equation, as provided by Equation (21.1), is defined as *parabolic* when the coefficients a, b, and c have values such that

$$b^2 - 4ac = 0 \tag{21.7}$$

Examples of parabolic PDEs include

$$\frac{\partial T}{\partial t} = a\frac{\partial^2 T}{\partial x^2} \tag{21.8}$$

and (the two-dimensional)

$$\frac{\partial T}{\partial t} = a\left[\frac{\partial^2 T}{\partial x^2} + \frac{\partial^2 T}{\partial y^2}\right] \tag{21.9}$$

Ketter and Prawler [2], as well as many others, have reviewed the finite difference approach to solving Equation (21.8). This is detailed below.

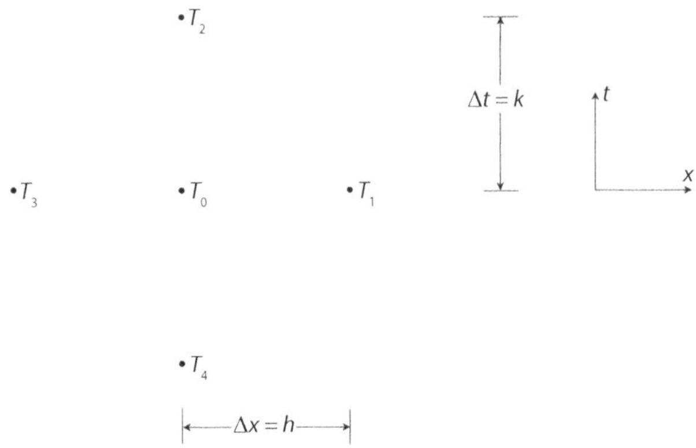

Figure 21.1 Parabolic grid.

Consider the (t, x) grid provided in Figure 21.1. The partial derivatives may be approximated by

$$\frac{\partial T}{\partial t} \cong \frac{\Delta T}{\Delta t} = \frac{-T_4 + T_2}{2(\Delta t)} = \frac{-T_4 + T_2}{2k}; \Delta t = k \tag{21.10}$$

and

$$\frac{\partial T^2}{\partial x^2} \cong \frac{\Delta}{\Delta x}\left(\frac{\Delta T}{\Delta x}\right) = \frac{T_3 - 2T_0 - T_1}{h^2}; \Delta x = h \tag{21.11}$$

Substituting Equations (21.10) and (21.11) into Equation (21.8) leads to

$$\frac{-T_4 - T_2}{2k} = \frac{T_3 - 2T_0 - T_1}{h^2} \tag{21.12}$$

Solving for T_2:

$$T_2 = T_4 + 2r(T_3 - 2T_0 - T_1) \tag{21.13}$$

where $r = k/h^2$.

Thus, T_2 may be calculated if T_0, T_1, T_3, and T_4 are known. Unfortunately, stability and error problems arise in employing the above approach. These can be removed by replacing the central difference term in Equation (21.10) by a forward difference term, i.e.,

$$\frac{\partial T}{\partial t} \cong \frac{\Delta T}{\Delta t} = \frac{-T_0 + T_2}{(\Delta t)} = \frac{-T_0 + T_2}{k} \tag{21.14}$$

With this substitution, Equation (21.13) becomes

$$T_2 = T_0 + r(T_3 - 2T_0 + T_1) \tag{21.15}$$

It can be shown that the problem associated with the central difference derivative is removed if $r \leq 0.5$.

Consider the following application. An unsteady-state heat conduction experiment conducted at Manhattan College's unit operations laboratory is concerned with the application of this parabolic equation. In the experiment, a 316 stainless steel rod is heated at each end with steam at temperature T_S. The initial temperature of the rod is T_A. The describing equation and *Ba/o* IC(s) for this system are

$$\frac{\partial T}{\partial t} = a\frac{\partial^2 T}{\partial x^2} \tag{21.16}$$

$$BC(1)\text{: } T = T_S \text{ at } x = 0$$

$$BC(2)\text{: } T = T_S \text{ at } x = L$$

$$\text{IC: } T = T_A \text{ at } t = 0,\ 0 \leq x \leq L$$

The analytical solution to this equation can be shown [3] to be:

$$T = T_S + (T_A - T_S)\left\{\sum_{n=1}^{\infty} 2\left[\frac{(-1)^{n+1}+1}{n\pi}\right] e^{-a\left(\frac{n\pi}{2}\right)^2 t} \sin\left(\frac{n\pi x}{2}\right)\right\} \tag{21.17}$$

The reader may choose to solve this problem numerically.

21.3 Parabolic PDE with Three Independent Variables

The previous section on parabolic PDEs was limited to systems involving only two independent variables. Problems involving three or more independent variables may be handled by analogous methods although the calculations become more tedious. For example, consider the simple unsteady-state heat-conduction equation for a solid in two linear dimensions:

$$\frac{\partial T}{\partial t} = a\left(\frac{\partial^2 T}{\partial x^2} + \frac{\partial^2 T}{\partial y^2}\right) \tag{21.18}$$

To generate a solution, divide the solid into a series of rectangles, each of length Δx and of width Δy. Let $m\Delta x$ denote the x distance of a point in the solid from a suitable reference y axis, $p\Delta y$ denote the y distance of the same point from a suitable reference x axis, and $n\Delta t$ denote the elapsed time since the start of the calculational process. The finite difference equivalent of Equation (21.18) is approximately given by

$$T_{m,p,n+1} - T_{m,p,n} = a\Delta t\left(\frac{T_{m+1,p,n} - 2T_{m,p,n} + T_{m-1,p,n}}{\Delta x^2} + \frac{T_{m,p+1,n} - 2T_{m,p,n} + T_{m,p-1,n}}{\Delta y^2}\right) \tag{21.19}$$

One again, the concept of a modulus may be employed:

$$R = a\frac{\Delta x^2}{\Delta t} \tag{21.20}$$

$$S = a\frac{\Delta y^2}{\Delta t} \tag{21.21}$$

The incremental changes in Δx and Δy need not be equal, but Δt must be independent of the position in space. Equation (21.19) with the moduli R and S is similar to the finite difference approximation to the one-distance coordinate case, and the solution of Equation (21.19) proceeds in a similar manner; however the boundary conditions are somewhat more troublesome.

21.4 Elliptical Partial Differential Equations

For this equation examine the grid in Figure 21.2 [2]. Using finite differences to replace the derivatives in Equation (21.3) ultimately leads to

$$T_0 = \frac{1}{4}(T_1 + T_2 + T_3 + T_4);\ \Delta x = \Delta y \tag{21.22}$$

In effect, each T value calculated reduces to the average of its four nearest neighbors in the square grid. This difference equation may then be written at each interior grid point, resulting in a linear system of N equations, where N is the number of grid points. The system can then be solved by one of several methods provided in the literature [1,2].

Another solution method involves applying the *Monte Carlo* approach, requiring the use of random numbers [1,2]. Consider the squares shown in Figure 21.3. If the describing equation for the variation of T within the grid structure is

$$\frac{\partial^2 T}{\partial x^2} + \frac{\partial^2 T}{\partial y^2} = 0 \tag{21.23}$$

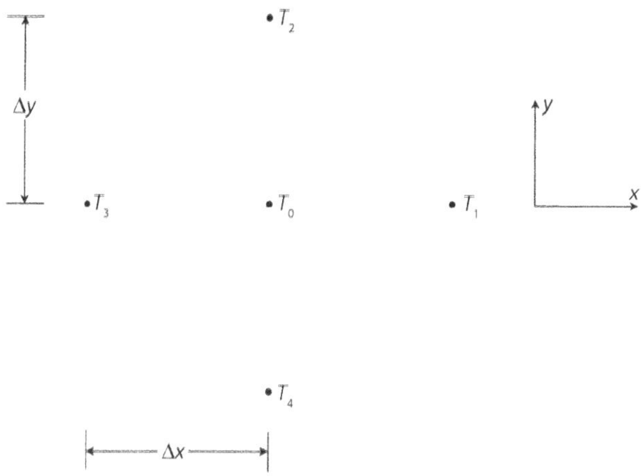

Figure 21.2 Partial differential equation elliptical grid.

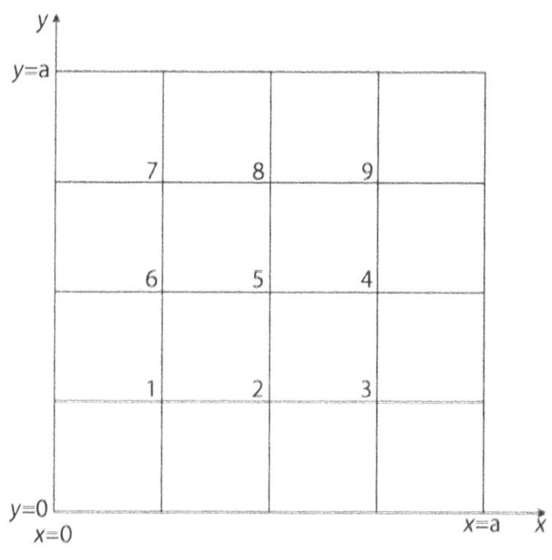

Figure 21.3 Monte Carlo grid approach.

with specified boundary conditions (BCs) for $T(x, y)$ of $T(0, y)$, $T(a, y)$, $T(x, 0)$, and $T(x, a)$, one may employ the following approach to generate a solution

1. Proceed to calculate T at point 1 (i.e., T_1).
2. Generate a random number between 00 and 99.
3. If the number is between 00 and 24, move to the left of point 1. For 25 to 49, 50 to 74, 75 to 99, move upwards, to the right, and downwards of point 1, respectively.
4. If the move in step 3 results in a new position that is at the outer surface (boundary), terminate the first calculation for point 1 and

record the T value of the boundary at that new position. However, if the move results in a new position that is not at a boundary, and is still at one of the other eight interval grid points, repeat steps 2 and 3. This process is continued until an outer surface boundary is reached.

5. Repeat steps 2 to 4 numerous times, for example, 1000 times.
6. After completing step 5, sum all the T values obtained and divide this value by the number of times steps (2 to 4) have been repeated. The resulting value provides a reasonable estimate of T_1.
7. Return to step 1 and repeat the calculation for the remaining eight grid points.

This method of solution is not limited to square systems. In addition, one of the authors [5] has applied this method of solution to numerous real-world problems.

Illustrative Example 21.1

Consider the system pictured in Figure 21.4. If the system is solid and the variable T is the temperature, the elliptical equation provided in Equation (21.3) applies. For this system, the temperature at each location is given by the average temperature of its four neighboring points. Using the Monte Carlo procedure provided above *and* in Part IV, Chapter 27, generate the temperature profile in the solid. The base temperature is maintained at 100°C and the other three surfaces at 0°C.

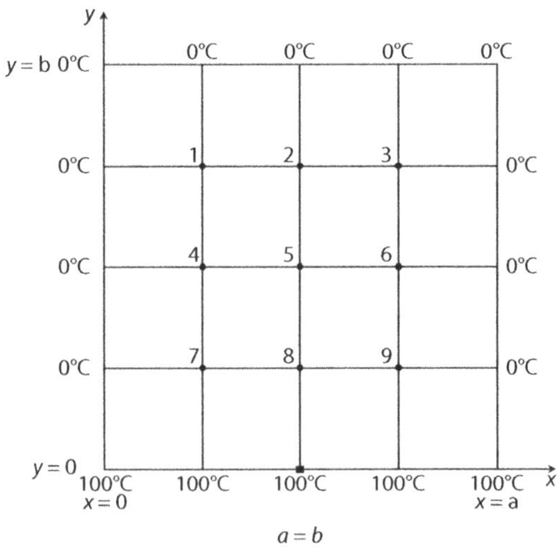

Figure 21.4 Temperature grid for a square.

Solution

Using the Monte Carlo algorithm and a random number generator leads to the following results:

$$T_1 = 7.07°C$$

$$T_2 = 9.80°C$$

$$T_3 = 7.16°C$$

$$T_4 = 18.67°C$$

$$T_5 = 25.12°C$$

$$T_6 = 18.77°C$$

$$T_7 = 42.93°C$$

$$T_8 = 52.57°C$$

$$T_9 = 42.80°C$$

This procedure can be extended to rectangles; it is not limited to squares. If a rectangle is subjected to the following (somewhat similar) boundary conditions,

$$T(x = 0,\ \text{any}\ y) = T_0$$

$$T(x = b,\ \text{any}\ y) = 0$$

$$T(\text{any}\ x,\ y = +a) = 0$$

$$T(\text{any}\ x,\ y = -a) = 0$$

The analytical solution for the above rectangle ($2a$ by b) [4] is

$$T = 2T_0 \sum_{n=0}^{\infty} \left\{ \frac{(-1)^n}{\lambda_n a} \right\} \left\{ \frac{\sinh[\lambda_n (b - x)]}{\sinh(\lambda_n b)} \right\} \cos(\lambda_n y)$$

where

$$\lambda_n = \frac{(2n+1)\pi}{2a};\ n = 0,\ 1,\ 2,\ 3,\ \ldots$$

Illustrative Example 21.2

Solve Illustrative Example 21.1 using the Gauss-Elimination method.

Solution [5]

The familiar finite difference approach will be employed. For example,

$$T_5 = \frac{1}{4}[T_2 + T_4 + T_6 + T_8]$$

This form of the equation may be applied to the nine points, which sets each T equal to the average of temperature of its four neighboring points. This produces a set of nine equations with nine unknowns. Employing Mathematica, Mathcad, Excel, or a similar program leads to

$$T_1 = 7.14°\text{C}$$

$$T_2 = 9.82°\text{C}$$

$$T_3 = 7.14°\text{C}$$

$$T_4 = 18.75°\text{C}$$

$$T_5 = 25.00°\text{C}$$

$$T_6 = 18.75°\text{C}$$

$$T_7 = 42.86°\text{C}$$

$$T_8 = 52.68°\text{C}$$

$$T_9 = 42.86°\text{C}$$

Illustrative Example 21.3

Comment on the results of the two previous examples.

Solution

As expected the results are in reasonable agreement with each other. In addition,

$$T_8 > T_7 > T_5 > T_4 > T_2 > T_1$$

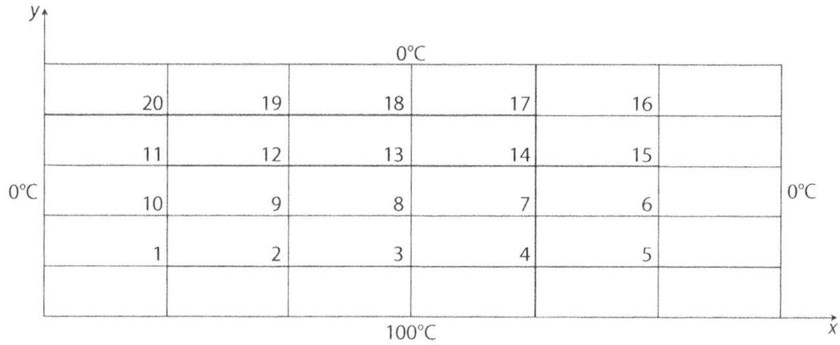

Figure 21.5 Rectangular grid.

with

$$T_1 \cong T_3$$

$$T_4 \cong T_6$$

$$T_7 \cong T_9$$

Illustrative Example 21.4

Outline how to solve the previous example if the face surface is a rectangular solid (see Figure 21.5).

Solution

The outline of the calculation presented in the previous examples remains the same. The reader should note that this illustrative example, as well as Illustrative Example 21.1, will be revisited in Part IV, Chapter 27.

References

1. B. Carnahan and J. Wilkes, *Digital Computing and Numerical Methods*, John Wiley & Sons, Hoboken, NJ, 1973.
2. R. L. Ketter and S. P. Prawler, *Modern Methods of Engineering Computations*, McGraw-Hill, New York City, NY, 1969.
3. L. Theodore, personal notes, East Williston, NY, 1965.
4. L. Theodore, personal notes, East Williston, NY, 1969.
5. L. Theodore, personal notes, East Williston, NY, 1985.

Part IV
STATISTICAL ANALYSIS

It is no secret that the teaching of probability and statistics is now considered essential in most curricula and is generally accepted as one of the fundamental branches of mathematics. This course, or its equivalent, is now a required undergraduate course at most colleges and universities and is slowly and justifiably finding its way into most environmental curricula.

Texts in this field are considered by some to be too advanced for the undergraduate student. This Part of the text is intended to overcome the difficulties and sometimes frightening experiences a beginning student encounters on being introduced to this subject. The authors' aim is to offer the reader the fundamentals of probability and statistics with appropriate environmental applications and to serve as an introduction to the specialized and more sophisticated texts in this area.

This Part has primarily evolved from a Theodore Tutorial prepared by Theodore and Taylor titled "*Probability and Statistics*." Material has also been drawn from notes prepared by one of the authors (many years ago) for a one-semester three-credit course given to senior students at Manhattan College; the course was also offered as an elective to other disciplines.

An important point needs to be made, and this applies to other parts of the book as well. There have been numerous occasions during one of the authors' tenure as an educator when students solved problems using packaged programs such as Excel, MathCAD, etc. For example, the problem could have involved a solution to a differential equation or the regression of some data. On being questioned how the packaged program performed the calculation, the student almost always responded with something to the effect of, "I don't know and I don't care." For this reason, the reader should note that no attempt was made to provide details on applied packaged computer programs that are presently available for this subject.

The author is deeply indebted to the aforementioned Frank Taylor who in a very real sense allowed this Part of the text to become a reality.

There are seven chapters in Part IV. The chapter numbers and accompanying titles are listed below.

Chapter 22: Basic Probability Concepts
Chapter 23: Estimation of the Mean and Variance
Chapter 24: Discrete Probability Distributions
Chapter 25: Continuous Probability Distributions
Chapter 26: Fault and Event Trees
Chapter 27: Monte Carlo Simulation
Chapter 28: Regression Analysis

22

Basic Probability Concepts

Probabilities are nonnegative numbers associated with the outcomes of so-called random experiments. A random experiment is an experiment whose outcome is uncertain. Examples include throwing a pair of dice, tossing a coin, counting the number of defectives in a sample from a lot of manufactured items, and observing the time to failure of a tube in a heat exchanger or a seal in a pump or a bus section in an electrostatic precipitator. The set of possible outcomes of a random experiment is called a *sample space* and is usually designated by S. The term $P(A)$, the probability of an event A, is the sum of the probabilities assigned to the outcomes constituting the subset A of the sample space S.

Consider, for example, tossing a coin twice. The sample space can be described as

$$S = \{HH,\ HT,\ TH,\ TT\} \tag{22.1}$$

If probability 1/4 is assigned to each element of S, and A is the event of at least one head, then

$$A = \{\mathrm{HH},\ \mathrm{HT},\ \mathrm{TH}\} \tag{22.2}$$

the sum of the probabilities assigned to the elements of A is 3/4. Therefore $P(A)=$ 3/4. The description of the sample space is not unique. The sample space S in the case of tossing a coin twice could be described in terms of the number of heads obtained. Then

$$S = \{0, 1, 2\} \tag{22.3}$$

Suppose probabilities 1/4, 1/2, and 1/4 are assigned to the outcomes 0, 1, and 2, respectively. Then A, the event of at *least* one head, would have for its probability,

$$P(\text{A}) = P\{1,2\} = \frac{3}{4} \tag{22.4}$$

Three sections complement the presentation of this chapter. Section numbers and subject titles follow:

22.1: Probability Definitions
22.2: Permutations and Combinations
22.3: Series and Parallel Systems

The bulk of the material in this chapter was drawn from the work of Theodore and Taylor [1].

22.1 Probability Definitions

How the probabilities are assigned to the elements of the sample space depends on the desired interpretation of the probability of an event. The aforementioned $P(A)$, can be interpreted as a *theoretical relative frequency*, i.e., a number about which the relative frequency of event A tends to cluster as n, the number of times the random experiment is preformed, increases indefinitely. This is the objective interpretation of probability. Under this interpretation, to say that $P(A)$ is 3/4 in the aforementioned example means that if a coin is tossed twice and this trial is repeated n times, the proportion of times one or more heads occurs clusters about 3/4 as n increases indefinitely.

As another example, consider a single valve that can stick in an open (O) or closed (C) position. The sample space can be described as follows:

$$S = \{O, C\} \tag{22.5}$$

Suppose that the valve sticks twice as often in the open position as it does in the closed position. Under the theoretical relative frequency interpretation, the probability assigned to element O in S would be 2/3, twice the probability assigned to element C. If two such valves are observed, the sample space S can be described as

$$S = \{OO,\ OC,\ CO,\ CC\} \tag{22.6}$$

Assuming that the two valves operate independently, a reasonable assignment of probabilities to the elements of S, as just listed above, should be 4/9, 2/9, 2/9, and 1/9 respectively. The reason for this assignment will become clear after consideration of the concept of independence. If A is the event of at least one valve sticking in the closed position, then

$$A = \{OC,\ CO,\ CC\} \tag{22.7}$$

the sum of the probabilities assigned to the elements of A is 5/9. Therefore, $P(A) = 5/9$.

Probability $P(A)$ can also be interpreted subjectively as a measure of degree of belief, on a scale from 0 to1, that event A occurs. This interpretation is frequently used in ordinary conversation. For example, if someone says, "The possibility that I will go to the casino tonight is 90%," then 90% is a measure of the person's belief that he or she will go to a casino. This interpretation is also used when, in the absence of concrete data needed to estimate an unknown probability based on an observed relative frequency, the opinion of an expert is sought. For example, an expert may be asked to estimate the probability that the seals in a newly designed pump will leak at high pressures. The estimate would be based on the expert's familiarity with pumps of a similar design.

Illustrative Example 22.1

Discuss the difference between a sample and a population.

Solution

When there is a set of n observations, one may wish to use the values of these observations to estimate certain characteristics of a larger number of observations that have not yet been made. In other words, the interest is not directly on a particular set of observations, but rather in using them to estimate something about a potentially larger set. One refers to the observations as a *sample*. The larger "supply," which one may have, is called a *population*. From the characteristics of a sample of observations, one can estimate similar characteristics for the population.

Illustrative Example 22.2

Discuss the difference between a statistic and a parameter.

Solution

When one calculates some measured value of a sample, it is referred to as a *statistic*. A statistic is any characteristic of a set of observations calculated from a sample. The measurement corresponding to a statistic obtained from a population is referred to as a *parameter*. In most cases of interest, the values of the parameters are unknown and must be estimated by the statistics derived from a sample. Thus, the sample value is an *estimate* of the population value.

22.2 Permutations and Combinations

The problem with calculating probabilities of *objects* or *events* in a finite group – defined above as the *sample space* – in which equal probabilities are assigned to the elements in the sample space requires counting the elements which make up the events. The counting of such events is often greatly simplified by employing the rules for permutations and combinations.

Permutations and combinations deal with the grouping and arrangement of objects or events. By definition, each different order or arrangement *with regard* to order of all or part of the objects is called a *permutation*. Alternately, each of the sets which can be made by using all or part of a given collection of objects *without regard* to order of the objects in the set is called a *combination*. Although, permutations or combinations can be obtained *with replacement* or *without replacement*, most analyses of permutations and combinations are based on sampling that is performed without replacement; i.e., each object or element can be used only once. For each of the two aforementioned *with/without* pairs (with/without regard to order and with/without replacement), four subsets if two may be drawn. These four are provided in Table 22.1 [2].

Each of the four paired subsets in Table 22.1 is considered in the following text with accompanying examples based on the letters A, B, and C. To personalize this, the reader could consider the options (games of chance) one of the authors faces while on a one-day visit to a casino. The only three options normally considered are dice (often referred to as craps), blackjack (occasionally referred to as 21), and pari-mutual (horses, trotters, dogs, and jai alai) simulcasting betting. All three of these may be played during a visit, although playing two or only one is also an option. In addition, the order may vary and the option may be repeated. Some possibilities include the following:

Dice, blackjack, and simulcast wagering
Blackjack, wagering, and dice
Wagering, dice, and wagering
Wagering and dice (one of the author's usual sequence)
Blackjack, blackjack (following a break), and dice

In order to simplify the four examples that follow, dice, blackjack, and wagering are referred to as *objects*, represented by the letters A, B, and C, respectively.

Table 22.1 Subsets of permutations and combinations.

Permutations (with regard to order)	**Combinations (without regard to order)**
Without replacement	Without replacement
With replacement	With replacement

1. Consider a scenario that involves three separate objects, A, B, and C. The arrangement of these objects is called a *permutation*. There are six different orders or permutations of these three objects possible, note that ABC ≠ CBA.

ABC BAC CAB

ACB BCA CBA

Thus, BCA would represent blackjack, wagering, and dice.

The number of different permutations of n objects is always equal to $n!$, where $n!$ is normally referred to as a factorial n. Factorial n or $n!$ is defined as the product of the n objects taken n at a time and denoted as $P(n, n)$. Thus,

$$P(n, n) = n! \tag{22.8}$$

With three objects, the number of permutations is $3! = 3 \times 2 \times 1 = 6$. Note that $0!$ is 1.

The number of different permutations of n objects taken r at a time is given by

$$P(n, r) = \frac{n!}{(n-r)!} \tag{22.9}$$

[**Note**: The permutation term P also appears in the literature as ${}^{n}P_{r}$ or $P(_{r}^{n})$.] For the three objects A, B, and C, taken two at a time, $n = 3$, and $r = 2$. Thus,

$$P(3, 2) = \frac{3!}{(3-2)!} = \frac{(3)(2)(1)}{1} = 6 \tag{22.10}$$

These possible different orders, noting once again that AB ≠ BC, are

AB BC AC

BA CB CA

Consider now a scenario involving n objects in which these can be divided into j sets with the objects within each set being alike. If $r_1, r_2, \ldots, r_j$ represents the number of objects within each of the respective sets, with $n = r_1 + r_2 + \ldots + r_j$, then the number of permutations of the n objects is given by

$$P(n; r_1, r_2, \ldots, r_j) = \frac{n!}{r_1!r_2!\ldots r_j!} \tag{22.11}$$

This represents the number permutations of n objects of which r_1 are alike, r_2 are alike, and so on. Consider, for example, 2 A's, 1B, and 1C; the number of permutations of these 4 objects is

$$P(4;2,1,1) = \frac{4!}{2!1!1!} = 12 \tag{22.12}$$

The 12 permutations for this scenario are as follows:

AABC ABAC ABCA BAAC BACA BCAA

AACB ACAB ACBA CAAB CABA CBAA

2. Consider the arrangement of the same three objects above, but obtain the number of permutations (*with* regard to order) and *with* replacement, PR [1]. There are 27 different permutations possible.

AAA BBB CCC

AAB BBA CCB

AAC BBC CCA

ABA BAB CBC

ABB BAA CBB

ACA BCB CAC

ACC BCC CAA

ABC BAC CBA

ACB BCA CAB

For this scenario,

$$PR(n,n) = (n)^n \tag{22.13}$$

so that

$$PR(3, 3) = (3)^3 = 27 \tag{22.14}$$

For n objects taken r at a time,

$$PR(n, r) = (n)^r \tag{22.15}$$

3. The number of different ways in which one can select r objects from a set of n *without* regard to order (i.e., the order does not count) and *without* replacement is defined as the number of *combinations*, C, of the n objects taken r at a time. The number of combinations of n objects taken r at a time is given by

$$C(n, r) = \frac{P(n, r)}{r!} = \frac{n!}{r!(n-r)!} \tag{22.16}$$

[Note: The combination term C also appears in the literature as C_r^n or $C(_r^n)$.] For the ABC example taken two letters at a time, one has

$$\begin{aligned} C(3, 2) &= \frac{3!}{2!1!} \\ &= \frac{(3)(2)(1)}{(2)(1)(1)} \\ &= 3 \end{aligned}$$

The number of combinations becomes

AB AC BC

Note that the combination BA is not included since AB = BA for combinations.

4. The arrangement of n objects, taken r at a time *without* regard to the order and *with* replacement is denoted by CR and given by [1]

$$CR(n, r) = C(n, r) + (n)^{n-1} \tag{22.17}$$

with

$$CR(n, n) = C(n, n) + (n)^{n-1}; \; r = n \tag{22.18}$$

For the ABC example above taken two letters at a time, one employs Equation (22.17).

$$\begin{aligned} CR &= C(3, 2) + (3)^{2-1} \\ &= 3 + 3 = 6 \end{aligned} \tag{22.19}$$

Table 22.2 Describing equations for permutations and combinations.

	Without replacement	With replacement	Type
With regard to order	$P(n,r)=\dfrac{n!}{(n-r)!}$ (Equation 22.9)	$PR(n,r)=(n)^r$ (Equation 22.15)	Permutation
Without regard to order	$C(n,r)=\dfrac{n!}{r!(n-r)!}$ (Equation 22.16)	$CR(n,r)=C(n,r)+(n)^{r-1}$ (Equation 22.17)	Combination

The number of combinations becomes

AA BB CC

AB AC BC

Table 22.2 may now be written to include the describing equation for each of the above four subsets [2].

Illustrative Example 22.3

Discuss the relative advantages or disadvantages of identifying an experimental laboratory sample as follows:

1. With three numbers
2. With three letters

Solution

This is an example of sampling with regard to order and with replacement because the numbers may be replaced (reused) or replicated. For this case, Equation (22.15) is employed.

1. Three numbers give the following solution:

$$PR(n,r)=(n)^r;\ n=10,\ r=3$$
$$=(10)(10)(10)=10^3=1{,}000$$

2. With three letters, the solution is as follows:

$$PR(26,3)=(26)(26)(26)=(26)^3$$
$$=17{,}576$$

Obviously, the latter choice (three letters) provides greater flexibility because numerous identification possibilities are available.

Illustrative Example 22.4

Determine the number of 4-element chemical compounds that can theoretically be generated from a pool of 112 elements. Assume each element counts only once in the chemical formula and that the order of the elements in the compound matters. An example of a three-element compound is H_2SO_4 (sulfuric acid) or CH_3OH (methanol). An example of a four-element compound is $NaNCO_3$.

Solution

As discussed earlier, each different ordering or arrangement of all or part of a number of symbols (or objects) in which the order matters is defined as a *permutation*. There are 112 elements to choose from in this application. For this problem, the describing equation is given by Equation (22.9).

$$P(n,r)=\frac{n!}{(n-r)!}$$

Based on the problem statement, $n = 112$ and $r = 4$. Therefore,

$$P(112,4)=\frac{112!}{(112-4)!}=\frac{112!}{108!}$$
$$=(112)(111)(110)(109)=1.49\times10^{8}$$

This is a large number. This number would be further increased if the number of a particular element appearing in the chemical formula was greater than one, for example, HCN (hydrogen cyanide) vs. C_3H_3N (acrylonitrile). However, a more realistic scenario would involve a calculation in which the order does not matter. This is left as an exercise for the reader.

Two points need to be made, one concerning the calculation and the other concerning the chemistry of the compound:

1. For calculations involving large factorials, it is often convenient to use an approximation known as *Stirling's formula*:

$$n!\approx(2\pi)^{0.5}\,e^{-n}\,n^{n+0.5} \tag{22.20}$$

2. For a real-world "viable" compound, the elements involved in this application must be capable of bonding.

Illustrative Example 22.5

Exposure to a nanoagent has four ways in which it can lead to a major health effect and twelve ways to a minor effect. How many ways may an individual be exposed to the following?

1. Two major and two minor health effects
2. One major and four minor health effects

Solution

Because the order in which the effects appear or impact on an individual does not matter, an analysis involving combinations without replacement is required. Employing Equation (22.16), the number of combined health effects (HE) is thus:

1. For two major and two minor effects:

$$HE = C(4,2)C(12,2) = \left(\frac{4!}{2!2!}\right)\left(\frac{12!}{10!2!}\right)$$

$$= \left(\frac{4\times 3}{2}\right)\left(\frac{12\times 11}{2}\right) = 396$$

2. For one major and four minor effects:

$$HE = C(4,1)C(12,4) = \left(\frac{4!}{3!1!}\right)\left(\frac{12!}{8!4!}\right)$$

$$= (4)\left(\frac{12\times 11\times 10\times 9}{4\times 3\times 2}\right) = 1{,}980$$

22.3 Series and Parallel Systems

Many systems in environmental practice consisting of several components can be classified as *series*, *parallel*, or a combination of both. However, many chemical and environmental industrial and process plants (units and systems) have a series of parallel configurations.

A *series system* is one in which the entire system fails to operate if any one of its components fails to operate. If such a system consists of n components that function independently, then the reliability of the system is the product of the reliabilities of the individual components. If R_s denotes the reliability of a series system and R_i denotes the reliability of the ith component $i = 1, \ldots, n$, then

$$R_s = R_1 R_2 \ldots R_n = \prod_{i=1}^{n} R_i \tag{22.21}$$

A *parallel* system is one that fails to operate only if all its components fail to operate. If R_i is the reliability of the ith component, then $(1 - R_i)$ is the probability that the ith component fails; $i = 1, \ldots, n$. Assuming that all n components function independently, the probability that all n components fail is $(1 - R_1)(1 - R_2) \ldots (1 - R_n)$. Subtracting this product from unity yields the following formula for R_p, the reliability of a parallel system.

$$R_p = 1-(1-R_1)(1-R_2)\ldots(1-R_n)$$

$$= 1-\prod_{i=1}^{n}(1-R_i) \tag{22.22}$$

The reliability formulas for series and parallel systems can be used to obtain the reliability of a system that combines features of a series and parallel system as shown in Figure 22.1. These calculations are illustrated in the illustrated examples that follow.

Illustrative Example 22.6

Consider the system diagrammed in Figure 22.2. Components A, B, C, and D have for their reliabilities 0.90, 0.80, 0.80, and 0.90, respectively. The system fails to operate if A fails, if B and C both fail, or if D fails. Determine the reliability of the system.

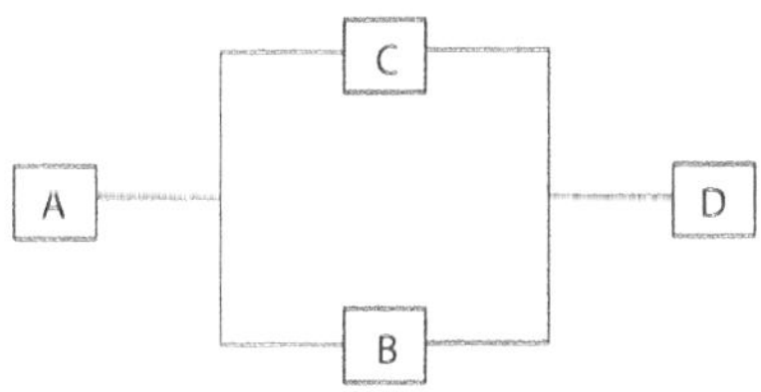

Figure 22.1 System with parallel and series components.

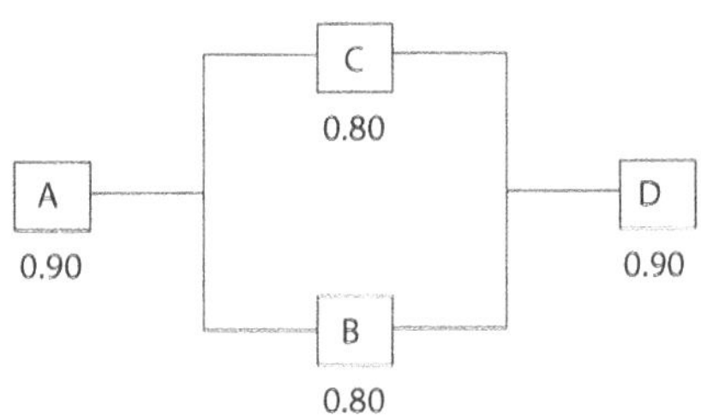

Figure 22.2 System with parallel and series components values.

Solution

Components B and C constitute a parallel subsystem connected in series to components A and D. The reliability of the parallel subsystem is obtained by applying Equation (22.22) which yields

$$R_p = 1-(1-0.80)(1-0.80) = 0.96$$

The reliability of the system is then obtained by applying Equation (22.21), which yields

$$R_s = (0.90)(0.96)(0.90) = 0.78$$

Illustrative Example 22.7

Determine the reliability of the electrical system shown in Figure 22.3 using the reliabilities indicated under the various components.

Solution

First identify the components connected in parallel. A and B are connected in parallel. D, E, and F are also connected in parallel. Therefore, compute the reliability of each subsystem of the components connected in parallel. The reliability of the parallel subsystem consisting of components A and B is

$$R_p = 1-(1-0.7)(1-0.7) = 0.91$$

The reliability of the parallel system consisting of components D, E, and F is

$$R_p = 1-(1-0.6)(1-0.6)(1-0.6) = 0.936$$

Multiply the product of the reliabilities of the parallel subsystems by the product of the reliabilities of the components to which the parallel subsystems are connected in series:

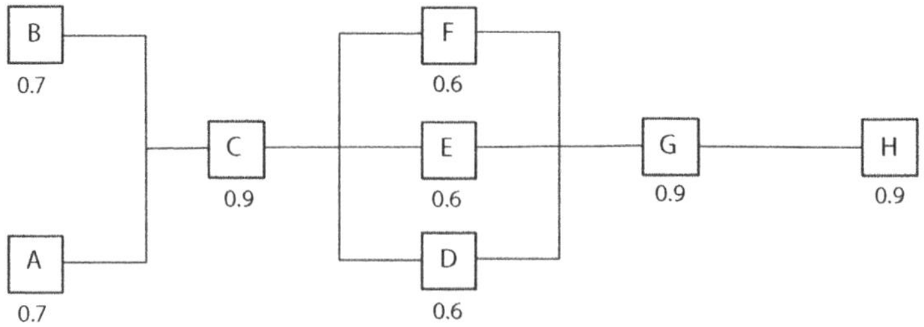

Figure 22.3 Diagram of electrical system I.

$$R_s = (0.91)(0.9)(0.936)(0.9)(0.9) = 0.621$$

The reliability of the whole system is therefore 0.621 or 62.1%

Illustrative Example 22.8

Determine the reliability of the components A, D, and G of the electrical system illustrated in Figure 22.4 using the reliabilities indicated under the various components. The overall reliability has been determined to be 0.42.

Solution

The reliability of the parallel subsystem consisting of components A and B is obtained once again by applying Equation (22.22), which yields

$$\begin{aligned} R_p &= 1-(1-0.7)(1-\text{A}) = 1-(-0.3-\text{A}+0.7\text{A}) \\ &= 0.7+0.3A \end{aligned}$$

The reliability of the parallel subsystem consisting of components D, E, and F is

$$\begin{aligned} R_p &= 1-(1-\text{D})(1-0.6)(1-0.6) \\ &= 1-0.16(1-\text{D}) = 0.84+0.16D \end{aligned}$$

The reliability of the entire system is obtained by applying Equation (22.21), which yields

$$R_s = 0.42 = (0.7+0.3\text{A})(0.9)(0.84+0.16\text{D})(\text{G})(0.9)$$

Since this single equation contains three unknowns, an infinite number of solutions are possible, including, for example,

$$A = 0;\ D = 0;\ G = 0.882$$

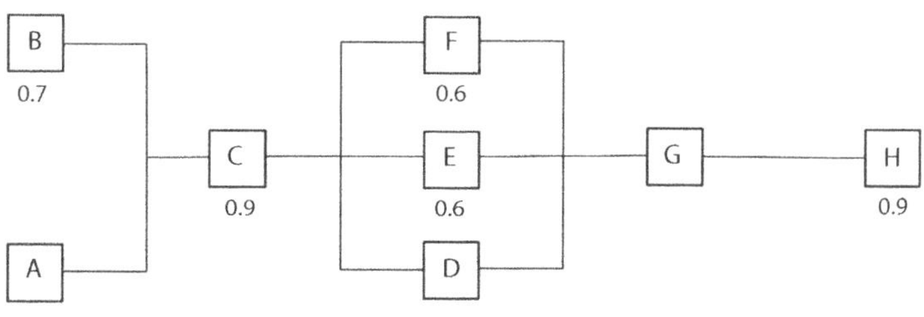

Figure 22.4 Diagram of electrical system II.

Figure 22.5 Exponential failure rate: series system.

Illustrative Example 22.9

Consider the system shown in Figure 22.5. Determine the reliability, R, if the operating time for each unit is 5000 hr. Components A and B have exponential failure rates, λ, of 3×10^{-6} and 4×10^{-6} failures per hour, respectively, where $R_i = e^{-\lambda_i t}$; t = time, hr. The term λ may be viewed as the reciprocal of the average time to failure. See Chapter 25 for additional details.

Solution

Because this is a series system,

$$R_s = R_A R_B$$

As indicated above, for an exponential failure rate

$$R = e^{-\lambda t};\ t = \text{time, } hr$$

so that

$$R_A = e^{(3\times10^{-6})(5000)} = e^{0.015} = 0.9851$$

and

$$R_B = e^{(4\times10^{-6})(5000)} = e^{0.02} = 0.9802$$

Therefore,

$$R_S = (0.9851)(0.9802) = 0.9656 = 96.56\%$$

References

1. L. Theodore and F. Taylor, *Probability and Statistics*, A Theodore Tutorial, Theodore Tutorials, East Williston, NY, originally published by USEPA/APTI, RTP, NC,1996.
2. L. Theodore, personal notes, East Williston, NY, 2006.

23

Estimation of Mean and Variance

By definition, the mean μ and the variance σ^2 of a random variable are constants characterizing the random variable's average value and its dispersion about its mean, respectively. The mean and variance can also be derived from the probability distribution function (pdf) of the random variable – a topic to be discussed in the next two chapters. If the pdf is unknown, however, the mean and the variance can be estimated on the basis of a random sample of some, but not all, observations on the random variable.

Three sections complement the presentation of this chapter. Section numbers and subject titles follow:

23.1: Estimation of the Mean
23.2: Estimation of the Variance
23.3: Interpretation of the Mean and Variance

Note that the bulk of the material in this chapter was drawn from the work of Theodore and Taylor [1].

23.1 Estimation of the Mean

Let $X_1, X_2, \ldots, X_n$ denote a random sample of n observations on X. The sample mean $\bar{X}$ is then defined by

$$\bar{X} = \sum_{i=1}^{n} \frac{X_i}{n} \tag{23.1}$$

Details on the following "*central tendency*" terms employed for the mean in equation form follow:

1. Arithmetic mean
2. Geometric mean
3. Median
4. Mode

These are defined below. Let $X_1, X_2, \ldots, X_n$ represents a set of n numbers.

1. For the arithmetic mean $\bar{X}$,

$$\bar{X} = \frac{X_1 + X_2 + \ldots + X_n}{n} \tag{23.2}$$

   If the numbers X_i have weighing factors W_i associated with them,

$$\bar{X} = \frac{W_1X_1 + W_2X_2 + \ldots + W_nX_n}{n} = \frac{\sum_{i=1}^{n} W_iX_i}{n} \tag{23.3}$$

   Weighing factors are often normalized, i.e., $\Sigma W_i = 1.0$. For this condition,

$$\bar{X} = \sum_{i=1}^{n} W_iX_i \tag{23.4}$$

2. The geometric mean, $\bar{X}_G$ is given by

$$\bar{X}_G = \left(X_1X_2, \ldots X_n\right)^{\frac{1}{n}}$$
$$\bar{X}_G = \left(\prod_{i=1}^{n} X_i\right)^{\frac{1}{n}} \tag{23.5}$$

3. The median is defined as the middle value (or arithmetic mean of the two middle values) of a set of numbers. Thus, the median of 4, 5, 9, 10, 15 is 9. It is also occasionally defined as the distribution's midpoint. Further, the median of a continuous probability distribution function *f(x)* is that value of *c* so that

$$\int_{-\infty}^{c} f(x)dx = 0.5 \tag{23.6}$$

4. The mode is the value that occurs with the greatest frequency in a set of numbers. Thus, it is the typical or most common value in a set.

Additional details follow. Other definitions also exist in the literature for the aforementioned four terms.

The reader should note that one basic way of summarizing data is by the computation of a central value. The most commonly used central value statistic is the arithmetic average, or the mean discussed above. This statistic is particularly useful when applied to a set of data having a fairly symmetrical distribution. The mean is an efficient statistic because it summarizes all the data in the set and each piece of data is taken into account in the computation. However, the arithmetic mean is not a perfect measure of the true central value of a given data set because arithmetic means can overemphasize the importance of one or two extreme data points.

When a distribution of data is asymmetrical, it is sometimes convenient to compute a different measure of central value. The second measure, known as the *median*, is simply the middle value of a distribution, or the quantity above which half of the data lie and below which the other half lie. If n data points are listed in their order of magnitude, the median is the $[(n + 1)/2]$th value. If the number of data points is even, then the numerical value of the median is the value midway between the two data points nearest the middle. The median, being a positional value, is less influenced by extreme values in a distribution than the mean. However, the median alone is usually not a good measure of central tendency. To obtain the median, the data values provided, i.e., 15, 8, 22, 18, 10, 13 must first be arranged in order of magnitude, such as

$$8, 10, 13, 15, 18, 22$$

Thus, the median for this data is 14, or the value halfway between 13 and 15 because this data set has an even number of measurements.

Another measure of central tendency used in specialized applications is the aforementioned *geometric mean*, $\overline{X}_G$.

Generally, the mean falls near the "middle" of the distribution. Actually, the mean may be thought of as the *center of gravity* of the distribution. The mean has another important property. If each measurement is subtracted from the mean, one obtains n "discrepancies" or differences; some of these are positive and some are negative, but the algebraic sum of all the differences is equal to zero.

23.2 Estimation of the Variance

Once again, let $X_1, X_2, \ldots, X_n$ denote a random sample of n observations on X. Then the *sample* variance s^2 is defined by

$$s^2 = \sum_{i=1}^{n} \frac{(X_i - \bar{X})^2}{n-1} \tag{23.7}$$

where $\bar{X}$ and s^2 are random variables in the sense that their values vary from sample to sample of the observations on X. It can be shown that the *expected* value of X is the population mean μ and that the *expected* value of s^2 is the population variance σ^2. Because of this, $\bar{X}$ and s^2 are called unbiased estimators of μ and σ^2, respectively.

The calculation of s^2 can be facilitated by use of the computation formula

$$s^2 = \frac{n\sum_{i=1}^{n} X_i^2 - \left(\sum_{i=1}^{n} X_i\right)^2}{n(n-1)} \tag{23.8}$$

For example, given the sample 5, 3, 6, 4, 7,

$$\sum_{i=1}^{5} X_i^2 = 135 \tag{23.9}$$

$$\sum_{i=1}^{5} X_i = 25 \tag{23.10}$$

and

$$n = 5 \tag{23.11}$$

Substituting in Equation (23.8) yields

$$s^2 = \frac{(5)(135) - (25)^2}{(5)(4)} = 2.5 \tag{23.12}$$

In the case of a random sample of observations of a continuous random variable assumed to have a so-called normal pdf (see also Chapter 25), the graph of which is a bell-shaped curve, the following statements give a more precise interpretation of the sample standard deviation s as a measure of spread or dispersion.

1. $\overline{X} \pm s$ includes approximately 68% of the sample observations.
2. $\overline{X} \pm 2s$ includes approximately 95% of the sample observations.
3. $\overline{X} \pm 3s$ includes approximately 99.7% of the sample observations.

The source of these percentages is the normal probability distribution, which is studied in more detail in Chapter 25.

Chebyshev's theorem provides an interpretation of the sample standard deviation (the positive square root of the sample variance) as a measure of the spread (dispersion) of sample observations about their mean. Chebyshev's theorem states that with $k > 1$, at least $(1 - 1/k^2)$ of the sample observations lie in the interval $(\overline{X}\text{-}ks,\ \overline{X} + ks)$. For $k = 2$, for example, this means that at least 75% of the sample observations lie in the interval $(\overline{X}\text{-}2s,\ \overline{X}_0 + 2s)$. The smaller the value of s, the greater the concentration of observations in the vicinity of $\overline{X}$.

The following is provided as a qualitative description of the standard deviation. The mean of a set of measurements provides some information about the aforementioned location of the "middle" or "center of gravity" of the set of measurements, but it gives no information about the *scatter* (or *dispersion* or amount of concentration) of the measurements. For example, the five measurements 14.0, 24.5, 25.0, 25.5, and 36.0 have the same mean as the five measurements 24.0, 24.5, 25.0, 25.5, and 26.0, but the two sets of measurements have different amounts of scatter.

One simple indication of the scatter of a set of measurements is the *range*, i.e., the largest measurement minus the smallest. In the two sets of measurements mentioned earlier, the ranges are 22 and 8, respectively. With fairly small sample sizes one would find the range to be very convenient. It is difficult, however, to compare a range for one sample size with that for a different sample size. For this and other reasons, the range, in spite of its simplicity, convenience, and importance, is used only in rather restricted situations.

One clearly needs a measure of scatter which can be used in samples of any size and in some sense, make use of all the measurements in the sample. There are several measures of scatter that can be used for this purpose, and the most common of these is the aforementioned *standard deviation*. The standard deviation may be thought of as the "natural" measure of scatter.

The two illustrative examples that follow will illustrate the procedure to obtain these values.

Illustrative Example 23.1

The average weekly wastewater temperatures (°C) for six consecutive weeks are

$$22, 10, 8, 15, 13, 18$$

Find the standard deviation of the temperature T of the water samples.

Solution

First calculate the arithmetic, $\overline{T}$,

$$\overline{T} = \frac{(8+10+13+15+18+22)}{6} = 14.33°\text{C}$$

As noted earlier, the most commonly used measure of dispersion, or variability, of sets of data is the standard deviation, s. Its defining formula is given by the expression in Equation (23.7).

$$s = \sqrt{\sum_{i=1}^{n} \frac{(X_1 - \overline{X})^2}{n-1}}$$

The expression $(X_i - \overline{X})$ shows that the deviation of each piece of data from the mean is taken into account by the standard deviation. Although the defining formula for the standard deviation gives insight into its meaning, the following algebraically equivalent formula (see Equation 23.8) makes computation much easier (now applied to the temperature, T):

$$s = \sqrt{\frac{\Sigma(T_i - \overline{T})^2}{n-1}} = \sqrt{\frac{n\Sigma T_i^2 - (\Sigma T_i)^2}{n-1}}$$

The standard deviation may now be calculated for the data at hand:

$$\Sigma T_i^2 = (8)^2 + (10)^2 + (13)^2 + (15)^2 + (18)^2 + (22)^2 = 1366$$

and

$$(\Sigma T_i)^2 = (8+10+13+15+18+22)^2 = 7396$$

Thus,

$$s = \sqrt{\frac{6(1366) - 7396}{(6)-(1)}} = 5.16°\text{C}$$

Illustrative Example 23.2

The following are SO_2 concentrations X (μg/m^3) for March, April, and May of 2017 at Floral Park, NY - home of Belmont Park racetrack (another place one of the authors regularly visits):

29 103 27 14 24 63 24

Calculate the standard deviation and variance.

Table 23.1 SO_2 calculations.

X_i	X_i^2
14	196
24	576
24	576
27	729
29	841
63	3969
103	10609
$\Sigma X_i = 284$	$\Sigma X_i^2 = 17496$

Solution

Once again, apply Equation (23.8),

$$s = \sqrt{\frac{\sum X_i^2 - \frac{(\sum X_i)^2}{n}}{n-1}}$$

The calculations are performed as follows based on Table 23.1. From the calculation of ΣX_i determine $(\Sigma X_i)^2$:

$$(\Sigma X_i)^2 = (284)^2 = 80{,}656$$

Substituting gives

$$s = \sqrt{\frac{17{,}496 - \left(\frac{80{,}656}{7}\right)}{6}} = \sqrt{995.62} = 31.55 \ \mu g/m^3$$

For the variance

$$\begin{aligned} s^2 &= (\text{standard deviation})^2 \\ &= (31.55)^2 = 995 \ (\mu g/m^3)^2 \end{aligned}$$

23.3 Interpretation of Mean and Variance

The mean and variance can be interpreted for both a *discrete* and *continuous* random variable. The following two illustrative examples review both terms.

Illustrative Example 23.3

A *discrete* random variable X assumes the values 49, 50, and 51 each with probability 1/3. A *discrete* random variable Y assumes the values 0, 50, and 100 each with probability 1/3. Compare the means and variances of X and Y, and interpret the results.

Solution

As indicated earlier, the pdf of a discrete random variable requires the probability assigned to each of the possible values of the random variable. For this problem,

$$f(x) = \frac{1}{3};\ x = 49, 50, 51$$

$$g(y) = \frac{1}{3};\ y = 0, 50, 100$$

To obtain μ_x, the mean of X, and μ_y, the mean of Y. Use the formulas [1, 2]:

$$\mu_x = E(X) = \Sigma x\, f(x) \tag{23.13}$$

$$\mu_y = E(Y) = \Sigma y\, g(y) \tag{23.14}$$

where $E(X)$ is the expected value of X. Substituting gives,

$$\mu_x = 49\left(\frac{1}{3}\right) + 50\left(\frac{1}{3}\right) + 51\left(\frac{1}{3}\right) = 50$$

$$\mu_y = 0\left(\frac{1}{3}\right) + 50\left(\frac{1}{3}\right) + 100\left(\frac{1}{3}\right) = 50$$

Obtain σ_x^2, the variance of X, and σ_y^2, the variance of Y, by applying the following equations [1, 2]:

$$\sigma_x^2 = E(X - \mu_x)^2 \tag{23.15}$$

$$\sigma_y^2 = E(X - \mu_y)^2 \tag{23.16}$$

The alternative formula for computing the variance of X is

$$E(X - \mu_x)^2 = E(X^2) - \mu_x^2 \tag{23.17}$$

Applying this form of the equation to both X and Y gives

$$\sigma_x^2 = \Sigma(x-\mu_x)^2 f(x) = \Sigma(x-50)^2 f(x)$$
$$= (49-50)^2\left(\frac{1}{3}\right) + (50-50)^2\left(\frac{1}{3}\right) + (51-50)^2\left(\frac{1}{3}\right) = \frac{2}{3} = 0.667$$

$$\sigma_y^2 = \Sigma(y-\mu_y)^2 g(y) = \Sigma(y-50)^2 g(y)$$
$$= (0-50)^2\left(\frac{1}{3}\right) + (50-50)^2\left(\frac{1}{3}\right) + (100-50)^2\left(\frac{1}{3}\right) = \frac{5000}{3} = 1667$$

Note that X and Y have the same mean but the variance of Y is greater than the variance of X, which reflects the greater dispersion (spread or variability) of the values of Y about its mean in contrast to the dispersion of the values of X about its mean.

Illustrative Example 23.4

Continuous random variables X and Y have pdfs specified by $f(x)$ and $g(y)$, respectively, as follows:

$$f(x) = \frac{1}{2};\ 1 < x < 1$$

$$g(y) = \frac{1}{4};\ 2 < y < 2$$

Compute the mean and variance of X and Y and compare the results.

Solution

Note that for a continuous random variable, integration replaces summation in the calculation of the mean and variance. Furthermore, the probability that a continuous random variable lies in a certain interval is obtained by integrating the pdf over that interval (See Chapter 25).

First compute μ_x, the mean of X, and μ_y, the mean of Y. Use the following formulas [1, 2]:

$$\mu_x = \int_{-\infty}^{\infty} x\, f(x)dx \tag{23.18}$$

$$\mu_y = \int_{-\infty}^{\infty} y\, g(y)dy \tag{23.19}$$

Substituting *f(x)* and *g(y)* gives

$$\mu_x = \int_{-1}^{1} x\left(\frac{1}{2}\right)dx = 0$$

$$\mu_y = \int_{-2}^{2} y\left(\frac{1}{4}\right)dy = 0$$

The terms $\sigma_x^{\ 2}$, the variance of *X*, and $\sigma_y^{\ 2}$, the variance of *Y*, are calculated as follows (1, 2):

$$\sigma_x^2 = E(X^2) - \mu_x^2 = \int_{-1}^{1} x^2\left(\frac{1}{2}\right)dx = \frac{1}{3} = 0.333$$

$$\sigma_y^2 = E(Y^2) - \mu_y^2 = \int_{-2}^{2} y^2\left(\frac{1}{4}\right)dy = \frac{4}{3} = 1.333$$

As with the previous illustrative example, *X* and *Y* have the same mean, i.e., 0. The variance of *Y* is greater than the variance of *X*, which reflects the greater dispersion of the values of *Y* about its mean.

Illustrative Example 23.5

A continuous random variable *X* has a pdf given by

$$f(x) = 4x^3;\ 0 < x < 1$$

Calculate μ and σ^2 for *x*.

Solution

By definition (See Chapter 25),

$$\mu = \int x\ f(x)dx \tag{23.20}$$

Substituting gives

$$\mu = \int_0^1 x(4x^3)dx = \int_0^1 4x^4\ dx$$

Integrating gives

$$\mu = 4\left(\frac{x^5}{5}\right)_0^1$$

$$= \frac{4}{5}$$

For σ^2, apply the methods described earlier.

$$\sigma^2 = E(X^2) - \mu^2 = \int x^2 f(x)dx - \mu^2$$

Substituting gives

$$\sigma^2 = \int_0^1 x^2(4x^3)dx - \mu^2$$

Integrating,

$$\sigma^2 = \frac{4x^6}{6} - \left(\frac{4}{5}\right)^2 = \frac{2}{3} - \frac{16}{25} = \frac{50-48}{75} = \frac{2}{75} = 0.0267$$

Reference

1. L. Theodore and F. Taylor, *Probability and Statistics*, A Theodore Tutorial, Theodore Tutorials, East Williston, NY, originally published by USEPA/APTI, RTP, NC,1996.
2. S. Shaefer and L. Theodore, *Probability and Statistics Applications for Environmental Science*, CRC Press/Taylor & Francis Group, Boca Raton, FL, 2007.

24

Discrete Probability Distributions

The probability distribution of a random variable concerns the distribution of the probability over the range of the random variable. The distribution of probability, i.e., the values of the discrete random variable together with their associated probabilities, is specified by the *probability distribution function* (pdf). This chapter is devoted to providing general properties of the pdf in the case of discrete random variables, as well as an introduction to the *cumulative distribution function* (cdf).

The pdf of a *discrete* random variable X is specified by $f(x)$, where $f(x)$ has the following three essential properties:

1. $f(x) = P(X = x) =$ probability assigned to the outcome corresponding to the number x in the range of X i.e., X is a specifically designed value of x (24.1)

2. $f(x) \geq 0$ (24.2)

3. $\sum_x f(x) = 1$ (24.3)

Property (24.1) indicates that the pdf of a discrete random variable generates probability by substitution. Property (24.2) and Property (24.3) restrict the values of $f(x)$ to nonnegative real numbers and numbers whose sum is 1, respectively.

Consider, for example, a box of 100 transistors containing 5 defectives. Suppose that a transistor is selected at random is to be classified as defective or nondefective. Then X is a discrete random variable with pdf specified by

$$f(x) = 0.05;\ x = 1 \tag{24.4}$$

$$= 0.95;\ x = 0 \tag{24.5}$$

For another example of the pdf of a discrete random variable, let X denote the number of the throw on which the first failure of an electrical switch occurs. Suppose the probability that a switch fails on any throw is 0.001 and that successive throws are independent with respect to failure. If the switch fails for the first time on throw x, it must have been successful on each of the preceding $x - 1$ trials. In effect, the switch survives up to $x - 1$ trials and fails at trial x. Therefore, the pdf of X is given by

$$f(x) = (0.999)^{x-1}(0.001);\ x = 1, 2, 3, \ldots, n, \ldots \tag{24.6}$$

Note that the range of X consists of a countable infinitude of values.

Another function used to describe the probability distribution of a random variable X is the aforementioned cdf. If $f(x)$ specifies the pdf of a random variable X, then $F(x)$ is used to specify the cdf. For *both* discrete and continuous random variables, the cdf of X is defined by

$$F(x) = P(X \geq x);\ -\infty < x < \infty \tag{24.7}$$

Note that the cdf is defined for all real numbers, not just the values assigned by the random variable.

To illustrate the derivation of the cdf from the pdf, consider a case of a *discrete random variable* X whose pdf is specified by

$$f(x) = 0.2;\ x = 2 \tag{24.8}$$

$$= 0.3;\ x = 5 \tag{24.9}$$

$$= 0.5;\ x = 7 \tag{24.10}$$

Applying the definition of the cdf in Equation (24.7), one obtains for the cdf of X (see also Figure 24.1)

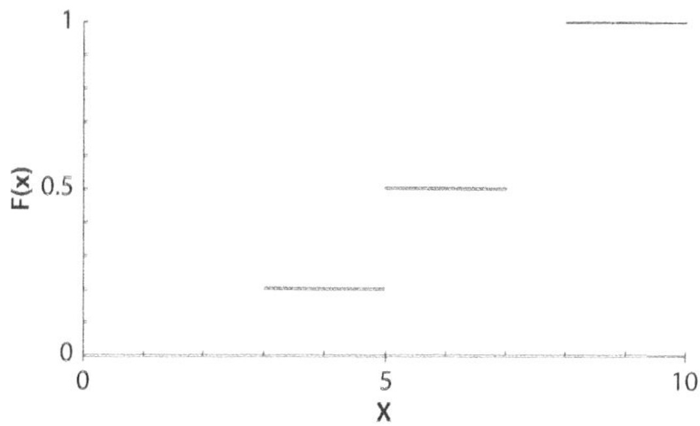

Figure 24.1 Graph of the cdf of a discrete random variable X.

$$\begin{aligned} F(x) &= 0;\ x < 2 \\ &= 0.5;\ 5 \le x \le 7 \\ &= 0.2;\ 2 \le x \le 5 \\ &= 7;\ x \ge 7 \end{aligned}$$

It is helpful to think of $F(x)$ as an accumulator of probability as x increases through all real numbers. In the case of a discrete random variable, the cdf is a step function increasing by finite jumps at the values of x in the range of X. In the above example, these jumps occur at the values 2, 5, and 7. The magnitude of each jump is equal to the probability assigned to the value at which the jump occurs. This is depicted in Figure 24.1.

Illustrative Example 24.1

Provide an example of a *discrete* random variable and a *continuous* random variable

Solution

A random variable may be *discrete*; i.e., it can take on only a finite or countable number of distinct values. Alternately, the random variable may be *continuous*, with values occurring anywhere in an interval. The outcome of the throw of a die is a discrete random variable. The exact height of an individual selected from a particular group is a continuous variable.

Illustrative Example 24.2

Let X denote the number of supply ships reaching a port on a given day. The pdf for X is given by:

$$f(x) = \frac{x}{21};\ x = 1, 2, 3, 4, 5, 6$$

Verify that $f(x)$ is a valid pdf and calculate the probability that at least three ships but less than six ships will arrive on a given day.

Solution

For $f(x)$ to be valid, the following two conditions must be satisfied:

$$0 \leq f(x) \leq 1; \; x \text{ between 1 and 6}$$

and

$$\sum_{1}^{6} f(x) = 1$$

Substituting into the second condition yields

$$\sum_{1}^{6} f(x) = \sum_{1}^{6}\left(\frac{x}{21}\right)$$
$$= \frac{1}{21} + \frac{2}{21} + \frac{3}{21} + \frac{4}{21} + \frac{5}{21} + \frac{6}{21} = \frac{21}{21} = 1.0$$

Thus, both conditions are satisfied and $f(x)$ is a valid pdf.

The probability that at least 3 ships but less than 6 ships will arrive on a given day can be determined by

$$P(3 \leq X < 6) = f(x=3) + f(x=4) + f(x=5)$$
$$= \frac{3}{21} + \frac{4}{21} + \frac{5}{21} = \frac{12}{21} = 0.429$$

Illustrative Example 24.3

Let X denote the annual number of floods in a certain region. The pdf of X is specified as

$$f(x) = 0.25; \; x = 0$$

$$f(x) = 0.35; \; x = 1$$

$$f(x) = 0.24; \; x = 2$$

$$f(x) = 0.11; \; x = 3$$

$$f(x) = 0.04; \; x = 4$$

$$f(x) = 0.01; \; x = 5$$

1. What is the probability of two or more floods in any year?
2. What is the probability of four or more floods in a year, given that two floods have already occurred?
3. What is the probability of at least one flood in a year, given that one has already occurred?

Solution

1. For two or more floods in any year, one may write

$$\begin{aligned} P(x \geq 2) &= \sum_{2}^{\infty} f(x) = \sum_{2}^{5} f(x) \\ &= f(2) + f(3) + f(4) + f(5) \\ &= 0.24 + 0.11 + 0.04 + 0.01 = 0.40 \end{aligned}$$

2. This involves a conditional probability calculation [1]. The describing equation and solution is

$$\begin{aligned} P(X \geq 4 \,/\, X \geq 2) &= \frac{P(X \geq 4 \text{ and } X \geq 2)}{P(X \geq 2)} \\ &= \frac{P(X \geq 4)}{P(X \geq 2)} = \frac{0.05}{0.40} = 0.125 \end{aligned}$$

3. This also involves a conditional probability calculation [1]. For this case,

$$\begin{aligned} P(X \geq 2 \,/\, X \geq 1) &= \frac{P(X \geq 2 \text{ and } X \geq 1)}{P(X \geq 1)} \\ &= \frac{P(X \geq 2)}{P(X \geq 1)} = \frac{0.40}{0.40 + 0.35} = \frac{0.40}{0.75} = 0.53 \end{aligned}$$

Three sections complement the presentation of this chapter. Section numbers and subject titles follow:

24.1: The Binomial Distribution
24.2: The Hypergeometric Distribution
24.3: The Poisson Distribution

Note that the bulk of the material in this chapter was drawn from the work of Theodore and Taylor [1].

24.1 The Binomial Distribution

Consider n independent performances of a random experiment with mutually exclusive outcomes that can be classified as *success* or *failure*. The words "success" and "failure" are to be regarded as labels for two mutually exclusive categories of outcomes of the random experiment and they do not necessarily have the ordinary connotation of success or failure. Assume that p, the probability of success on any performance of the random experiment, is constant. Let Q be the probability of failure, so that

$$Q = 1 - p \tag{24.11}$$

The probability distribution of X, the number of successes in n performances of the random experiment, is the binomial distribution, with a probability distribution function (pdf) specified by

$$f(x) = \frac{n!}{x!(n-x)!} p^x Q^{n-x};\ x = 0, 1, \ldots, n \tag{24.12}$$

where $f(x)$ is the probability of x successes in n performances. One can show that the expected value of the random variable X is np and its variance is npQ [2].

As a simple example of the binomial distribution, consider the probability distribution of the number of defectives in a sample of 5 items drawn *with replacement* from a lot of 1000 items, 50 of which are defective. Associate "success" with drawing a defective item from the lot. The result of each drawing can then be classified as success (defective item) or failure (nondefective item). The sample of five items is drawn with replacement (i.e., each item in the sample is returned before the next is drawn from the lot; therefore, the probability of success remains constant at 0.05). Substituting the values $n = 5$, $p = 0.05$, and $Q = 0.95$ in Equation (24.12) yields

$$f(x) = \frac{5!}{x!(5-x)!}(0.05)^x (0.95)^{5-x};\ x = 0, 1, 2, 3, 4, 5 \tag{24.13}$$

as the pdf for X, the number of defectives in the sample. The probability that the sample contains exactly three defectives is given by

$$P(X = 3) = \frac{5!}{3!2!}(0.05)^3 (0.95)^2 = 0.0011 \tag{24.14}$$

The binomial distribution can be used to calculate the reliability of a *redundant system*. A redundant system consisting of n identical components is a system that fails only if more than r components fail. Familiar examples include single-usage equipment such as missile engines, short-life batteries, and flash bulbs which are required to operate for one time period and are not reused. Once again, associate "success" with the failure of a component. Assume that the n components are independent with respect to failure, and that the reliability of each is $1 - p$. Then X, the number of failures, has the binomial pdf in Equation (24.12) and the reliability of the redundant system is

$$P(X \leq r) = \sum_{x=0}^{r} \frac{n!}{x!(n-x)!} p^x Q^{n-x} \tag{24.15}$$

The binomial distribution has been applied in problems in which samples are drawn from a large population with specified "*success*" or "*failure*" probabilities and there is a desire to evaluate instances of obtaining a certain number of successes in the sample. It has applications in quality control, reliability, environmental engineering, environmental management, consumer sampling, and many other cases.

Suppose a trial or event can result in one and only one of mutually exclusive events E_1, E_2, ..., E_j with probabilities p_1, p_2, ..., p_j respectively, where $p_1 + p_2 + \ldots + p_j = 1$. If n independent trials are made, then one can show that the probability function of obtaining x_1 of E_1s, x_2 of E_2s, ..., x_j of E_js is given by

$$f(x_1, \ldots, x_j) = \frac{n!}{x_1! \ldots x_j!} p_1^{x_1} \ldots p_j^{x_j} \tag{24.16}$$

where $0 \leq x_j \leq n$, $i = 1, \ldots, j$, and $x_1 + \ldots + x_j = n$. This is the probability function of the *multinominal distribution*. Note that if $j = 2$ this function reduces to the binomial distribution. Hence, the multinominal distribution is essentially an extension of the binomial distribution for just two possible outcomes - success and failure.

Illustrative Example 24.4

A sample of five transistors is drawn *with replacement* from a lot which is 5% defective. Once again, "with replacement" means that each transistor drawn is returned to the lot before the next is drawn. What is the probability that the number of defective transistors and the sample is:

1. Exactly 2
2. At most 2
3. At least 2

Solution

This random experiment consists of drawing a transistor at random with replacement from a lot. The random experiment is preformed five times because a sample of five transistors is drawn with replacement from the lot. Therefore, $n = 5$. Also, note that the performances are independent because each transistor is replaced before the next is drawn. Therefore, the composition of the lot is exactly the same before each drawing.

1. For this problem, associate "success" with drawing a defective transistor. Associate "failure" with drawing a nondefective. Refer once again to Equation (24.12). Because 5% of the lot is defective, $p = 0.05$. Therefore, $Q = 0.95$. These values (n, p, and Q) may be substituted in the binomial pdf.

$$f(x) = \frac{5!}{x!(5-x)!}(0.05)^x(0.95)^{5-x};\ x = 0, 1, \ldots, 5 \tag{24.13}$$

2. Substitute the appropriate value of X to obtain the required probabilities.
3. For this problem,

$$\begin{aligned} P(\text{exactly 2 defectives}) &= P(X = 2) \\ &= \frac{5!}{2!3!}(0.05)^2(0.95)^3 = 0.214 \end{aligned}$$

$$\begin{aligned} P(\text{at most 2 defectives}) &= P(X \le 2) \\ &= \sum_{x=0}^{2} \frac{5!}{x!(5-x)!}(0.05)^x(0.95)^{5-x} \\ &= 0.7738 + 0.2036 + 0.0214 = 0.998 \end{aligned}$$

$$\begin{aligned} P(\text{at least 2 defectives}) &= \mathrm{P}(X \ge 2) \\ &= 1 - P(X \le 1) \\ &= 1 - \sum_{x=0}^{1} \frac{5!}{x!(5-x)!}(0.05)^x(0.95)^{5-x} \\ &= 1 - (0.7748 + 0.2036) = 0.0226 \end{aligned}$$

Note that this problem involved summing the binomial pdf over the appropriate values of X; this procedure applies if the probability of more than a single value of X is required.

Illustrative Example 24.5

The probability that an exposure to a nanocarcinogen will be fatal is 0.80. Find the probability of the following events for a group of 15 workers:

1. At least 9 will die.
2. From 4 to 8 will die.

Solution

1. For event 1,

$$
\begin{aligned}
P(\text{at least 9 will die}) = P(X \geq 9);\ p = 0.8,\ Q = 0.2,\ m{=}15 \\
&= \sum_{x=9}^{15} \frac{15!}{x!(15-x)!}(0.8)^x(0.2)^{15-x} \\
&= 0.0430 + 0.1032 + 0.1876 + 0.2501 + 0.2309 + \\
&\quad 0.1319 + 0.0352 \\
&= 0.982
\end{aligned}
$$

This calculation can be performed by longhand or obtained directly from binomial tables.

2. For event 2,

$$P(4 \leq X \leq 8) = 1.0 - P(X \geq 9) - P(0 \leq X \leq 4)$$

One notes, almost immediately that

$$P(0 \leq X \leq 4) \cong 0$$

Therefore,

$$P(4 \leq X \leq 8) = 1.0 - 0.982 - 0.0 = 0.018$$

Illustrative Example 24.6

A redundant system consisting of three operating pumps can survive two pump failures. Assume that the pumps are independent with respect to failure and each has a probability of failure of 0.10. What is the reliability of the system?

Solution

The system consists of three pumps. Therefore, $n = 3$. The system can survive the failure of two pumps. Equation (24.12) should be employed with $r = 2$. The probability of a pump failure is 0.10. Therefore, $p = 0.10$. Substitute these values for n and p (with $Q = 0.90$) in the binomial pdf:

$$f(x) = \frac{3!}{x!(3-x)!}(0.10)^x(0.90)^{3-x};\ x = 0, 1, \ldots, 3$$

Sum the binomial pdf from 0 to r to find the reliability, R:

$$R = \sum_{x=0}^{2} \frac{3!}{x!(3-x)!}(0.10)^x(0.90)^{3-x} = 0.999$$

Illustrative Example 24.7

The probabilities of hospitals A, B, and C obtaining a particular type of serum are 0.5, 0.3, and 0.2, respectively. Four such serums are to be provided. What is the probability that a single hospital receives all four serums?

Solution

This involves the use of the multinomial distribution. The desired probability is (see Equation (24.16))

$$\begin{aligned}
&P(\text{one hospital alone getting serum})\\
&= f(4,0,0) + f(0,4,0) + f(0,0,4)\\
&= \frac{4!}{4!0!0!}(0.5)^4(0.3)^0(0.2)^0 + \frac{4!}{0!4!0!}(0.5)^0(0.3)^4(0.2)^0\\
&\quad + \frac{4!}{0!0!4!}(0.5)^0(0.3)^0(0.2)^4\\
&= 0.0625 + 0.0081 + 0.0016 = 0.0722
\end{aligned}$$

Note that for this calculation a hospital receives all four serum shipments.

24.2 Hypergeometric Distribution

The hypergeometric distribution is applicable to situations in which a random sample of r items is drawn *without replacement* from a set of n items. Without replacement means that an item is *not* returned to the set after it is drawn. Recall that the binomial distribution is frequently applicable in cases where the item is drawn *with replacement*.

Suppose that it is possible to classify each of the n items as a *success* or *failure*. Again, the words "success" and "failure" do not have the usual connotation. They are merely labels for two mutually exclusive categories into which n items have been classified. Thus, each element of the population may be dichotomized as belonging to one of two disjointed classes.

Let a be the number of items in the category labeled *success*. Then $n - a$ will be the number of items in the category labeled *failure*. Let X denote the number of successes in a random sample of r items drawn without replacement from the set of n items. Then the random variable X has a hypergeometric distribution whose probability distribution function (pdf) is specified as follows [1]:

$$f(x) = \frac{\dfrac{a!}{x!(a-x)!}\,\dfrac{(n-a)!}{(r-x)!(n-a-r+x)!}}{\dfrac{n!}{r!(n-r)!}};\ x = 0, 1, \ldots, \min(a,r) \qquad (24.17)$$

The term $f(x)$ in Equation (24.17) is the probability of x successes in a random sample of n items drawn without replacement from a set of n items, a of which are classified as successes and $n - a$ as failures. The term $\min(a, r)$ represents the smaller of the two numbers a and r, i.e., $\min(a, r) = a$ if $a < r$ and $\min(a, r) = r$ if $r \leq a$.

The hypergeometric distribution can also be arrived at through the use of combinations. Using this approach, the total number of possible selections of r elements out of n is $C(n, r)$. Furthermore, x "good" elements may be chosen out of the total of a good elements in a total of $C(a, x)$ combinations. For each such combination, it is also possible to select $r - x$ of $n - a$ "bad" elements in $C(a, x)\ C(n - a, r - x)$. Because each selection possibility is equally likely, the probability of picking exactly x good elements is

$$f(x) = \frac{C(a,x)C(n-a,r-x)}{C(n,r)} \qquad (24.18)$$

which, upon substitution of C is the expression given in Equation 24.17 for the probability density function of the hypergeometric distribution. As with the binomial distribution, a tabulation of cumulative and individual terms of the hypergeometric distribution is available in the literature [2].

The hypergeometric distribution is applicable in situations similar to those when the binomial density is used, except that samples are taken from a small population. Examples arise in the sampling of small numbers of chemical, medical, and environmental samples, as well as from manufacturing lots. The hypergeometric distribution is obviously a special case of the binomial distribution when applied to finite populations. In particular, the hypergeometric distribution approaches the binomial distribution when the population size approaches infinity. Note, however, that others have claimed that the binomial distribution is a special case of a hypergeometric distribution.

Illustrative Example 24.8

A sample of 5 transistors is drawn at random *without replacement* from a lot of 1000 transistors, 50 of which are defective. What is the probability that the sample contains exactly 3 defectives?

Solution

The number of items in the set from which the sample is drawn is the number of transistors in the lot. Therefore, $n = 1000$. Associate once again "success" with drawing a defective transistor, and "failure" with drawing a nondefective one. Determine a, the number of "successes" in the set of n items. Because 50 of the transistors in the lot are defective, $a = 50$. Also, note that the sample is drawn without replacement and that the size of the sample is $r = 5$.

Substituting the values of n, r, and a in a hypergeometric pdf provided in Equation (24.17) gives

$$f(x) = \frac{\dfrac{50!}{x!(50-x)!}\,\dfrac{950!}{(5-x)!(945+x)!}}{\dfrac{1000!}{5!995!}};\ x = 0, 1, \ldots, 5$$

Also, substitute the appropriate value of X above to obtain the required probability.

$$P(\text{sample contains exactly 3 defectives}) = P(X = 3)$$

Therefore,

$$P(X = 3) = \frac{\dfrac{50!}{x!47!}\,\dfrac{950!}{2!(945+3)!}}{\dfrac{1000!}{5!995!}} = 0.0011$$

Illustrative Example 24.9

A pillbox contains 24 drug tablets, 3 of which are contaminated. If a sample of six is chosen at random from the pillbox, what is the probability that:

1. Zero will be contaminated
2. One will be contaminated
3. Two will be contaminated
4. Three will be contaminated

Solution

First note that this once again involves sampling without replacement. Employing the combination equation provided above, Equation (24.18) yields

$$f(x) = \frac{C(a, x)C(n-a, r-x)}{C(n, r)} : \ a = 3, r = 6, n = 24$$

Therefore,

$$f(x) = \frac{C(3, x)C(21, 6-x)}{C(24, 6)}; \ 0 \leq x \leq 3$$

Note that there are only four values X can assume: 0, 1, 2, and 3.

1. Substituting $X = 0$ gives

$$P(X=0) = \frac{C(3,0)C(21,6)}{C(24,6)}$$
$$= \frac{\left(\frac{3!}{0!3!}\right)\left(\frac{21!}{6!15!}\right)}{\left(\frac{24!}{6!18!}\right)} = \frac{(1)(54,264)}{(134,596)} = 0.40316$$

2. For one pill,

$$P(X=1) = \frac{C(3,1)C(21,5)}{C(24,6)} = 0.45356$$

3. For two pills,

$$P(X=2) = \frac{C(3,2)C(21,4)}{C(24,6)} = 0.13340$$

4. For three pills,

$$P(X=3) = \frac{C(3,3)C(21,3)}{C(24,6)} = 0.00988$$

Illustrative Example 24.10

An environmental quality control engineer inspects a random sample of three vacuum pumps at a water treatment plant drawn *without replacement* from each incoming lot of 25 vacuum pumps. A lot is accepted only if the sample contains no defectives. If any defectives are found, the entire lot is inspected, at the cost charged to the vendor. What is the probability of accepting a lot containing 10 defective vacuum pumps?

Solution

Each lot consists of 25 vacuum pumps. Therefore, $n = 25$. Once again, associate "success" with a defective vacuum pump. Determine, a, the number of "successes" in the set of n items. For this problem, $a = 10$. A sample of three vacuum pumps is drawn without replacement from each lot; therefore, $r = 3$.

Substitute the values of n, r, and a in the pdf of the hypergeometric distribution.

$$f(x) = \frac{\dfrac{10!}{x!(10-x)!}\dfrac{15!}{(3-x)!(12+x)!}}{\dfrac{25!}{3!22!}};\ x = 0, 1, 2, 3$$

Substitute the appropriate values of X to obtain the required probability. The probability of accepting a lot containing 10 defective vacuum pumps is

$$P(X=0) = \frac{\dfrac{10!}{0!10!}\dfrac{15!}{(3-0)!12!}}{\dfrac{25!}{3!22!}} = \frac{(1)(455)}{2300} = 0.20$$

Illustrative Example 24.11

An order was recently received to manufacture a special type of water pollution monitor. The total order consisted of 25 monitors. Five of these monitors are to be selected at random and initially life tested. The contract specifies that if not more than 1 of the 5 monitors fails a specified life test, the remaining 20 will be accepted; otherwise, the complete lot will be rejected. What is the probability of lot acceptance if exactly 4 of the 25 submitted monitors are defective?

Solution

The lot will be accepted if the sample of 5 contains either 0 or 1 of the 4 defective monitors. The probability of this happening is given by the hypergeometric distribution as

$$P(X \le 1) = \frac{C(4,x)C(21,5-x)}{C(24,5)}$$

$$= P(X=0) + P(X=1) = \frac{\frac{4!}{4!0!}\frac{21!}{5!16!}}{\frac{25!}{20!5!}} + \frac{\frac{4!}{3!1!}\frac{21!}{4!17!}}{\frac{25!}{20!5!}}$$

$$= \frac{(1)(20349)}{(53130)} + \frac{(4)(5985)}{(53130)} = \frac{20349}{53130} + \frac{23940}{53130}$$

$$= 0.383 + 0.451 = 0.834$$

24.3 Poisson Distribution

The probability distribution function (pdf) of the Poisson distribution can be derived by taking the limit of the binomial as $n \to \infty$, $P \to \infty$, and $nP = \mu$ remains constant. The Poisson pdf is given by (1, 2)

$$f(x) = \frac{e^{-\mu}\mu^x}{x!};\ x = 0, 1, 2, \ldots \tag{24.19}$$

Here, $f(x)$ is the probability of x occurrences of an event that occurs on the average μ times per unit of space or time. Both the mean and the variance of a random variable X having a Poisson distribution are μ.

The Poisson pdf can be used to approximate probabilities obtained from the binomial pdf given earlier when n is large and p is small. In general, good approximations will result when n exceeds 10 and p is less than 0.05. When n exceeds 100 and np is less than 10, the approximation will generally be excellent. Table 24.1

Table 24.1 Binomial and Poisson Comparison.

x	Binomial	Poisson	Binomial	Poisson
	(n = 20, p = 0.05)	(n = 20, p = 0.05)	(n = 100, p = 0.01)	(n = 100, p = 0.01)
0	0.3585	0.3679	0.3660	0.3679
1	0.3774	0.3679	0.3697	0.3679
2	0.1887	0.1839	0.1849	0.1839
3	0.0596	0.0613	0.0610	0.0613
4	0.0133	0.0153	0.0149	0.0153
5	0.0022	0.0031	0.0029	0.0031
6	0.0003	0.0005	0.0005	0.0005
≥ 7	0.0000	0.0001	0.0001	0.0001

compares binomial and Poisson probabilities for the case of ($n = 20, p = 0.05$) and ($n = 100, p = 0.01$).

If λ is the failure rate (per unit of time) of each component of a system, then λt is the average number of failures for a given unit of time. The probability of x failures in the specified unit of time is obtained by substituting $\mu = \lambda t$ in Equation (24.19) to obtain

$$f(x) = \frac{e^{-\lambda t}(\lambda t)^x}{x!}; \; x = 0, 1, 2, \ldots \tag{24.20}$$

Suppose, for example, that in a certain country the average number of airplane crashes per year is 2.5. What is the probability of 4 or more crashes during the next year? Substituting $\lambda = 2.5$ and $t = 1$ in Equation 24.20 yields

$$f(x) = \frac{e^{-2.5}(2.5)^x}{x!}; \; x = 0, 1, 2, \ldots \tag{24.21}$$

as the pdf of X, the number of airplane crashes in a year. The probability of 4 or more airplane crashes next year is then

$$\begin{aligned} P(X \geq 4) &= 1 - \sum_{x=0}^{3} \frac{e^{-2.5}(2.5)^x}{x!} \\ &= 1 - (0.0821 + 0.205 + 0.257 + 0.214) \\ &= 1 - 0.76 = 0.24 \end{aligned} \tag{24.22}$$

As another example, suppose that the average number of breakdowns of personal computers during 1000 h of operation of a computer center is 3. What is the probability of no breakdowns during a 10-h work period? Note that the given average is the average number of breakdowns during 1000 h, so the unit of time associated with the given average is 1000 h. The probability required is the probability of no breakdowns during a 10-h period, and the unit of time connected with the required probability is 10 h. If 10 is divided by 1000, the result is 0.01 so that in a 10-h period there are 0.01 time periods of 1000-h duration. The given average of breakdown is 3; multiplication by 0.01 yields 0.03, the average number of occurrences during a 10-h time period μ. This value of μ may be substituted in the Poisson pdf:

$$f(x) = \frac{e^{-0.03}(0.03)^x}{x!}; \; x = 0, 1, 2, \ldots \tag{24.23}$$

One may now substitute for x, the number of occurrences whose probability is required. The probability of no breakdowns in a 10-h period is:

$$P(X=0)=\frac{e^{-0.03}(0.03)^{0}}{0!}=e^{-0.03}=0.97 \tag{24.24}$$

In addition to the applications cited, the Poisson distribution can be used to obtain the reliability R of a standby redundancy system in which one unit is in the operating mode and n identical units are in standby mode. Unlike parallel systems, the standby units are inactive. The reliability of the standby redundancy system may be calculated by employing the Poisson distribution under the following conditions:

1. all units have the same failure rate in the operating mode
2. unit failures are independent
3. standby units have 0 failure rate in the standby mode
4. there is perfect switch over to a standby when the operating unit fails

The Poisson distribution is a distribution in its own right and arises in many different situations. For instance, it provides probabilities of specified numbers of telephone calls per unit interval of time, of environmental sampling procedures, of given numbers of defects per unit area of glass, textiles, or papers, and of various numbers of bacterial colonies per unit volume.

Illustrative Example 24.12

Microscopic slides of a certain culture of microorganisms contain on the average of 20 microorganisms per square centimeter. After treatment by a chemical, one cm^2 is found to contain only 10 such microorganisms. If the treatment had no effect, what would be the probability of finding 10 or fewer microorganisms in a given square centimeter?

Solution

Let X denote the number of microorganisms in one cm^2. If the chemical treatment had no effect, X has a Poisson pdf with $\mu = 20$. The probability of 10 or fewer microorganisms in a given square centimeter is

$$\begin{aligned}P(X\le 10)&=\sum_{x=0}^{10}e^{-20}(20)^{x}\\&=P(X=0)+P(X=1)+\ldots+P(X=10)\end{aligned}$$

Longhand calculation leads to

$$P(X\le 10)=0.0128$$

Illustrative Example 24.13

The average number of defective welds detected at the final examination of the tail section of an aircraft is 5. What is the probability of detecting at least one defective well during the final examination of the tail section?

Solution

The given average is the average number of defective welds detected at the final inspection of an aircraft tail section. The unit of space connected with the given average is the space occupied by the tail section of the aircraft. Therefore, the associated unit of space is the same as the space occupied by the tail section of the aircraft. Because both the unit of space connected with the given average and the unit of space connected with the required area are the same, their quotient is one. The probability required is the probability of detecting at least one defective weld at the final inspection of the tail section. The given average number of defectives is 5; multiplication by 1 yields the value of μ as 5. This value of μ is to be substituted in the Poisson pdf.

$$f(x) = \frac{e^{-5}(5)^x}{x!};\ x = 0, 1, 2, \ldots$$

Substitute for x, the number of occurrences whose probability is required. The probability of detecting at least one defective weld is therefore

$$P(X \geq 0) = 1 - P(X = 0)$$

$$= 1 - \frac{e^{-5}(5)^0}{0!} = 0.9933$$

Illustrative Example 24.14

Over the last 10 years, a local hospital reported that the number of deaths per year due to temperature inversions (air pollution) was 0.5 [3]. What is the probability of exactly 3 deaths in a given year?

Solution

For this problem,

$$P(X = 3) = \frac{e^{-0.5}(0.5)^3}{3!} = 0.0126$$

Illustrative Example 24.15

The number of hazardous waste trucks arriving daily at a certain hazardous waste incineration facility has a Poisson distribution with parameter $n = 2.5$. Present facilities can accommodate 3 trucks a day. If more than 3 trucks arrive in a day, the trucks in excess of 3 must be sent elsewhere.

1. On a given day, what is the probability of having to send a truck elsewhere?
2. How much must the present waste facilities be increased to permit handling of all trucks on about 95% of the days?

Solution

1. Let X denote the number of trucks arriving on a given day. Then

$$P(\text{sending trucks elsewhere}) = P(X \geq 4)$$

This may also be written as

$$P(X \geq 4) = 1 - P(X < 4)$$

$$P(X \geq 4) = 1 - P(X=0) - P(X=1) - P(X=2) - P(X=3)$$

Employing the Poisson pdf with $\mu = 2.5$ gives

$$\begin{aligned} P(X \geq 4) &= 1 - \left(e^{-2.5} + 2.5e^{-2.5} + \frac{(2.5)^2 e^{-2.5}}{2} + \frac{(2.5)^3 e^{-2.5}}{6}\right) \\ &= 1 - (0.0821 + 0.2052 + 0.2565 + 0.2138) \\ &= 0.2424 \end{aligned}$$

2. Note that the number of trucks has not been specified. The general solution is given by

$$P(\text{sending trucks elsewhere}) = 0.05 = 1 - \sum_{i=0}^{n} P(X = i)$$

A trial-and-error solution is required because a different equation is obtained for each value of n. For example, if $n = 5$.

$$0.05 = 1 - [P(X=0) + P(X=1) + P(X=2) + P(X=3) + P(X=4) + P(X+5)]$$

Substituting gives

$$\begin{aligned}0.05 &= 1 - e^{-2.5}\left[1 + 2.5 + \frac{(2.5)^2}{2} + \frac{(2.5)^3}{6} + \frac{(2.5)^4}{24} + \frac{(2.5)^5}{120}\right]\\ &= 1 - 0.0821[1.0 + 2.5 + 3.125 + 2.604 + 1.628 + 0.814]\\ &= 1 - 0.958\\ 0.05 &\approx 0.042\end{aligned}$$

The increase is from 3 to 5; therefore, the facilities must be increased by approximately 67%.

References

1. L. Theodore and F. Taylor, *Probability and Statistics*, A Theodore Tutorial, Theodore Tutorials, East Williston, NY, originally published by USEPA/APTI, RTP, NC,1996.
2. S. Shaefer and L. Theodore, *Probability and Statistics Applications for Environmental Science*, CRC Press/Taylor & Francis Group, Boca Raton, FL, 2007.
3. Theodore, *Air Pollution Control Equipment Calculations*, John Wiley & Sons, Hoboken, NJ, 2008.

25

Continuous Probability Distributions

The probability distribution of a continuous random variable concerns the distribution of probability over the range of the random variable. The distribution probability, i.e., the values of the variables together with their associated probabilities, is specified by the aforementioned *probability distribution function* (pdf). This chapter is devoted to providing general properties of the pdf for the case of *continuous* random variables as well as an introduction to the *cumulative distribution function* (cdf).

The pdf of a continuous random variable X has the following properties:

$$\int_a^b f(x)dx = P(a < X < b) \tag{25.1}$$

$$f(x) \geq 0 \tag{25.2}$$

$$\int_{-\infty}^{\infty} f(x)dx = 1 \tag{25.3}$$

Equation (25.1) indicates that the pdf of a continuous random variable generates probability by integration of the pdf over the interval whose probability is required. When this interval contracts to a single value, the integral over the interval becomes zero. Therefore, the probability associated with any particular value of a continuous random variable is *zero*. Consequently, if X is continuous

$$\begin{aligned} P(a \le X \le b) &= P(a < X \le b) \\ &= P(a < X < b) \\ &= P(a \le X < b) \end{aligned} \tag{25.4}$$

Equation (25.2) restricts the value of $f(x)$ to nonnegative numbers. Equation (25.3) follows from the fact that

$$P(-\infty < X < \infty) = 1 \tag{25.5}$$

As an example of the pdf of a continuous random variable, consider the pdf of the time X in hours between successive failures of an aircraft air conditioning system. Suppose the pdf of X is specified by

$$f(x) = 0.01e^{-0.01x};\ x > 0 \tag{25.6}$$

$$= 0;\ \text{elsewhere} \tag{25.7}$$

A plot of $f(x)$ vs. X for positive values of X is provided in Figure 25.1. Inspection of the graph indicates that intervals in the lower part of the range of X are assigned greater probabilities than intervals of the same length in the upper part of the range of X because the areas over the former are greater than the areas over the latter.

The expression $P(a < X < b)$ provided in Equation (25.1) can be interpreted geometrically as the area under the pdf curve over the interval (a, b). Integration

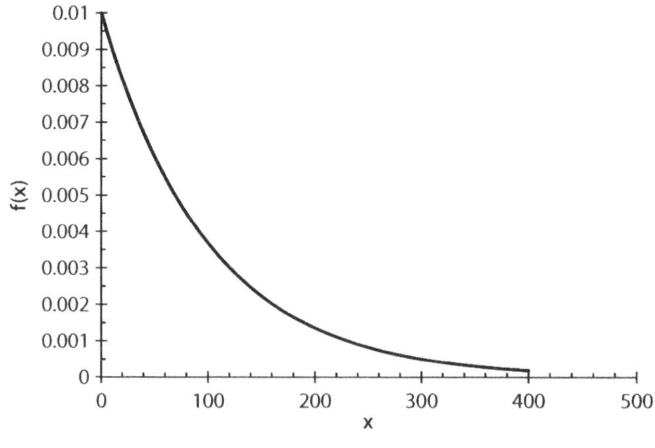

Figure 25.1 The pdf for a continuous variable.

of the pdf over the interval yields the probability assigned to the interval. For example, the probability that the time in hours between successive failures of the aforementioned aircraft air conditioning system is greater than 6 but less than 10 is

$$\begin{aligned} P(6 < X < 10) &= \int_{6}^{10} 0.01e^{-0.01}dx = 0.01\int_{6}^{10} e^{-0.01}dx \\ &= 0.01\left[-\left(\frac{1}{0.01}\right)e^{-0.01x}\right]_{6}^{10} \\ &= -e^{-0.01x}\Big|_{6}^{10} \\ &= -e^{-(0.01)10} - (-e^{-(0.01)6}) \\ &= -e^{-0.1} + e^{-0.06} \\ &= -0.9048 + 0.9418 = 0.037 = 3.7\% \end{aligned} \tag{25.8}$$

Another function used to describe the probability distribution of a random variable X is the aforementioned cdf. If $f(x)$ specifies the pdf of a random variable X, then $F(x)$ is used to specify the cdf. For both discrete and continuous random variables, the cdf of X is defined by

$$F(x) = P(X \geq x); \quad -\infty < x < \infty \tag{25.9}$$

In the case of a continuous random variable, the cdf is a continuous function.

Suppose, for example, that x is a continuous random variable with a pdf specified by

$$f(x) = 2x; \quad 0 \leq x < 1 \tag{25.10}$$

$$= 0; \text{ elsewhere} \tag{25.11}$$

Applying the definition in Equations (25.1)–(25.3), one obtains

$$F(x) = 0; \ x < 0 \tag{25.12}$$

$$F(x) = \int_{0}^{x} 2x\,dx = x^2; \quad 0 \leq x < 1 \tag{25.13}$$

$$F(x) = 1; \ x \geq 1 \tag{25.14}$$

Figure 25.2 displays the graph of this cdf, which is simply a plot of $F(x)$ vs. x. Differentiating the cdf and setting the pdf equal to zero where the derivative of the cdf does not exist can provide the pdf of a continuous random variable. For

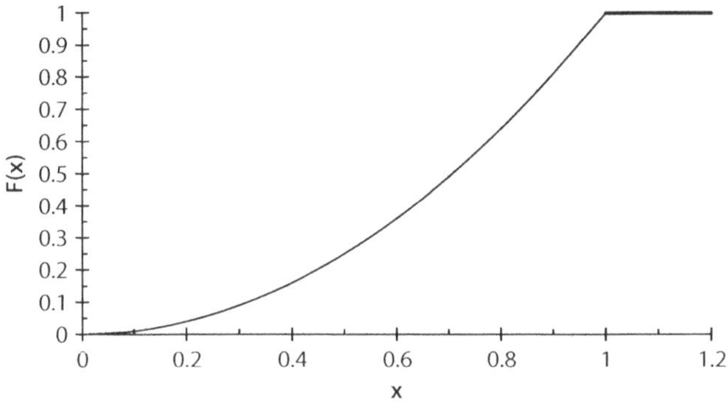

Figure 25.2 Graph of the cdf of a continuous random variable x.

example, differentiating the cdf of x^2 yields the specific pdf of $2x$. In this case, the derivative of the cdf does not exist for $x = 1$.

The following properties of the cdf of a random variable X can be deduced directly from the definition of $F(x)$.

$$F(b) - F(a) = P(a < X \leq b) \tag{25.15}$$

$$F(+\infty) = 1 \tag{25.16}$$

$$F(-\infty) = 0 \tag{25.17}$$

$$F(x) \text{ is a nondecreasing function of } x \tag{25.18}$$

Interestingly, these properties apply to the cases of both discrete and continuous random variables.

Illustrative Example 25.1

The difference between the magnitude of a large earthquake, as measured on the Richter scale, and the threshold value of 3.25 is a random variable X having the pdf

$$f(x) = 1.7e^{-1.7x};\ x > 0$$

$$= 0;\ \text{elsewhere}$$

Calculate $P(2 < X < 6)$.

Solution

Applying the definition of cdf leads to

$$P(2 < X < 6) = \int_2^6 f(x)dx$$

Substituting and integrating yields

$$P(2 < X < 6) = \int_2^6 1.7e^{-1.7x}dx$$
$$= e^{-3.4} - e^{-10.2}$$
$$= 0.0334 - 0.0$$
$$= 0.0334$$
$$= 3.34\%$$

Illustrative Example 25.2

A random variable X denoting the useful life of a battery in years has the pdf

$$f(x) = \left(\frac{3}{8}\right)x^2;\ 0 < x < 2$$
$$= 0;\ \text{elesewhere}$$

Find the cdf of X.

Solution

As before, the cdf of X is given by

$$f(x) = P(X \leq x)$$
$$= \int_{-\infty}^{\infty} f(x)dx$$

Therefore,

$$F(x) = 0;\ x \leq 0$$

And,

$$F(x) = \int \left[\left(\frac{3}{8}\right)x^2\right]dx$$
$$= \frac{x^3}{8};\ 0 < x < 2$$

In addition,

$$F(x) = 1;\ x \geq 2$$

Illustrative Example 25.3

The probability that a light switch fails is 0.0001. Let X denote the trial number on which the first failure occurs. Find the probability that X exceeds 1000.

Solution

The probability of no failures on the first $x - 1$ trials and a failure on trial x is $(0.9999)^{x-1}(0.0001)$. Thus, the pdf of X is

$$f(x) = (0.9999)^{x-1}(0.0001)$$

which is a geometric series with the first term equal to

$$f(x) = (0.9999)^{1000}(0.0001)$$

and common ratio equal to 0.9999. This reduces to

$$P(X > 1000) = \frac{(0.9999)^{1000}(0.0001)}{1 - 0.9999} = 0.9048 = 90.48\%$$

Four sections complement the presentation of this chapter. Section numbers and subject titles follow.

25.1: Exponential Distribution
25.2: Weibull Distribution
25.3: Normal Distribution
25.4: Log-Normal Distribution

Note: The bulk of the material in this chapter was drawn from the work of Theodore and Taylor [1].

25.1 Exponential Distribution

The exponential distribution is an important distribution in that it represents the distribution of the time required for a single event from a Poisson process to occur [1]. In particular, in sampling from a Poisson distribution with parameter μ, the probability that no event occurs during (0, t) is $e^{-\lambda t}$. Consequently, the probability that an event will occur during (0, t). is

$$F(t) = 1 - e^{-\lambda t} \tag{25.19}$$

This represents the cumulative distribution function (cdf) of t. One can therefore show that the probability distribution function (pdf) is

$$f(t) = e^{-\lambda t} \tag{25.20}$$

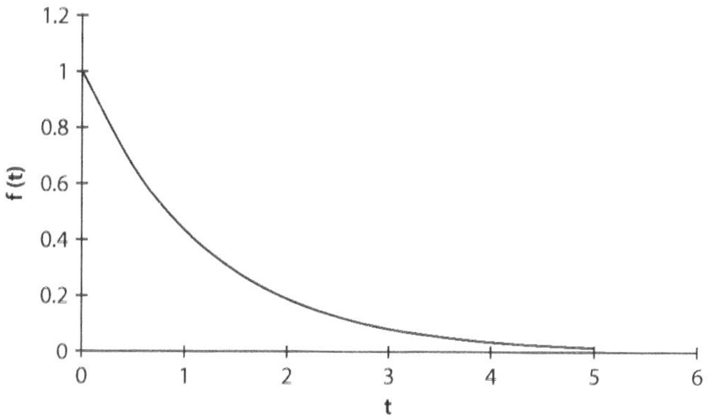

Figure 25.3 Exponential distribution.

Note that the parameter $1/\lambda$ (sometimes denoted as μ) is the expected value. Normally, the reciprocal of this value is specified and represents the expected value of $f(x)$. Because the exponential function appears in the expression for both the pdf and cdf, the distribution is justifiably called the *exponential distribution*. A typical pdf of t plot is provided in Figure 25.3.

Alternatively, the cumulative exponential distribution can be obtained from the pdf (with x now replacing t):

$$F(x) = \int_0^x \lambda e^{-\lambda x} dx = 1 - e^{-\lambda x} \tag{25.21}$$

All that remains is a simple evaluation of the negative exponent in Equation (25.21).

In statistical and reliability applications one often encounters a random variable's conditional failure density or hazard function, $g(x)$. In particular, $g(x)dx$ is the probability that a "product" will fail during $(x, x + dx)$ under the condition that it had not failed before (time) x. Consequently,

$$g(x) = \frac{f(x)}{1 - F(x)} \tag{25.22}$$

If the probability density function $f(x)$ is exponential, with parameter λ, it follows from Equation (25.20) and Equation (25.21) that

$$g(x) = \frac{\lambda e^{-\lambda x}}{1-(1-e^{-\lambda x})} = \frac{\lambda e^{-\lambda x}}{e^{-\lambda x}} \tag{25.23}$$
$$g(x) = \lambda$$

Equation (25.23) indicates that the failure probability is constant, irrespective of time. It implies that the probability that a component whose time-to-failure distribution is exponential fails in an instant during the first hour of its life is the same as its failure probability during an instant in the thousandth hour - presuming it has survived up to that instant. It is for this reason that the parameter λ is usually referred to in life-test applications as a *failure rate*. This definition generally has meaning only with an exponential distribution.

This natural association with life-testing and the fact that it is very tractable mathematically makes the exponential distribution attractive as representing the life distribution of a complex system or several complex systems. In fact, the exponential distribution is as prominent in reliability analysis as the normal distribution is in other branches of statistics.

Illustrative Example 25.4

Discuss some of the advantages and limitations of the exponential distribution.

Solution

It has been shown theoretically that this distribution provides a reasonable model for systems designed with a limited degree of redundancy and made up of many components, none of which has a high probability of failure. This is especially true when low component failure rates are maintained by periodic inspection and replacement, or in situations in which failure is a function of outside phenomena rather than a function of previous conditions. On the other hand, the exponential distribution often cannot represent individual component life (because of its "infant mortalities" and wearout patterns), and it is sometimes questionable even as a system model.

Illustrative Example 25.5

Estimate the probability that a pump at a nuclear power plant will survive at least:

1. 3 times its expected life
2. 5 times its expected life
3. 10 times its expected life

Assume an exponential distribution applies.

Solution

The exponential distribution gives

$$P(T) = e^{-\lambda t};\ t = time$$

with $\lambda = 1/a$ where a = expected life of the pump. Thus for 3 times its expected life $t = 3a$,

$$P(T > 3a) = e^{-\left(\frac{1}{a}\right)(3a)} = e^{-3}$$
$$= 0.0498 = 4.98\%$$

Therefore, there is a 5% chance that the pump will survive past 3 times its expected life.

For 5 times its expected life $t = 5a$

$$P(T > 5a) = e^{-\left(\frac{1}{a}\right)(5a)} = e^{-5}$$
$$= 0.0067 = 0.67\%$$

and for 10 times its expected life $t = 10a$

$$P(T > 10a) = e^{-\left(\frac{1}{a}\right)(10a)} = e^{-10}$$
$$= 4.54 \times 10^{-5} = 4.54 \times 10^{-3}\%$$

As expected, the probability decreases with increasing survival time.

Illustrative Example 25.6

The time to failure for a battery is presumed to follow an exponential distribution with $\lambda = 0.1$ (per year). What is the probability of a failure within the first year?

Solution

Refer to Equation (25.20)

$$F(x) = \int_0^1 \lambda e^{-\lambda x} dx = 1 - e^{-\lambda x} \tag{25.21}$$

For this case (with x representing time),

$$P(X \le 1) = \int_0^1 (0.1)e^{-(0.1)x} dx$$
$$= -\frac{0.1}{0.1} e^{-(0.1)x} \Big|_0^1 = -e^{-0.1} + e^0$$
$$= 1 + e^{-0.1} = 0.095 = 9.5\%$$

Therefore, there is nearly a 10% probability that the battery will fail within the first year.

Illustrative Example 25.7

An electronic system consists of 3 components (1, 2, 3) connected in parallel. If the time to failure for each component is exponentially distributed and mean times of failures for components 1, 2, and 3, are 200, 300, and 600 d, respectively, determine the system reliability for a year (365 d).

Solution

For this series system, the probability of failure is

$$\begin{aligned} P(F) &= P(\text{all the components fail}) \\ &= (1-P_1)(1-P_2)(1-P_3) \end{aligned}$$

where P_i is the probability of surviving 365 d. The system reliability is

$$\begin{aligned} R &= 1-P(F) \\ &= 1-(1-P_1)(1-P_2)(1-P_3) \end{aligned}$$

Based on the data provided:

$$P_1 = e^{-\left(\frac{365}{200}\right)} = 0.161$$

$$P_2 = e^{-\left(\frac{365}{300}\right)} = 0.296$$

$$P_3 = e^{-\left(\frac{365}{600}\right)} = 0.544$$

Therefore, the system reliability is

$$\begin{aligned} R &= 1-(1-0.161)(1-0.296)(1-0.544) \\ &= 0.731 = 73.1\% \end{aligned}$$

Illustrative Example 25.8

A pumping system at a nuclear facility consists of 4 components. 3 are connected in parallel, which in turn are connected downstream (in series) with the other component. The arrangement is schematically shown in Figure 25.4.

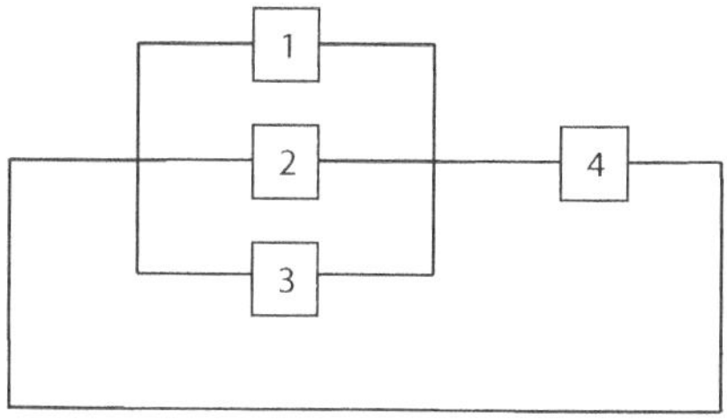

Figure 25.4 Pumping system.

If the pumps have the same exponential failure rate, λ, of 0.5 $(\text{year})^{-1}$, estimate the probability that the system will not survive for more than one year.

Solution

Based on the information provided, the pumping system fails when the 3 parallel components fail *or* when the downstream component fails. This is a combination of a parallel and a series system. The reliability is

$$R(t) = 1 - F(t) \tag{25.24}$$

where $F(t)$ is the probability of failure between 0 and t. The reliability of the parallel system is

$$\begin{aligned} R_p &= 1-(1-R_1)(1-R_2)(1-R_3);\ R_i = R \\ &= 1-(1-R)^3 \end{aligned}$$

For a series system

$$R_s = R_p R_4 = R_p R$$

where R_s also represents the overall system reliability.

Applying the exponential model gives

$$R(t) = 1 - F(t) = 1 - (1 - e^{-\lambda t}) = e^{-\lambda t}$$

$$\begin{aligned} &R_1 = R_2 = R_3 = R_4 = R = e^{-(0.5)(1)} \\ &R = 0.6065 \end{aligned}$$

Thus,

$$R_p = 1 - (1 - 0.6065)^3 = 1 - 0.0609 = 0.9391 = 93.9\%$$

and

$$R_s = (0.9391)(0.6065)$$
$$= 0.57 = 57\%$$

Illustrative Example 25.9

The probability that a thermometer placed in three different locations of a hazardous waste incinerator will not survive for more than 36 months is 0.925, 0.95, and 0.99 respectively. How often should the thermometer be replaced in each one of these situations? Assume the time to failure is exponentially distributed and that the replacement time should be based on the thermometer's expected life.

Solution

This requires the calculation of μ in the exponential model with units of $(\text{month})^{-1}$. Once again,

$$F(t) = 1 - e^{-\lambda t}$$

Based on the information provided

$$P(T \leq 36) = 0.925$$

or

$$0.925 = 1 - e^{-(\lambda)(36)}$$

Solving gives

$$\lambda = -\frac{1}{36}\ln(1 - 0.925) = 0.07195$$

The expected time (or life), $E(T)$, is [1, 2]

$$E(T) = \frac{1}{\lambda} = \frac{1}{0.07195} = 13.9 \text{ months}$$

Therefore, for the first location the thermometer should be replaced in approximately 14 months.

For the second location, the describing equation remains

$$F(t) = 1 - e^{-\lambda t}$$

For $F(t) = 0.95$,

$$0.95 = 1 - e^{-(\lambda)(36)}$$

Solving for μ gives

$$\lambda = -\frac{1}{36}\ln(1-0.95) = 0.0832$$

The expected life is then

$$E(T) = \frac{1}{\lambda} = \frac{1}{0.0832} = 12 \text{ months}$$

For the third location

$$\lambda = -\frac{1}{36}\ln(1-0.99) = 0.128$$

and

$$E(T) = \frac{1}{\lambda} = \frac{1}{0.128} = 7.82 \text{ months}$$

25.2 Weibull Distribution

Unlike the exponential distribution, the failure rate of equipment frequently exhibits 3 stages: a break-in stage with a declining failure rate, a useful life stage characterized by a fairly constant failure rate, and a wear out period characterized by an increasing failure rate. Many industrial parts and components follow this behavior. A failure rate curve exhibiting these 3 phases (see Figure 25.5) is called a *bathtub* curve.

In the case of the bathtub curve, the failure rate during useful life is constant. Letting this constant be α yields

$$F(t) = \alpha \exp\left(-\int_0^t \alpha dt\right) = \alpha \exp(-\alpha t);\ t > 0 = \alpha e^{-\alpha t} \qquad (25.25)$$

as the probability distribution function (pdf) of time to failure during the useful life stage of the bathtub curve. Equation (25.25) defines an exponential (pdf) that is a special case of the pdf defining the Weibull distribution.

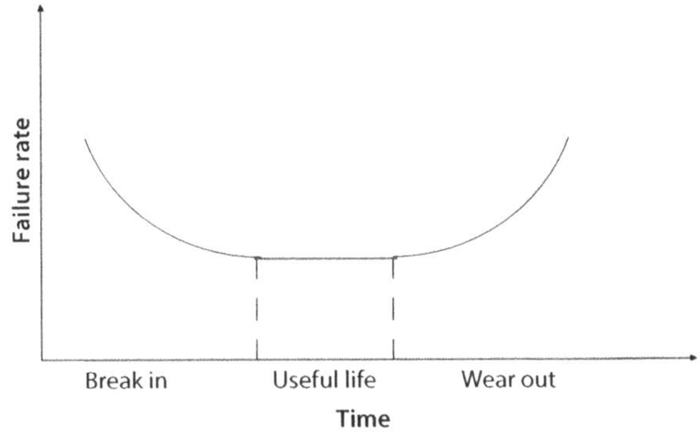

Figure 25.5 Bathtub curve.

Weibull introduced the distribution, which bears his name principally on empirical grounds, to represent certain life-test data. The Weibull distribution provides a mathematical model of all 3 stages of the bathtub curve. This is now discussed. An assumption about the failure rate $Z(t)$ that reflects all 3 stages of the bathtub curve is

$$Z(t) = \alpha\beta t^{\beta-1};\ t > 0 \tag{25.26}$$

where α and β are constants. For $\beta < 1$, the failure rate $Z(t)$ decreases with time. For $\beta = 1$, the failure rate is constant and equal to α. For $\beta > 1$, the failure rate increases with time. Using this equation again to translate the assumption about failure rate into a corresponding assumption about the pdf of T, time to failure, one obtains

$$f(t) = \alpha\beta t^{\beta-1} \exp\left(\int_0^1 \alpha\beta t^{\beta-1} dt\right) = \alpha\beta t^{\beta-1} \exp(-\alpha t^{\beta});\ t > 0;\ \alpha > 0, \beta > 0 \tag{25.27}$$

Equation (25.27) defines the pdf of the Weibull distribution. As noted, the exponential distribution discussed in the preceding section, whose pdf was given in Equation (25.20) earlier, is a special case of the Weibull distribution with $\beta = 1$.

A variety of assumptions about failure rate and the probability distribution of time to failure that can be accommodated by the Weibull distribution make it especially attractive in describing failure time distributions in industrial and process plant applications

To illustrate probability calculations involving the exponential and Weibull distributions introduced in conjunction with the bathtub curve of failure rate, consider first the case of a transistor having a constant rate of failure of 0.01 per thousand hours. To find the probability that a transistor will operate for at least 25,000 h, substitute the failure rate

$$Z(t) = 0.01 \tag{25.28}$$

which yields

$$f(t) = \exp\left(-\int_0^t 0.01dt\right) \\ = 0.01e^{-0.01t};\ t > 0 \tag{25.29}$$

as the pdf of T, the time to failure of the transistor. Because t is measured in thousands of hours, the probability that the transistor will operate for at least 25,000 h is given by

$$P(T > 25) = \int_{25}^{\infty} -0.01e^{-0.01t}dt \tag{25.30}$$

$$= -e^{-\infty} + e^{-0.01(25)}$$

$$= 0 + 0.78$$

$$= 0.78 = 78\%$$

Now suppose it is desired to determine the 10,000 h reliability of a circuit of 5 such transistors connected in series. The 10,000 h reliability of one transistor is the probability that it will last at least 10,000 h. This probability can be obtained by integrating the pdf of T, time to failure, which gives

$$P(T > 10) = \int_{10}^{\infty} -0.01e^{-0.01t}dt \tag{25.31}$$

$$= -e^{-\infty} + e^{-0.01(10)}$$

$$= 0 + 0.90 = 90\%$$

The 10,000 h reliability of a circuit of 5 transistors connected in series is obtained by applying the formula for the reliability of series systems to obtain

$$R_s = [R(10)]^5 = (0.9)^5$$

$$= 0.59 = 59\% \tag{25.32}$$

As another example of probability calculations, consider a component whose time to failure T, in hours, has a Weibull pdf with parameters $\alpha = 0.01$ and $\beta = 0.50$ in Equation (25.27). This gives

$$f(t) = (0.01)(0.5)t^{0.5-1}e^{-(0.01)t^{0.5}};\ t > 0 \tag{25.33}$$

As the Weibull pdf of the failure time of the component under consideration, the probability that the component will operate at least 8100 h is then given by

$$P(T > 8100) = \int_{8100}^{\infty} f(t)dt = \int_{8100}^{\infty} 0.005t^{-0.5}e^{-(0.01)t^{0.5}}dt \tag{25.34}$$

$$= e^{-0.01t^{0.5}}\Big|_{8100}^{\infty} = 0 + e^{-(0.01)(8100)^{0.5}} = 0.41 = 41\%$$

Illustrative Example 25.10

Discuss some Weibull distribution applications.

Solution

A variety of conditional failure distributions, including wear out patterns, can be accommodated by the Weibull distribution. Therefore, this distribution has been frequently recommended – instead of the exponential distribution – as an appropriate failure distribution model. Empirically, satisfactory fits have been obtained for failure data on electron tubes, relays, ball bearings, metal fatigue, and even business mortality.

Illustrative Example 25.11

The life (time to failure) of a machine component has a Weibull distribution. Outline how to determine the probability that the component lasts a given period of time if the failure rate is $t^{-1/2}$.

Solution

Identify the failure rate, $Z(t)$, from Equation (25.26).

$$Z(t) = t^{-\frac{1}{2}}$$

Also, identify the values of and appearing in the failure rate. If the failure rate is $t^{-1/2}$,

$$\beta - 1 = -\frac{1}{2}$$

and

$$\alpha\beta = 1$$

Therefore, $\beta = 1/2$, and $\alpha = 2$.

For these values of α and β obtained above, determine the Weibull pdf

$$f(t) = t^{-\frac{1}{2}} e^{-2t^{\frac{1}{2}}};\ t > 0$$

Integration of this pdf will yield the required probability. This is demonstrated in the next illustrative example.

Illustrative Example 25.12

Refer to Illustrative Example 25.11. Determine the probability that the component lasts at least 25,000 h if t is measured in thousands of hours.

Solution

For this case,

$$t = 25$$

Because time is measured in thousands of hours the probability that the component lasts at least 25,000 h is

$$P(T > 25) = \int_{25}^{\infty} t^{-\frac{1}{2}} e^{-2t^{\frac{1}{2}}}\, dt$$

$$= -e^{-2\sqrt{t}} \Big|_{25}^{\infty}$$

This may be integrated to give

$$P(T > 25) = -0 - (-e^{-2(25)^{0.5}})$$

$$= 4.5 \times 10^{-5} \approx 0\%$$

Illustrative Example 25.13

The life of a gasket has a Weibull distribution with failure rate

$$Z(t) = \frac{1}{\sqrt{t}} = t^{-1/2}$$

where t is measured in years. What is the probability that the gasket will last at least 4 years?

Solution

The pdf specified by $f(t)$ in terms of the failure rate, $Z(t)$, is as follows:

$$f(t) = Z(t)\exp\left(-\int_0^t Z(t)dt\right)$$

Substituting $1/t^{1/2}$ for $Z(t)$ yields

$$f(t) = t^{-\frac{1}{2}}\exp\left(-\int_0^{\infty} t^{-\frac{1}{2}}dt\right) = t^{-\frac{1}{2}}\exp\left(-2t^{\frac{1}{2}}\right)dt;\ t > 0$$

Once again, employ the integration procedure provided earlier. The probability that the seal lasts at lasts 4 years is

$$P(T \geq 4) = \int_4^{\infty} t^{-\frac{1}{2}}\exp\left(-2t^{\frac{1}{2}}\right)dt = [-e^{-2t^{0.5}}]_4^{\infty}$$

$$= -0 - (-e^{-4}) = 0.0183 = 1.83\%$$

25.3 Normal Distribution

The initial presentation in this section on normal distributions will focus on failure rate, but can be simply applied to all other applications involving normal distributions. When T, time to failure, has a normal distribution, its probability distribution function (pdf) is given by

$$f(T) = \frac{1}{\sqrt{2\pi\sigma}}\exp\left[-\frac{1}{2}\left(\frac{T-\mu}{\sigma}\right)^2\right];\ -\infty < t < \infty \qquad (25.35)$$

where μ is the mean value of T and σ is its standard deviation. The graph of $f(T)$ is the familiar bell-shaped curve shown in Figure 25.6.

The reliability function corresponding to the normal distributed failure time is given by

$$f(T) = \frac{1}{\sqrt{2\pi\sigma}}\int_t^{\infty}\exp\left[-\frac{1}{2}\left(\frac{T-\mu}{\sigma}\right)^2\right]dt \qquad (25.36)$$

If T is normally distributed with mean μ and standard deviation σ, then the random variable $(T - \mu)/\sigma$ is normally distributed with mean 0 and standard deviation 1. The term $(T - \mu)/\sigma$ is called a *standard normal curve* that is represented by

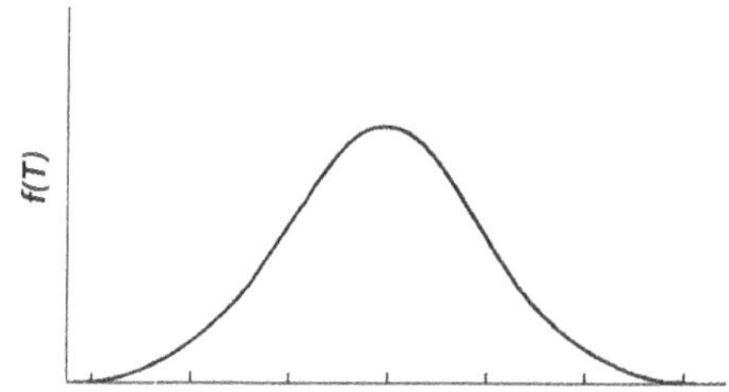

Figure 25.6 Normal pdf of time to failure.

Z, not to be confused with the failure rate $Z(t)$. This is discussed in more detail later in this section.

Table 25.1 is a tabulation of areas under a standard normal curve to the right of z_0 for nonnegative values of z_0. Probabilities about the normal variable Z can be determined from this table. For example,

$$P(Z > 1.54) = 0.062 \tag{25.37}$$

is obtained directly from Table 25.1 as the area to the right of 1.54. As presented in Figure 25.7, the symmetry of the standard normal curve about zero implies that the area to the right of zero is 0.5.

Plots demonstrating the effect of μ and σ on the bell-shaped curve are provided in Figure 25.8 and Figure 25.9. Consequently, one can deduct from Table 25.1 and Figure 25.7 that

$$P(0 < Z < 154) = 0.5 - 0.062 = 0.438 \tag{25.38}$$

Also, because of symmetry

$$P(-1.54 < Z < 0) = 0.438 \tag{25.39}$$

and

$$P(Z < -1.54) = 0.062 \tag{25.40}$$

The following probabilities can also be deduced by noting that the area to the right of 1.54 is 0.062 in Figure 25.7.

$$P(-1.54 < Z < 1.54) = 0.876 \tag{25.41}$$

$$P(Z < 1.54) = 0.938 \tag{25.42}$$

Table 25.1 Standard normal cumulative probability; right-hand tail.

Second decimal place of *z*										
z	**.00**	**.01**	**.02**	**.03**	**.04**	**.05**	**.06**	**.07**	**.08**	**.09**
0.0	.5000	.4960	.4920	.4880	.4840	.4801	.4761	.4721	.4681	.4641
0.1	.4602	.4562	.4522	.4483	.4443	.4404	.4364	.4325	.4286	.4247
0.2	.4207	.4168	.4129	.4090	.4052	.4013	3971	.3936	.3897	.3859
0.3	.3821	.3783	1745	.3707	.3669	.3632	.3594	.3557	.3520	.3483
0.4	.3446	.3409	.3372	.3336	.3300	.3264	.3228	.3192	.3156	.3121
0.5	.3085	.3050	.3015	.2981	.2946	.2912	.2877	.2843	.2810	.2776
0.6	.2743	.2709	.2676	.2643	.2611	.2578	.2546	.2514	.2483	.2451
0.7	.2420	.2389	.2358	.2327	.2296	2266	.2236	.2206	.2177	.2148
0.8	.2119	.2090	.2061	.2033	.2005	.1977	.1949	.1922	1894	.1867
0.9	.1841	.1814	.1788	.1762	.1736	.1711	.1685	.1660	.1635	.1611
1.0	.1587	.1562	.1539	.1515	.1492	.1469	.1446	.1423	.1401	.1379
1.1	.1357	.1335	.1314	.1292	.1271	.1251	.1230	.1210	.1190	.1170
1.2	.1151	.1131	.1112	.1093	.1075	.1056	.1038	.1020	.1003	.0985
1.3	.0968	.0951	.0934	.0918	.0901	.0885	.0869	.0853	.0838	.0823
1.4	.0808	.0793	.0778	076-1	.0749	.0735	1721	.0708	0694	.0681
1.5	.0668	.0655	.06-13	.0630	.0618	.0606	.0591	.0582	.0571	.0559
1.6	.0548	.0537	.0526	.0516	.0505	.0495	.0485	.0475	.0465	.0455
1.7	.0446	.0436	.0127	.0118	.0109	.0401	.0392	.0384	.0375	.0367
1.8	.0359	.0351	.0344	.0336	.0329	.0322	.0314	.0307	.0301	.0294
1.9	.0287	.0281	.0274	.0268	.0262	.0256	.0250	.0244	.0239	.0233
2.0	.0228	.0222	.0217	.0212	.0207	.0202	.0197	.0192	.0188	.0183
2.1	.0179	.0174	.0170	.0166	.0162	.0158	.0154	.0150	.0146	.0143
2.2	.0139	.0136	.0132	.0129	.0125	.0122	.0119	.0116	.0113	.0110
2.3	.0107	.0104	.0102	.0099	.0096	.0094	.0091	.0089	.0087	.0084
2.4	.0082	.0080	.0078	.0075	.0073	.0071	.0069	.0068	.0066	.0064
2.5	.0062	0060	.0059	.0057	.0055	.0054	.0052	.0051	.0049	.0048
2.6	.0047	.0045	.0044	.0043	.0041	.0040	.0039	.0038	.0037	.0036
2.7	.0035	0034	.0033	.0032	.0031	.0030	.0029	.0028	.0027	.0026
2.8	.0026	.0025	.0024	.0023	.0023	.0022	.0021	.0021	.0020	.0019
2.9	.0019	.0018	.0018	.0017	.0016	.0016	.0015	.0015	.0014	.0014

$$P(Z > -1.54) = 0.938 \tag{25.43}$$

Table 25.1 can also determine probabilities concerning normal random variables that are not standard normal variables. The required probability is first converted to an equivalent probability about a standard normal variable. For example,

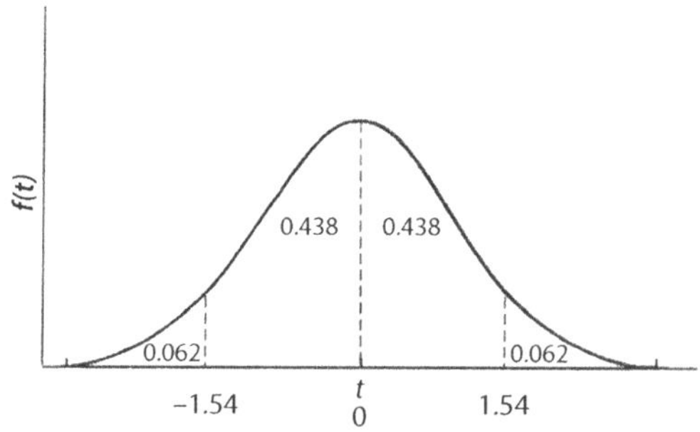

Figure 25.7 Areas under a standard normal curve.

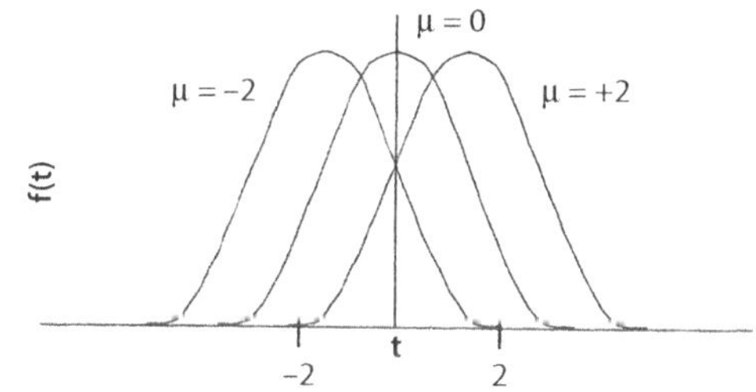

Figure 25.8 Normal pdf – varying μ.

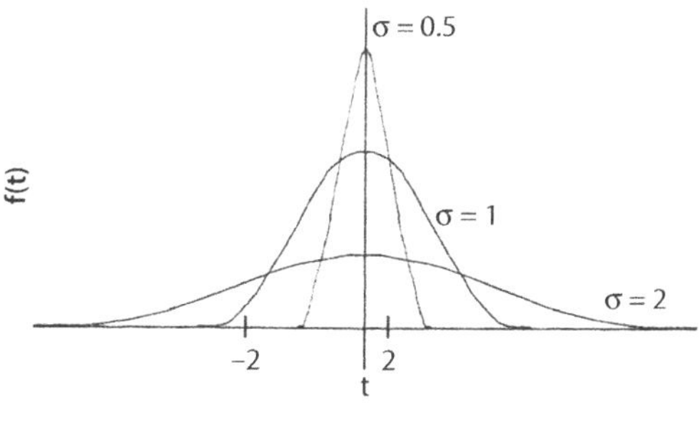

Figure 25.9 Normal pdf – varying σ.

if T, the time to failure, is normally distributed with mean $\mu = 100$ and standard deviation $\sigma = 2$ then $(T - 100)/2$ is a standard normal variable and one may write

$$P(T_1 < T < T_2) = P\left(\frac{T_1 - \mu}{\sigma} < \frac{T - \mu}{\sigma} < \frac{T_2 - \mu}{\sigma}\right) \tag{25.44}$$

where

$$\frac{T - \mu}{\sigma} = Z = \text{standard normal variable} \tag{22.45}$$

Therefore, if $T_1 = 98$ and $T_2 = 104$ as an example, the describing equation becomes

$$\begin{aligned} P(98 < T < 104) &= \left(\frac{98 - \mu}{\sigma} < \frac{T - \mu}{\sigma} < \frac{104 - \mu}{\sigma}\right) \\ &= P\left(\frac{98 - 100}{2} < \frac{T - \mu}{2} < \frac{104 - 100}{2}\right) \\ &= P\left(-1 < \frac{T - 100}{2} < 2\right) \\ &= P(-1 < Z < 2) \\ &= 0.341 + 0.477 \\ &= 0.818 = 81.8\% \end{aligned} \tag{25.46}$$

For any random variable X - where X has now replaced T - that is normally distributed with mean μ and standard deviation σ, one may now write the following:

$$\begin{aligned} P(\mu - \sigma < X < \mu + \sigma) &= P\left(-1 < \frac{X - \mu}{\sigma} < 1\right) \\ &= P(-1 < Z < 1) = 0.68 \end{aligned} \tag{25.47}$$

$$\begin{aligned} P(\mu - 2\sigma < X < \mu + 2\sigma) &= P\left(-2 < \frac{X - \mu}{\sigma} < 2\right) \\ &= P(-2 < Z < 2) = 0.95 \end{aligned} \tag{25.48}$$

$$\begin{aligned} P(\mu - 3\sigma < X < \mu + 3\sigma) &= P\left(-3 < \frac{X - \mu}{\sigma} < 3\right) \\ &= P(-3 < Z < 3) = 0.997 \end{aligned} \tag{25.49}$$

The probabilities given above are the sources of the percentages cited earlier [2]. These can be used to interpret the standard deviation s of a sample of observations on a normal random variable as a measure of dispersion about the sample mean $\overline{X}$.

The normal distribution can be used to obtain probabilities concerning the mean $\overline{X}$ of a sample of n observations on a random variable X if X is normally distributed with mean μ and standard deviation $\sigma/\sqrt{n}$. For example, suppose X is normally distributed with mean 100 and standard deviation 2. Then $\overline{X}$, of a sample of 16 observations on X, is normally distributed with mean 100 and standard deviation 0.5. To calculate the probability that X is greater than 101, one would write

$$P(X > 101) = P\left[\frac{\overline{X} - 100}{0.5} > \frac{101 - 100}{0.5}\right] \tag{25.50}$$

$$\begin{aligned} P(X > 101) &= P\left[\frac{\overline{X} - 100}{0.5} > \frac{101 - 100}{0.5}\right]; \; Z = \frac{\overline{X} - 100}{0.5} \\ &= P(Z > 2) = 0.023 = 2.3\% \end{aligned} \tag{25.51}$$

If X is not normally distributed, then $\overline{X}$, the mean of a sample of n observations on X, is *approximately* normally distributed with mean μ and standard deviation $\sigma/\sqrt{n}$, provided the sample size n is large (>30). This result is based on an important theorem and probability called the *central limit theorem*.

Suppose for example, that the mean and variance for a random variable are 1 and 1/3 respectively. If $\overline{X}$ is the mean of a random sample of 48 observations on X, $\overline{X}$ is approximately normally distributed with mean 1 and standard deviation $\sigma/\sqrt{n}$. The latter term is therefore given by

$$\frac{\sigma}{\sqrt{n}} = \frac{\sqrt{1/3}}{\sqrt{48}} = \sqrt{\frac{1}{(3)(48)}} = \sqrt{\frac{1}{144}} = \frac{1}{12} \tag{25.52}$$

The following example is now provided for this case.

$$P\left(X > \frac{9}{8}\right) = P\left[\frac{\overline{X} - 1}{\frac{1}{12}} > \frac{\frac{9}{8} - 1}{\frac{1}{12}}\right] = P(Z > 1.5) = 0.067 \tag{25.53}$$

One of the principal applications of the normal distribution in reliability calculations and hazard risk analysis is the distribution of time to failure due to "wear out." Suppose, for example, that a production lot of thermometers, to be employed in an incinerator especially designed to withstand high temperatures and intense vibrations, has just come off the assembly line. A sample of 25 thermometers from the lot is tested under the specified heat and vibration conditions. Time to failure, in hours, is recorded for each of the 25 thermometers. Application of the equations for the sample mean X and sample variance s^2 yields

$$\overline{X} = 125; \ s^2 = 92 \tag{25.54}$$

Past experience indicates that the "wear out" time of this "unit," like that of a large variety of products in many different industries, tends to be normally distributed. Using the above values of $\overline{X}$ and s as best estimates of μ and σ, one can obtain the 110-h reliability of the thermometers.

$$\begin{aligned} P(X > 110) &= P\left[\frac{X-125}{\sqrt{92}} > \frac{110-125}{\sqrt{92}}\right] \\ &= P(Z > -1.56) \end{aligned}$$

From values in the standard normal table,

$$P(X > 110) = 0.94 = 94\% \tag{25.55}$$

As indicated earlier, the normal distribution is symmetric. The data from a normal distribution could be plotted on special graph paper, known as normal probability paper. The resulting plot appears as a straight line. The parameters μ and σ, can be estimated for such a plot. A nonlinear plot on this paper is indicative of nonnormality.

Actual (experimental) data have shown many physical variables to be normally distributed. Examples include physical measurements on living organisms, molecular velocities in an ideal gas, scores on an intelligence test, the average temperatures in a given locality, etc. Other variables, though not normally distributed *per se*, sometimes approximate a normal distribution after an appropriate transformation, such as taking the logarithm or square root of the original variable. The normal distribution also has the advantage that it is tractable mathematically. Consequently, many of the techniques of statistical inference have been derived under the assumption of underlying normal variants.

Illustrative Example 25.14

The parts-per-million concentration of a particular toxic substance in a wastewater stream is known to be normally distributed with mean $\mu = 100$ and standard deviation $\sigma = 2.0$. Calculate the probability that the toxic concentration, C, is between 98 and 104.

Solution

Because C is normally distributed with $\mu = 100$ and a standard deviation $\sigma = 2.0$, then $(C - 100)/2$ is a standard normal variable and

$$P(98 < C < 104) = P\left(-1 < \left[\frac{C-100}{2}\right] < 2\right)$$
$$= P(-1 < Z < 2)$$

From the values in the standard normal table,

$$P(98 < C < 104) = 0.341 + 0.477 = 0.818 = 81.8\%$$

The reader should note that the calculation for this illustrative example is identical to one presented earlier for a time to failure application.

Illustrative Example 25.15

Acceptance limits require the plate-to-plate spacing in an electrostatic precipitator to be between 24 and 25 cm [3]. Spacing from other installations suggests that it is normally distributed with an average length of 24.6 cm and a standard deviation of 0.4 cm. What proportion of the plate spacing in a typical unit can be assumed unacceptable [3]?

Solution

The problem is equivalent to determining the probability that an observation or sample from a normal distribution with parameters $\mu = 0$ and $\sigma = 1$ either exceeds

$$Z_1 = \frac{25.0 - 24.6}{0.4} = 1$$

or falls below

$$Z_2 = \frac{24.0 - 24.6}{0.4} = -1.5$$

These probabilities are 0.159 and 0.067, respectively, from the table of the standard normal distribution. Consequently, the proportion of unacceptable spacing is

$$P(\text{unacceptable spacing}) = P(Z > 1) + P(Z < -1.5)$$
$$= 0.159 + 0.067$$
$$= 0.226 = 26.6\%$$

Illustrative Example 25.16

The temperature of a polluted estuary during the summer months is normally distributed with mean 56°F and standard deviation 3.0°F. Calculate the probability that the temperature is between 55 and 62°F.

Solution

Normalizing the temperature T gives

$$Z_1 = \frac{55-56}{3.0} = -0.333$$

$$Z_2 = \frac{62-56}{3.0} = 2.0$$

Thus,

$$P(0.333 < Z < 2.0) = P(0.0 < Z < 2.0) + P(0.0 < Z < 0.333)$$
$$= 0.4722 + 0.1293 = 0.6015 = 60.15\%$$

Illustrative Example 25.17

The regulatory specifications on a toxic substance in a solid waste ash calls for a concentration level of 1.0 ppm or less [4]. Earlier observations of the concentration of the ash, C, indicate a normal distribution with a mean of 0.60 ppm and a standard deviation of 0.20 ppm. Estimate the probability that ash will exceed the regulatory limit.

Solution

The problem requires the calculation of $P(C > 1.0)$. Normalizing the variable C,

$$P\left(\left[\frac{C-0.6}{0.2}\right] > \left[\frac{1.0-0.6}{0.2}\right]\right)$$

$$P(Z > 2.0)$$

From the standard normal table

$$P(Z > 2.0) = 0.0228 = 2.28\%$$

For this situation, the area to the right of the 2.0 is 2.28% of the total area. This represents the probability that ash will exceed the regulatory limit of 1.0 ppm.

Illustrative Example 25.18

The diameter D of wires installed in an electrostatic precipitator (ESP) has a standard deviation of 0.01 in [3]. At what value should the mean be if the probability of its exceeding 0.21 in is to be 1%?

Solution

For this example,

$$P\left(\left[\frac{D-\mu}{\sigma}\right]>\left[\frac{0.21-\mu}{0.01}\right]\right)=0.01$$

where

$$Z=\frac{D-\mu}{\sigma}$$

Therefore,

$$P\left(Z>\left[\frac{0.21-\mu}{0.01}\right]\right)=0.01$$

For a *one-tailed* test at the 1% (0.01) level [1, 2],

$$Z=2.326$$

so that

$$2.326=\frac{0.21-\mu}{0.01}$$
$$\mu=0.187 \text{ in}$$

Illustrative Example 25.19

The lifetime, T, of a circuit board employed is normally distributed with a mean of 2500 d. What is the largest lifetime variance the installed circuit boards can have if 95% of them need to last at least 365 d?

Solution

For this application,

$$P(T>365)=0.95$$

Normalizing gives

$$P\left(\left[\frac{T-\mu}{\sigma}\right]>\left[\frac{365-2500}{\sigma}\right]\right)=0.95$$

$$P\left(Z>-\frac{2135}{\sigma}\right)=-1.645$$

The following equation must apply for this condition:

$$-\frac{2135}{\sigma}=-1.645$$

Solving,

$$\sigma=1298\ \mathrm{d}$$

In addition,

$$\sigma^2=1.68\times10^6\ \mathrm{d}^2$$

Illustrative Example 25.20

Let X denote the coded quality of bag fabric used in a particular utility baghouse [3]. Assume that X is normally distributed with mean 10 and standard deviation 2.

1. Find c such that $P(|X-10|<c)=0.90$

2. Find k such that $P(X>k)=0.90$

Solution

1. Because X is normally distributed with mean 10 and standard deviation 2, $(X-10)/2$ is a standard normal variable. For this *two-sided* test [1, 2],

$$\begin{aligned}P(|X-10|<c)&=P(-c<(X-10)<c)\\&=P\left(\frac{-c}{2}<\frac{X-10}{2}<\frac{c}{2}\right)=P\left(\frac{-c}{2}<Z<\frac{c}{2}\right)\\&=2P\left(0<Z<\frac{c}{2}\right);\ \text{due to symmetry}\end{aligned}$$

Because

$$P(|X-10|<c)=0.90$$

one may write

$$2P\left(0<Z<\frac{c}{2}\right)=0.90$$

$$P\left(0<Z<\frac{c}{2}\right)=0.45$$

For this condition to apply,

$$\frac{c}{2}=1.645$$
$$c=3.29$$

2. Based on the problem statement,

$$P(X>k)=0.90$$
$$P\left(\frac{X-\mu}{\sigma}>\frac{k-10}{2}\right)=0.90$$
$$P\left(Z>\frac{k-10}{2}\right)=0.90$$

For this equation to be valid

$$\frac{k-10}{2}=-1.28$$

Solving for k gives

$$k=7.44$$

25.5 Log-Normal Distribution

A nonnegative random variable X has a log-normal distribution whenever $\ln X$, i.e., the natural logarithm of X, has a normal distribution. The probability distribution function (pdf) of a random variable X having a log-normal distribution is specified by

$$f(x) = \frac{1}{\sqrt{2\pi\beta}} x^{-1} \exp\left[-\frac{(\ln x - a)^2}{2\beta^2}\right]; \; x > 0 \tag{25.56}$$
$$= 0; \text{ elsewhere}$$

The mean and variance of a random variable X having a log-normal distribution are given by

$$\mu = e^{a + \frac{\beta^2}{2}} \tag{25.57}$$

$$\sigma^2 = e^{2a+\beta^2}(e^{\beta^2} - 1) \tag{25.58}$$

Figure 25.10 plots the pdf of the log-normal distribution for $a = 0$ and $\beta = 1$. Probabilities concerning random variables having a log-mean distribution can be calculated from the previously employed tables of the normal distribution. If X has a log-normal distribution with parameters a and β, then the ln X has a normal distribution with $\mu = a$ and $\sigma = \beta$. Probabilities concerning X can therefore be converted into equivalent probabilities concerning ln X.

For example, suppose that X has a log-normal distribution with $a = 2$ and $\beta = 0.1$. Then

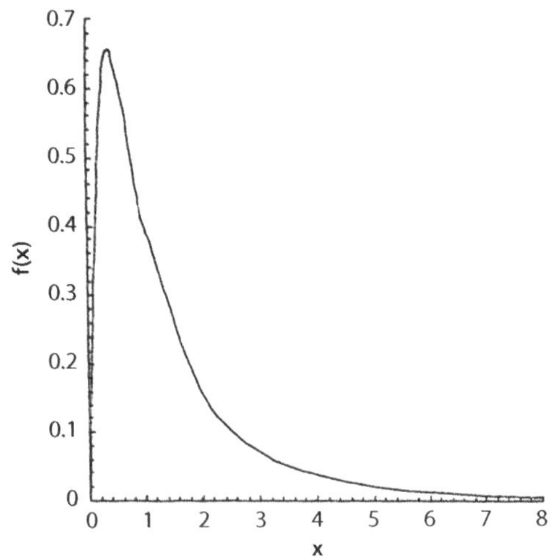

Figure 25.10 Log-normal pdf for $a = 0, \beta = 1$.

$$\begin{aligned} P(6 < X < 8) &= P(\ln 6 < \ln X < \ln 8) \\ &= \left[\frac{\ln 6 - 2}{0.1} < \frac{X - 2}{0.1} < \frac{\ln 8 - 2}{0.1}\right] \\ &= P(-2.08 < Z < 0.79) \\ &= (0.5 - 0.019) + (0.5 - 0.215) \\ &= 0.481 + 0.285 \\ &= 0.78 \end{aligned} \tag{25.59}$$

Estimates of the parameters α and β in the pdf of a random variable X having a log-normal distribution can be obtained from a sample of observations on X by making use of the aforementioned fact that $\ln X$ is normally distributed with a mean α and a standard deviation β. Therefore, the mean and standard deviation of the natural logarithms of the sample observations on X furnish estimates of α and β. To illustrate this procedure, suppose the time to failure T, in thousands of hours, was observed for a sample of five pumps at a water treatment plant. The observed values of T were 8, 11, 16, 22, and 34. The natural logarithms of these observations are 2.08, 2.40, 2.77, 3.09, and 3.53. Assuming that T has a log-normal distribution, the estimates of the parameters α and β in the pdf are obtained from the mean and standard deviation of the natural logs of the observations on T. Applying Equation (23.1) and Equation (23.7) developed earlier yields 2.77 as the estimate of α and 0.57 as the estimate of β. See also Illustrative Example 25.21.

The log-normal distribution has been employed as an appropriate model in a wide variety of situations from environmental management to biology to economics. Additional applications include the distributions of personal incomes, inheritances, bank deposits as well as the distribution of organism growth subject to many small impurities. Perhaps the primary application of the log-normal distribution has been to represent the distribution for particle sizes in gaseous emissions from many industrial processes, details of which are provided in the literature [3].

Illustrative Example 25.21

The time to failure, T, in thousands of hours was observed for a sample of 5 electric motors as follows:

$$8, 11, 16, 22, 34$$

Assuming T has a log-normal distribution estimate the probability that an electric motor last less than 5000 h.

Solution

Under the assumption that T has a log-normal distribution, the natural logs of the observation constitute a *sample* from; a normal population. First, obtain the natural logs of the given observations on T, time to failure:

$$\ln 8 = 2.08, \quad \ln 16 = 2.77, \quad \ln 34 = 3.53, \quad \ln 11 = 2.40, \quad \ln 22 = 3.09$$

Compute the mean, μ, and the standard deviation, σ, of these results.

$$\mu = \sum_{i=1}^{5} \frac{\ln T_i}{5} = 2.77$$

$$\sigma = \sqrt{\sum_{i=1}^{5} \frac{[\ln T_i - 2.77]^2}{4}} = 0.57$$

Estimate α and β in the pdf of T from μ and σ, respectively.

$$\text{Estimate of } \alpha = 2.77$$

$$\text{Estimate of } \beta = 0.57$$

Convert the required probability concerning T into a probability about ln T:

$$P(T < 5) = P(\ln T < \ln 5) = P(\ln T < 1.61)$$

The required probability using the standard normal table can now be obtained; by treating ln T as a random variable that is normally distributed with mean α and standard deviation β.

$$P(\ln T < 1.61) = P\left(\left[\frac{\ln T - \alpha}{\beta}\right] < \left[\frac{1.61 - \alpha}{\beta}\right]\right)$$

$$= P\left(Z < \frac{1.61 - 2.77}{057}\right) = P(Z < -2.04)$$

$$= 0.021 = 2.1\%$$

Illustrative Example 25.22

Resolve Illustrative Example 25.21 by calculating the probability that an electric motor lasts more than

1. 5,000 h
2. 10,000 h

Solution

1. One can immediately conclude that

$$P(T > 5) = P(\ln T > 1.61) = 1 - P(T < 5)$$
$$= 1.0 - 0.0021 = 0.979 = 97.9\%$$

2. The describing equation becomes

$$P(T > 10) = P(\ln T > \ln 10) = P(\ln T > 2.303)$$

Converting to a standard normal variable,

$$P(\ln T > 2.303) = P\left(\left[\frac{\ln T - a}{\beta}\right] > \left[\frac{2.303 - a}{\beta}\right]\right)$$
$$= P\left[Z > \left(\frac{2.303 - 2.77}{0.57}\right)\right] = P(Z > -2.434)$$
$$= 0.9926 = 99.26\%$$

Illustrative Example 25.23

The failure rate per year, Y, of a coolant recycle pump in a wastewater treatment plant has a log-normal distribution. If ln Y has a mean of 2.0 and variance of 1.5, find $P(0.175 < Y < 1)$.

Solution

If Y has a log-normal distribution, ln Y has a normal distribution with mean 2 and standard deviation $\sigma = 1.5^{1/2} = 1.22$. Therefore,

$$P(0.175 < Y < 1) = P(\ln 0.175 < \ln Y < \ln 1)$$
$$= P\left(\left[\frac{\ln 0.175 - 2}{1.22}\right] < \frac{\ln Y - 2}{1.22} < \left[\frac{\ln 1 - 2}{1.22}\right]\right)$$
$$= P(-3.07 < Z < -1.64) = 0.1587 - 0.0011$$
$$= 0.1576 = 15.76\%$$

Illustrative Example 25.24

Normalized biological oxygen demand (BOD) levels in an estuary during the past 10 years are summarized in Table 25.2 [4]. If the BOD levels are assumed to follow a log-normal distribution, predict the level that would be exceeded only once in 100 years [2].

Table 25.2 Estuary BOD data.

Year	1	2	3	4	5	6	7	8	9	10
BOD level	23	38	17	210	62	142	43	29	71	31

Note: BOD = biological oxygen demand

Table 25.3 BOD calculations.

Year (*Y*)	BOD Level	*X* = ln BOD	X^2
1	23	3.13	9.83
2	38	3.64	13.25
3	17	2.83	8.01
4	210	5.35	28.62
5	62	4.13	17.06
6	142	4.96	24.60
7	43	3.76	14.14
8	29	3.37	11.36
9	71	4.26	18.15
10	31	3.43	11.67
Total	–	38.86	156.78

Solution

For this case, refer to Table 25.3. Based on the data provided

$$\bar{X} = \frac{\sum X^2}{n} = \frac{38.86}{10} = 3.886$$

$$s^2 = \frac{\sum X^2 - (\sum X)^{\frac{2}{n}}}{n-1}$$

$$= \frac{156.78 - \dfrac{(38.86)^2}{10}}{10-1}$$

$$= 0.64$$

and

$$s = 0.80$$

For this test, with $Z = 2.327$ for the 99% value,

$$Z = \frac{X - \bar{X}}{s}$$

$$2.327 = \frac{X - 3.886}{0.80}$$

Solving for X yields

$$X = 5.748$$

For this log-normal distribution,

$$X = \ln(\text{BOD})$$

$$X = 5.748 = \ln(\text{BOD})$$

$$\text{BOD} = 313$$

Reference

1. L Theodore and F. Taylor, *Probability and Statistics*, A Theodore Tutorial, Theodore Tutorials, East Williston, NY, originally published by USEPA/APTI, RTP, NC, 1997.
2. S. Shaefer and L. Theodore, *Probability and Statistics Applications for Environmental Science*, CRC Press/Taylor & Francis Group, Boca Raton, FL, 2007.
3. L. Theodore, *Air Pollution Control Equipment Calculations*, John Wiley & Sons, Hoboken, NJ, 2008.
4. R. Dupont, K. Ganesan, and L. Theodore, *Pollution Prevention: Sustainability, Industrial Ecology, and Green Engineering*, 2nd ed., CRC Press/Taylor & Francis Group, Boca Raton, FL, 2017.

26

Fault Tree and Event Tree Analysis [1]

Tree analysis has become an integral part of probability analysis for environmental engineering and science practitioners. There are two tree analyses that are employed: fault trees and event trees. Fault trees attempt to determine the cause of failure in a system while event trees attempt to determine the consequences associated with a system that has failed.

Two sections compliment the presentation of this chapter. Section numbers and subject titles follow.

26.1: Fault Trees
26.2: Event Trees

26.1 Fault Trees

A *fault tree* is a graphical technique used to analyze some complex systems. The objective is to spotlight faulty conditions that have caused a system to fail. Fault tree analysis attempts to describe how and why an accident or other undesirable event has occurred. It may also be used to describe how and why an accident or other undesirable event could take place. A fault tree analysis also finds wide application in environmental management as it applies to hazard analysis and risk assessment of process and plant systems.

Fault tree analysis seeks to relate the occurrence of an undesired event, the "top event," to one or more antecedent events, called *basic events*. The top event may be, and usually is, related to the basic events via certain intermediate events. A fault tree diagram exhibits the usual chain, linking the basic events to the intermediate events and the latter to the top event. In this chain, the logical connection between events is indicated by so-called *logic gates*. The principal logic gates on the fault tree are the "AND" gate, symbolized by , and the "OR" gate, symbolized by .

As a simple example of a fault tree, consider a water pumping system consisting of 2 pumps A and B, where A is the pump ordinarily operating and B is a standby unit that automatically takes over if A fails. A control valve in both cases regulates flow of water through the pump. Suppose that the top event is no water flow, resulting from the following basic events: failure of pump A and failure of pump B or failure of the control valve. The full tree diagram for the system is shown in Figure 26.1.

Because one of the purposes of a fault tree analysis is the calculation of the probability of the top event, let A, B, and C represent the failure of pump A, the failure of pump B, and the failure of the control valve, respectively. Then, if T represents the top event of no water flow, one can write

$$T = AB + C \tag{26.1}$$

This equation indicates that T occurs if both A and B occur or if C occurs.

Assume that A, B, and C are independent and $P(A) = 0.01$, $P(B) = 0.005$, and $P(C) = 0.02$. Then, application of the addition theorem [1] yields

$$P(T) = P(AB) + P(C) - P(ABC) \tag{26.2}$$

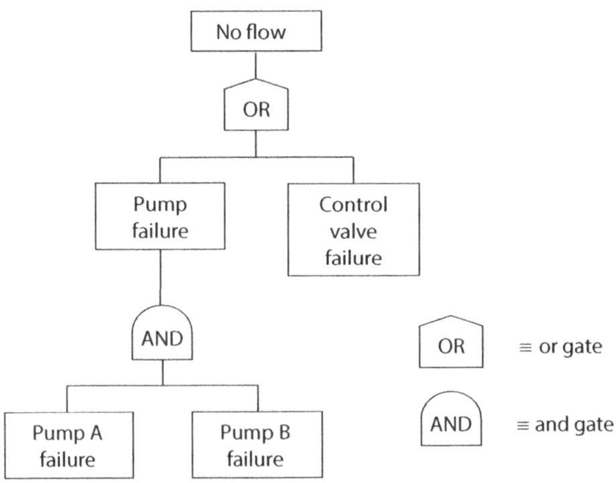

Figure 26.1 Fault tree for a water pumping system.

The independence of A, B, and C implies

$$P(AB) = P(A)P(B) \tag{26.3}$$

and

$$P(ABC) = P(A)P(B)P(C) \tag{26.4}$$

Therefore, the proceeding equation can be written as

$$P(T) = P(A)P(B) + P(C) - P(A)P(B)(C) \tag{26.5}$$

Substituting the proceeding data gives

$$\begin{aligned} P(T) &= (0.01)(0.005) + 0.02 - (0.01)(0.005)(0.02) \\ &= 0.020049 \end{aligned} \tag{26.6}$$

In connection with fault trees, cut sets and minimal cut sets are defined as follows. A *cut set* is a basic event or intersection of basic events that will cause the top event to occur. A *minimal cut set* is a cut set that is not a subset of any other cut set. In the example presented above, AB and C are cut sets because if AB occurs then T occurs, and if C occurs, then T occurs. Also, AB and C are minimal cut sets, because neither is a subset of the other. The event T describes what can be regarded as a top event represented as a union of cut sets.

The reader should note the behavior of a fault tree includes the following:

1. A fault tree works backwards from an undesirable event or ultimate consequence to the possible causes and failures.
2. It relates the occurrence of an undesired event to one or more preceding events.
3. It "*chain-links*" basic events to intermediate events that are in turn connected to the top event.
4. It is used in the calculation of the probability of the top event.
5. It is based on the most likely or credible events that lead to a particular failure or accident.
6. Its analysis includes *human error* as well as *equipment failure.*

Illustrative Example 26.1

A runaway chemical reaction can lead to a fire, an explosion, or massive pollutant emissions. It occurs if coolers fail (A) *or* there is a bad chemical batch (B). Coolers fail only if both cooler #1 fails (C) *and* cooler #2 fails (D). A bad chemical batch occurs if there is a wrong mix (E) *or* there is a process upset (F). A wrong mix occurs only if there is an operator error (G) *and* instrument failure (H). Construct a fault tree.

Solution

Begin with the top event for this problem shown in Figure 26.2. Generate the first branch of the fault tree (see Figure 26.3), applying the logic gates seen in Figure 26.1. Then generate the second and third branches using the same process. The addition of the second and third branches can be seen in Figure 26.4 and Figure 26.5, respectively.

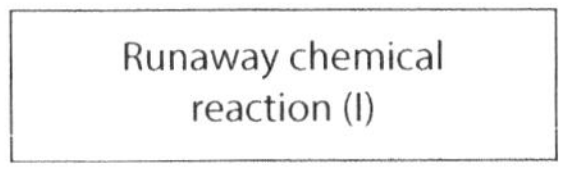

Figure 26.2 Top event of fault tree.

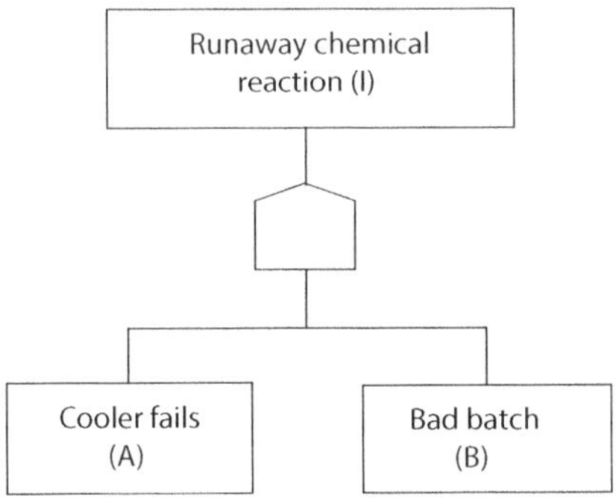

Figure 26.3 Fault tree with first branch.

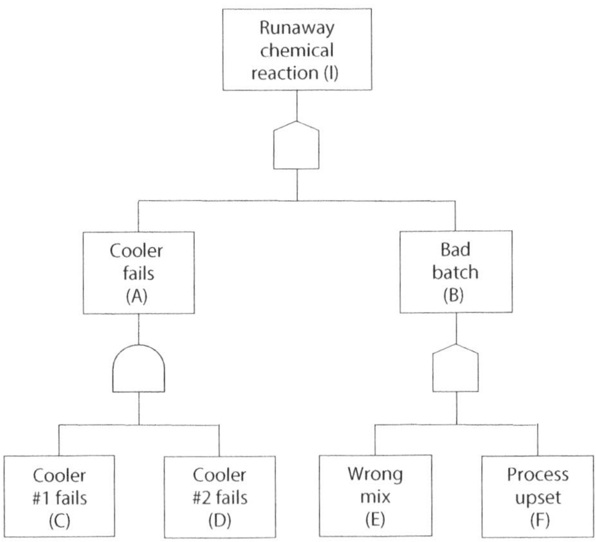

Figure 26.4 Fault tree with second branch.

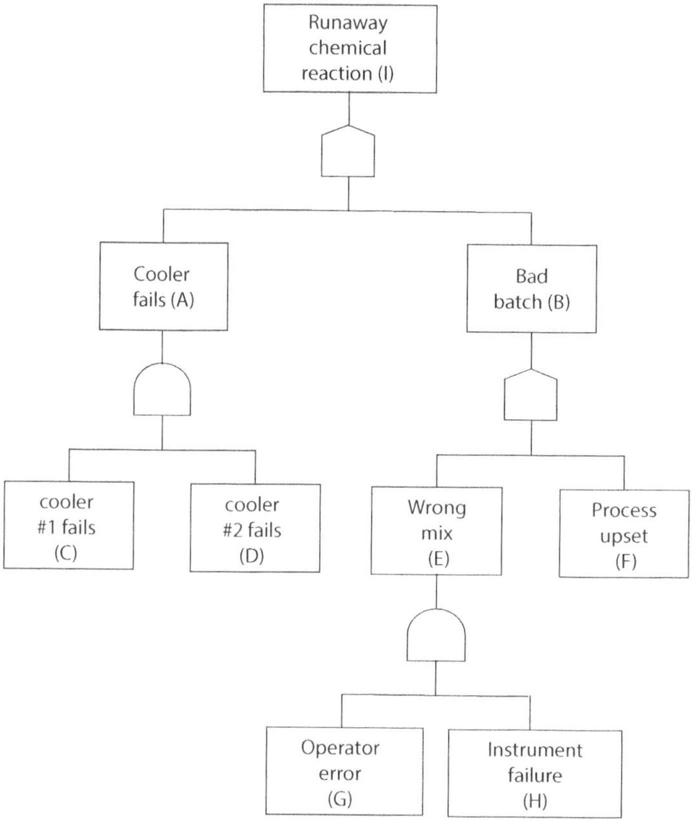

Figure 26.5 Fault tree with third branch.

Illustrative Example 26.2

Refer to Illustrative Example 26.1. If the following annual probabilities are provided by the plant engineer, calculate the probability of a runaway chemical reaction occurring in a year's time given the following annual probabilities:

$$P(C) = 0.050$$

$$P(D) = 0.080$$

$$P(F) = 0.060$$

$$P(G) = 0.030$$

$$P(H) = 0.010$$

Solution

Redraw Figure 26.5 and insert the given probabilities (see Figure 26.6). Calculate the probability that the runaway reaction will occur

$$
\begin{aligned}
P &= (0.05)(0.08) + (0.01)(0.03) + 0.06 \\
&= 0.0040 + 0.0003 + 0.06 \\
&= 0.064
\end{aligned}
$$

Note that the process upset, F, is the major contributor to the probability. The reader should also note that if A and B are not mutually exclusive then

$$
\begin{aligned}
P &= 0.004 + 0.0603 - (0.004)(0.0603) \\
&= 0.06406
\end{aligned}
$$

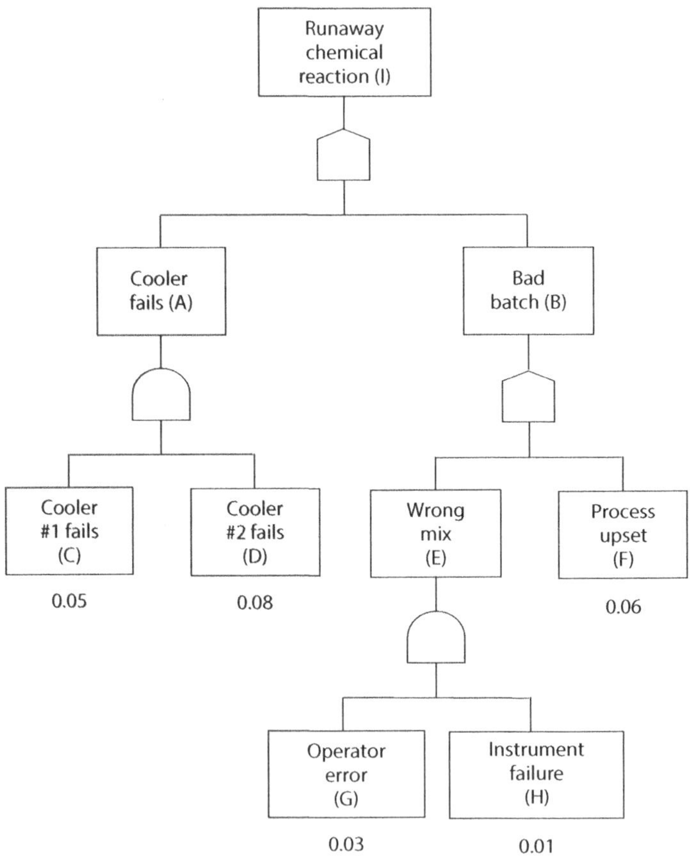

Figure 26.6 Runaway chemical reaction probabilities.

Illustrative Example 26.3

A distillation column explosion can occur if the overhead cooler fails (*OC*) *and* the condenser fails (*CO*) *or* there is a problem with the reboiler (*RB*) [2]. The overhead unit fails (*OUC*) only if both the cooler fails (*OC*) *and* the condenser fails (*CO*). Reboiler problems develop if there is a power failure (*PF*) *or* there is a failed tube (*FT*) [3]. A power failure occurs only if there are both operator error (*OE*) *and* instrument failure (*IF*). Construct a fault tree with the information below.

$$P(OC) = 0.02$$

$$P(CO) = 0.05$$

$$P(FT) = 0.04$$

$$P(OE) = 0.60$$

$$P(IF) = 0.005$$

Solution

The fault tree is given in Figure 26.7. Based on the data provided,

$$\begin{aligned} P(\text{explosion}) &\approx (0.02)(0.05) + (0.005)(0.6) + 0.04 \\ &\approx 0.001 + 0.003 + 0.04 \approx 0.0413 \end{aligned}$$

Note that the major contributor to an explosion is the failed tube (*FT*), whose probability is 0.04.

26.2 Event Trees

An event tree provides a diagrammatic representation of event sequences that begin with a so-called initiating event and terminate in one or more undesirable consequences. In contrast to a fault tree, which works backwards from an undesirable consequence to possible causes, an event tree looks forward from the initiating event to possible undesirable consequences. The initiating event may be equipment failure, human error, power failure, or some other event that has the potential for adversely affecting an ongoing process or environment.

The following illustration of an event tree analysis is based on one reported in the classic work by Lees [4]. Consider a situation in which the probability of an external power outage in any given year is 0.1. A backup generator is available, and the probability that it will fail to start on any given occasion is 0.02. If the generator starts, the probability that it will not supply sufficient power for the duration of the

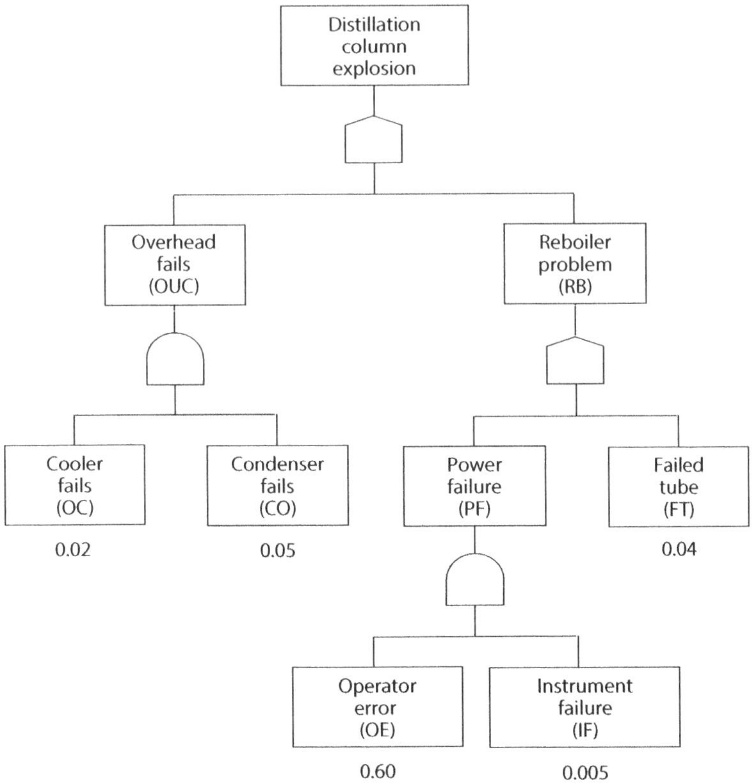

Figure 26.7 Distillation column explosion fault tree.

external power outage is 0.001. An emergency battery power supply is available; the probability that it will be inadequate is 0.01.

Figure 26.8 shows the event tree with the initiating event-external power outage, denoted by *I*. Labels for the other events on the event tree are also indicated. The event sequences *I S G B* and *I S B* terminate in the failure of emergency power supply. Applying the multiplication theorem, one obtains

$$P(IS\overline{G}\overline{B}) = (0.1)(0.98)(0.001)(0.01) = 9.8\times10^{-7}$$

$$P(I\overline{SB}) = (0.1)(0.02)(0.01) = 2\times10^{-5}$$

Therefore, the probability of emergency power supply failure in any given year is 2.098×10^{-5}, the sum of these two probabilities.

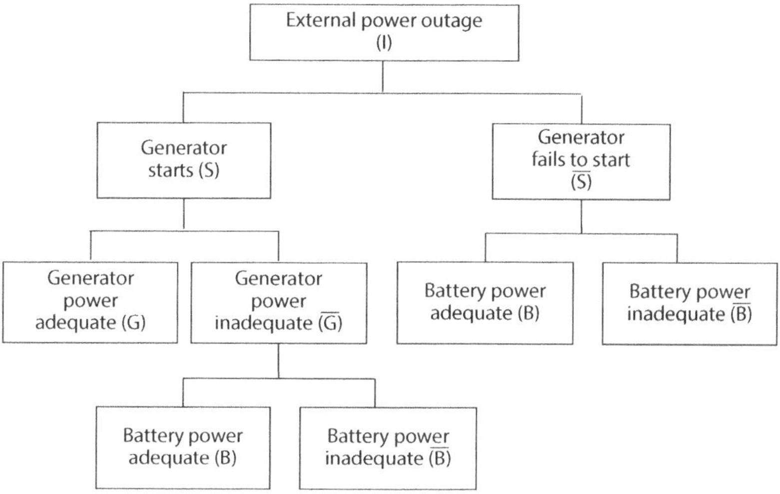

Figure 26.8 Event tree for power outage.

The reader should note the behavior of an event tree includes the following:

1. An event tree works forward from an initial event, or an event that has the potential for adversely affecting an ongoing process, and ends at one or more undesirable consequences.
2. It is used to represent the possible steps leading to a failure or accident.
3. It uses a series of branches that relate the proper operation and/or failure of a system with the ultimate consequences.
4. It is a quick identification of the various hazards that could result from a single initial event.
5. It is beneficial in examining the possibilities and consequences of a failure.
6. It does not usually quantify (although it can) the potential of the event occurring.
7. It can be incomplete if all the initial events are not identified.

Thus, the use of event trees is sometimes limited for hazard analysis because they usually do not quantify the potential of the event occurring. They may also be incomplete if all the initial occurrences are not identified. Their use is beneficial on examining, rather than evaluating, the possibilities and consequences of a failure. For this reason, a fault tree analysis should supplement this model to establish the probabilities of the event tree branches.

Illustrative Example 26.4

Construct a decision tree given the following information:

> Date: a couple's anniversary
> Decision of husband: buy flowers or do not buy flowers
> Consequences (buy flowers): domestic bliss or suspicious wife
> Consequences (do not buy flowers): status quo, or wife in tears/ husband in doghouse

Solution

Begin by setting the initial event "tree" format which is the top event in Figure 26.9. Also note that this is an event tree. Set up the first branch from the initial event. This is the first decision point (Figure 26.10).

One may now complete the tree. Note that the bottom four events evolve from what may be defined as resolution of uncertainty points (Figure 26.11).

This is an example of an event tree. As indicated above, in contrast to a fault tree, which works backwards from a consequence to possible causes, an event tree works forward from the initiating (or top) event to all possible consequences. Thus, this type of tree provides a diagrammatic representation of sequences that

Figure 26.9 Top event.

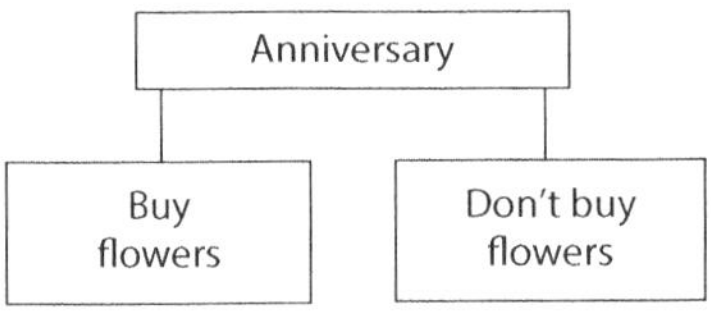

Figure 26.10 First decision point.

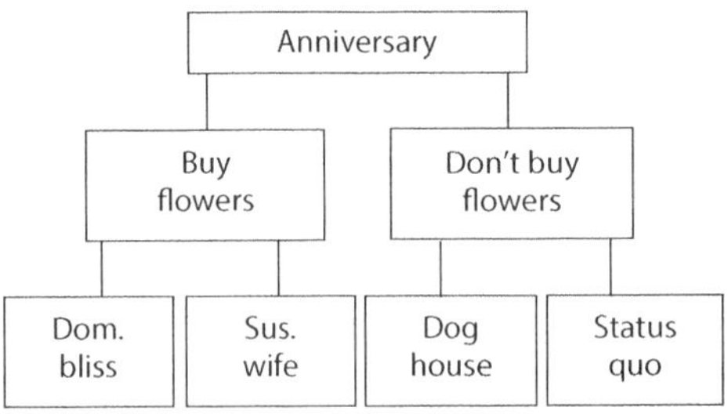

Figure 26.11 Marital bliss event tree.

begin with a so-called initiating event and terminate in one or more consequences. It primarily finds application in hazard analysis.

Illustrative Example 26.5

If a plant fire occurs, a smoke alarm sounds with probability 0.9. The sprinkler system functions with probability 0.7 whether or not the smoke alarm sounds. The consequences are minor fire damage (alarm sounds, sprinkler works), moderate fire damage with few injuries (alarm fails, sprinkler works), moderate fire damage with many injuries (alarm sounds, sprinkler fails), and major fire damage with many injuries (alarm fails, sprinkler fails). Construct an event tree, and indicate the probabilities for each of the four consequences.

Solution

The first set of consequences of the plant fire with their probabilities are shown in Figure 26.12. The second set of consequences of the plant fire and their probabilities are shown in Figure 26.13. The final set of consequences and the probabilities of minor fire damage, moderate fire damage with few injuries, moderate fire damage with many injuries, and major fire damage with many injuries are shown in Figure 26.14. Note that for each new branch in an event tree, the sum of probabilities must equal 1.0.

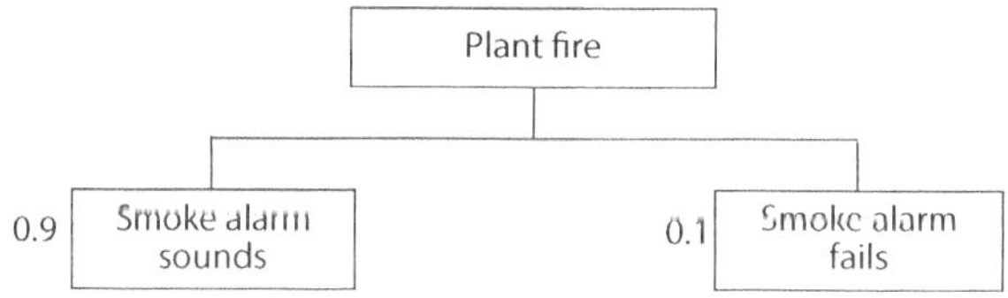

Figure 26.12 Event tree with first set of consequences.

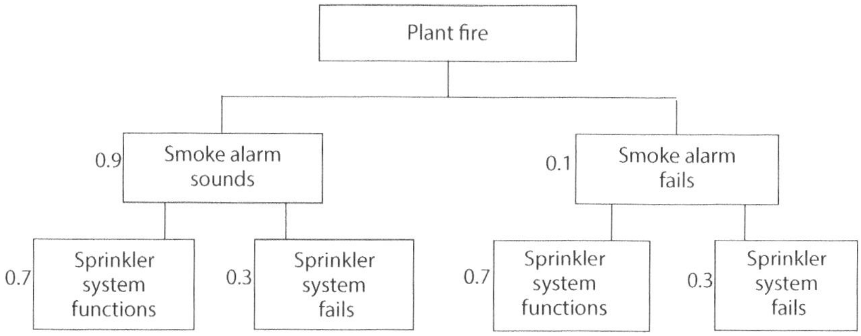

Figure 26.13 Event tree with second set of consequences.

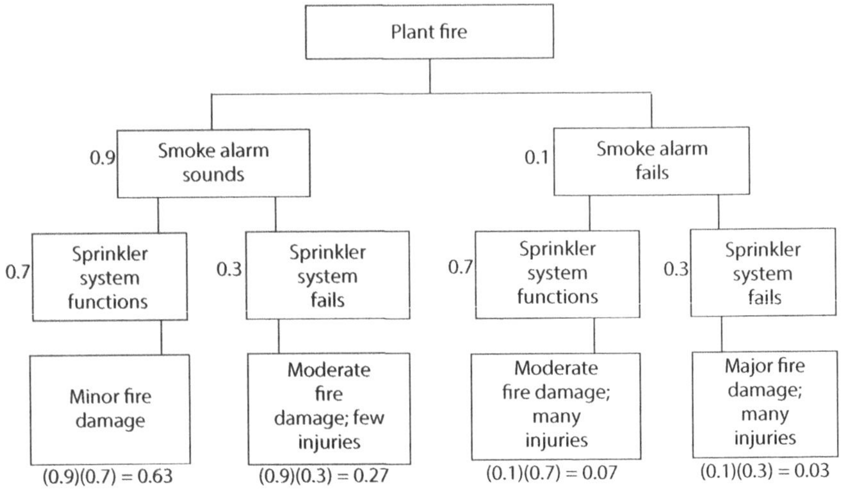

Figure 26.14 Event tree with final set of consequences.

Illustrative Example 26.6

An explosion followed immediately by an emission occurs in a plant. An alarm (A) then sounds with probability 0.7. The emission control system (E) functions with probability 0.9 whether or not the alarm sounds. The consequences are minor damage (A sounds, E works), moderate damage with few injuries (A sounds, E fails), moderate damage with many injuries (A fails, E works), and major damage with many injuries (A fails, E fails). Construct an event tree and indicate the probabilities for each of the four consequences.

Solution

This is a "takeoff" on illustrative example 26.5. The event tree for this application is shown in Figure 26.15.

Illustrative Example 26.7

A risk assessment is being conducted at a chemical plant to determine the consequences of two incidents, with the initiating event being an unstable chemical. The incidents are as follows:

Incident I: an explosion resulting from detonation of an unstable chemical
Incident II: a release of a flammable toxic gas

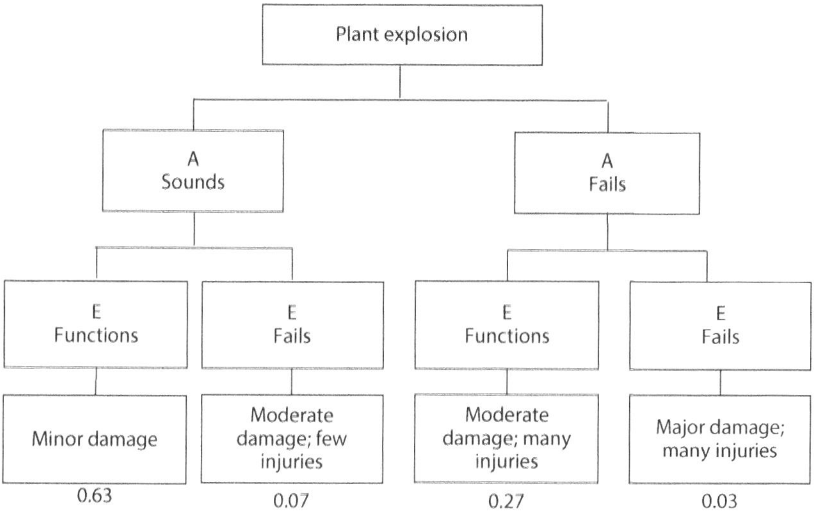

Figure 26.15 Event tree for plant explosion.

Incident I has one possible outcome, the explosion, the consequences of which are assumed to be unaffected by weather conditions. Incident II has several possible outcomes, which, for purposes of simplification, are reduced to the following:

Outcome IIA: vapor-cloud explosion, caused by ignition of the gas released, centered at the release point, and unaffected by weather conditions

Outcome IIB: toxic cloud extending downwind and affected by weather conditions.

For purposes of further simplification only two weather conditions are envisioned, a northeast wind and a southwest wind. Associated with these two wind directions are events IIB1 and IIB2:

Event IIB1: toxic cloud to the southwest

Event IIB2: toxic cloud to the Northeast

Prepare an event tree for incident I and incident II. (Note: this was an actual analysis conducted at a chemical plant.)

Solution

The event tree is provided in Figure 26.16. Additional details, including numerical data and calculations, are available in the literature [5].

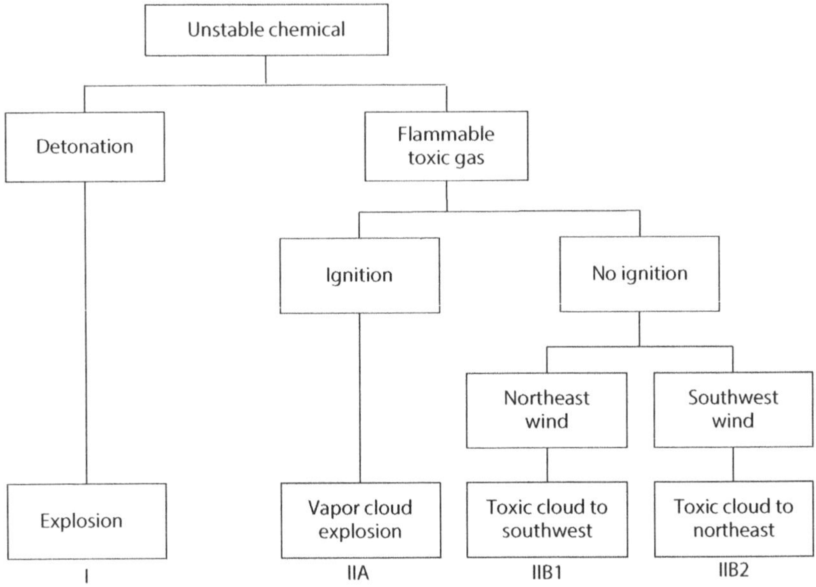

Figure 26.16 Event tree showing outcomes of two incidents (I and II) involving an unstable chemical.

References

1. L. Theodore and F. Taylor, *Probability and Statistics*, A Theodore Tutorial, Theodore Tutorials, East Williston, NY, originally published by USEPA/APTI, RTP, NC,1996.
2. L. Theodore and F. Ricci, *Mass Transfer Operations for the Practicing Engineer*, John Wiley & Sons, Hoboken, NJ, 2009.
3. L. Theodore, *Heat Transfer Applications for the Practicing Engineer*, John Wiley & Sons, Hoboken, NJ, 2012.
4. Lees, *Loss Prevention in the Chemical Industries*, Butterworth, Boston, MA, 1980.
5. L. Theodore and R. Dupont, *Environmental Health and Hazard Risk Assessment: Principles and Calculations*, CRC Press / Taylor & Francis Group, Boca Raton, FL, 2012.

27

Monte Carlo Simulation

Monte Carlo simulation is a procedure for mimicking observations on a random variable that permits verification of results that would ordinarily require difficult mathematical calculations and/or extensive experimentation. The method normally uses computer programs called *random number generators*. A random number is a number selected from the interval (0, 1) in such a way that the possibilities that the number comes from any two subintervals of equal "length" are equal. For example, the probability that the number is in the subinterval (0.1, 0.3) is the same as the probability that the number is in the subinterval (0.5, 0.7). Thus, random numbers are observations on a random variable X having a uniform distribution on the interval (0, 1). This means that the probability distribution function (pdf) of x is specified by

$$f(x) = 1;\ 0 < x < 1 \tag{27.1}$$

$$= 0;\ \text{elsewhere} \tag{27.2}$$

The above pfd assigns equal probability to subintervals of equal length in the interval (0, 1). Using random number generators, Monte Carlo simulation can generate observed values of a random variable having any specified pfd. For example, to generate observed values of T, the time to failure, when T is assumed to have a pdf specified by $f(t)$, one first uses the random number generator to generate the value of T between 0 and 1. The solution is an observed value of the random variable T having a pdf specified by $f(t)$.

One of the authors' first academic introductions to Monte Carlo methods involved the evaluation of a simple integral [1]. Consider the following integral, I

$$I = \int_{0}^{99} I(x)dx \tag{27.3}$$

where $I(x)$ is an integrand which is a function of x. A solution to this integral can be obtained in the following manner:

1. Generate by any means a random number between 00 and 99.
2. Set x to the value of the random number.
3. Calculate $I(x)$ at that value of x, i.e., the random number.
4. Repeat steps (1) – (3) for N times (a large number of random numbers).
5. Sum all the N values of $I(x)$.
6. Divide the sum obtained in (5) by N. This represents the average value $\bar{I}(x)$ for $I(x)$ over the 00 – 99 range of x.
7. Multiply $\bar{I}(x)$ by the range of x, i.e.,

$$I = \bar{I}(x)[00-99] = 99\bar{I}(x) \tag{27.4}$$

Monte Carlo methods find application in a host of applied mathematical and engineering areas. Most of these involve the solution of problems that cannot be solved by analytical or ordinary numerical methods. Solving these types of problems using Monte Carlo methods will be explored in the illustrative examples to follow.

Three sections compliment the presentation of this chapter. Section numbers and subject titles follow.

27.1: Exponential Distribution Applications
27.2: Normal Distribution Applications
27.3: Heat Conduction Applications

27.1 Exponential Distribution Applications

Three illustrative examples are provided below for the chapter's section involving the exponential distribution.

Illustrative Example 27.1

A pump has time to failure, T, measured in years, with an exponential pdf specified by

$$f(t) = e^{-t}; t > o$$
$$= 0; \text{elsewhere}$$

Generate 15 simulated values of T using a Monte Carlo procedure, i.e., the generation of random numbers.

Solution

Calculate the cumulative distribution function (cdf), $F(t)$, from the given pdf specified by $f(t)$.

$$f(t) = e^{-t}; t > 0$$
$$= 0; t \leq 0$$

From Chapter 25 one may write

$$F(t) = P(T \leq t)$$
$$= \int_0^t f(t)dt$$
$$= \int_0^t e^{-t}dt = 1 - e^{-t}; t > 0$$
$$= 0; t \leq 0$$

Fifteen random numbers are now generated in the interval or range of 0 to 1:

0.93, 0.06, 0.53, 0.56, 0.41, 0.38, 0.78, 0.54,

0.49, 0.89, 0.77, 0.85, 0.17, 0.34, 0.56

For each random number generated, solve the equation obtained by setting the random number equal to $F(t)$. The first random number generated is 0.93. Setting this equal to the cdf, $F(t)$, produces the equation

$$0.93 = 1 - e^{-t}$$

Table 27.1 Simulated time to failure.

Random number	Simulated time to failure (years)
0.93	2.66
0.06	0.06
0.53	0.76
0.41	0.82
0.38	0.48
0.78	1.52
0.54	0.78
0.49	0.67
0.89	2.21
0.77	1.47
0.85	1.90
0.17	0.19
0.34	0.42
0.56	0.82

Solving yields

$$e^{-t} = 0.07$$

$$t = -\ln 0.07 = 2.66$$

Therefore, the first simulated value of T, time to failure, is 2.66 years. The other simulated values of T obtained in the same manner are shown in Table 27.1.

Illustrative Example 27.2

Use the results of the previous example to estimate the average life of the pump.

Solution

The average value of T is the sum of all 15 simulated values divided by the number of simulated values (N). in this case $N = 15$. The average value of T is 1.02 years; this represents the Monte Carlo estimate of the average life of the pump.

Illustrative Example 27.3

Outline at least two other methods that could be employed to estimate the average life of the pump in Illustrative Example 27.1.

Solution

The exact value of the average life of the pump can be calculated by finding the expected value of T from its pdf [2].

$$E(T) = \int_{-\infty}^{\infty} tf(t)dt$$

$$= \int_{0}^{\infty} te^{-t}dt = 1.0$$

A more accurate estimate of the true value of the average life of the pump can be obtained by increasing the number of simulated values on which the estimate is based. The expected value, or mean, can also be shown to equal the coefficient associated with the exponential term, i.e., t, for an exponential distribution. Therefore, the mean for this distribution is 1.0 year.

27.2 Normal Distribution Applications

Two illustrative examples are provided for different applications involving the normal distribution.

Illustrative Example 27.4

A series system consists of two electrical components, A and B. Component A has a time to failure, T_A, and assumed to be normally distributed with mean (μ) of 100 h and standard deviation (σ) of 20 h. Component B has a time to failure, T_B, and assumed to be normally distributed with mean 90 h and standard deviation 10 h. The system fails whenever either component A or component B fails. Therefore, T_S, the time to failure of the system is the minimum of the times to failure of components A and B.

Estimate the average value of T_s on the basis of the simulated values of 10 simulated values of T_A and 10 simulated values of T_B.

Solution

First, generate 20 random numbers in the range 0 to 1:

0.10, 0.54, 0.42, 0.02, 0.81, 0.07, 0.06, 0.27, 0.57, 0.80; A

0.92, 0.86, 0.45, 0.38, 0.88, 0.21, 0.26, 0.51, 0.73, 0.71; B

Use the table of the standard normal distribution (see Table 25.1) and obtain the simulated value of Z corresponding to each of the above companion random numbers for A+B. The first random number is 0.10. the corresponding simulated value of Z is −1.28 because the area under a standard normal curve to the left of −1.28 is 0.10. The remaining simulated values of *Z* are obtained in similar fashion. The 20 simulated values of *Z* are provided in Table 27.2.

Using the first 10 simulated values of *Z*, obtain 10 simulated values of T_A by multiplying each simulated value of *Z* by 20 (the standard deviation σ) and adding 100 (the mean, μ), i.e.,

$$T_A = \sigma Z + 100$$

Note that

$$Z = \frac{T_A - 100}{\sigma}$$

Table 27.2 Simulated values of Z

Random number	Simulated values of Z
0.10	−1.28
0.54	0.10
0.42	−0.20
0.02	−2.05
0.81	0.88
0.07	−1.48
0.06	−1.56
0.27	−0.61
0.57	0.18
0.80	0.84
0.92	1.41
0.86	1.08
0.45	−0.13
0.38	−0.31
0.88	1.17
0.21	−0.81
0.26	−0.64
0.51	0.03
0.73	0.61
0.71	0.56

The life time or time to failure of each component, T, is calculated using this equation. Thus, multiplying each of the first 10 simulated values of Z by 20 and adding 100 yields the following simulated values of T_A:

74, 102, 96, 59, 118, 70, 69, 88, 104, 117

Multiplying each of the second 10 simulated values of Z by 10 and adding 90 yields the following simulated values of T_B:

104, 101, 89, 87, 102, 82, 84, 90, 96, 96

Simulated values of T_S corresponding to each companion pair of simulated values of T_A and T_B are obtained by recording the minimum of each pair. The values are shown in Table 27.3.

The average of the 10 simulated values of T_S is 84, the estimated time to failure of the system.

Illustrative Example 27.5

According to state regulations, three thermometers (A, B, C) are positioned near the outlet of an afterburner. Assume that the individual thermometer component lifetimes are normally distributed with means and standard deviations given in Table 27.4.

Using the 10 random numbers from 0 to 1 provided in Table 27.5 for each thermometer, simulate the lifetime (time to thermometer failure) of the temperature recording system, estimate its mean and standard deviation, and the estimated

Table 27.3 Minimum simulated values.

Simulated time to failure		
Component *A* (T_A)	**Component *B* (T_B)**	**System (T_S)**
74	104	74
102	101	101
96	89	89
59	87	59
118	102	102
70	82	70
69	84	69
88	90	88
104	96	96
117	96	96

Table 27.4 Thermometer data.

Thermometer	A	B	C
Mean (weeks)	100	90	80
Standard deviation (weeks)	30	20	10

Table 27.5 Thermometer random numbers.

For A		For B		For C	
0.52	0.01	0.77	0.67	0.14	0.90
0.80	0.50	0.54	0.31	0.39	0.28
0.45	0.29	0.96	0.34	0.06	0.51
0.68	0.34	0.02	0.00	0.86	0.56
0.59	0.46	0.73	0.48	0.87	0.82

time to failure for this system. The lifetime is defined as the time (in weeks) for one of the thermometers to "fail."

Solution

Let T_A, T_B, and T_C denote the lifetimes of thermometer components, A, B, and C, respectively. Let T_s denote the lifetime of the system. The random number generated in Table 27.5 may be viewed as the cumulative probability, and the cumulative probability is the area under the standard normal distribution curve. Because the standard normal distribution curve is symmetrical, the negative values of Z and the corresponding area are once again found by symmetry. For example,

$$P(Z < -1.54) = 0.062$$
$$P(Z > 1.54) = 0.062$$
$$P(0 < Z < 1.54) = 0.5 - P(Z > 1.54)$$
$$= 0.5 - 0.062 = 0.438$$

Recall that the lifetime or time to failure of each component, T, is calculated using the equation

$$T = \mu + \sigma Z$$

where μ is the mean, σ, the standard deviation, and Z, the standard normal variable. First; determine the values of the standard normal variable, Z, for component A using the 10 random numbers given in the problem statement and the standard normal table. Then calculate the lifetime of thermometer component, T_A, using

Table 27.6 Lifetime of thermometer A (T_A).

Random number	***Z* (standard normal table)**	**$T_A = 100 + 30\,Z$**
0.52	0.05	102
0.80	0.84	125
0.45	–0.13	96
0.68	0.47	114
0.59	0.23	107
0.01	–2.33	30
0.50	0.00	100
0.29	–0.55	84
0.34	–0.41	88
0.46	0.10	97

Table 27.7 Lifetime of thermometer B (T_B).

Random number	***Z* (standard normal table)**	**$T_B = 90 + 20\,Z$**
0.77	0.74	105
0.54	0.10	92
0.96	1.75	125
0.02	–2.05	49
0.73	0.61	102
0.67	0.44	99
0.31	–0.50	80
0.34	–0.41	82
0.00	–3.90	12
0.48	–0.05	89

the above equation for *T*. Results are provided in Table 27.6. Next, determine the values of the standard normal variable and the lifetime of the thermometer component of components B and C (see also Table 27.7 and 27.8, respectively)

For each random value of each component, determine the system lifetime, T_S. Since this is system a series, the system lifetime is limited by the component with the minimum lifetime (see Table 27.9).

The mean value, μ, of T_S, is

$$\mu = \frac{635}{10} = 63.5 \text{ years}$$

Table 27.8 Lifetime of thermometer C (T_C).

Random number	***Z* (standard normal table)**	$T_C = 80 + 10\,Z$
0.14	−1.08	69
0.39	−0.28	77
0.06	−1.56	64
0.86	1.08	91
0.87	1.13	91
0.90	1.28	93
0.28	−0.58	74
0.51	0.03	80
0.56	0.15	81
0.82	0.92	89

Table 27.9 Thermometer system lifetime.

T_A	T_B	T_C	T_S
102	105	69	69
125	92	77	77
96	125	64	64
114	49	49	49
107	102	91	91
30	99	30	30
100	80	74	74
84	82	80	80
88	12	12	12
97	89	89	89
Total			635

Calculate the standard deviation, σ, of T_S using the equation [2]

$$\sigma^2 = \frac{1}{n}\Sigma(T_S - \mu)^2$$

where n is 10, the number in the population. Note that this is not a sample, so that a modified equation applies for σ^2 (see Table 27.10). Therefore,

$$\sigma = \left(\frac{5987}{10}\right)^{0.5} = 24.5 \text{ years}$$

Table 27.10 Thermometer standard deviation calculation.

System lifetime (T_S)	$(T_S-\mu)^2$
69	30.25
77	182.25
64	0.25
49	210.25
91	756.25
30	1122.25
74	110.25
80	272.25
12	2652.25
89	650.25
Total	5987

Monte Carlo simulation is an extremely powerful tool available to the environmental engineer/scientist that can be used to solve multivariable systems, ordinary and partial differential equations, numerical integrations, etc. The application to solving differential equations is provided in the next section.

27.3 Heat Conduction Applications

Three illustrative examples are provided in this section, two of which were introduced in Part III, Chapter 21.

Illustrative Example 27.6

A new radiant heat transfer rod has been proposed for use in the incineration of hazardous waste. One of the first steps in determining the usefulness of this form of "hot" rod, is to analytically estimate the temperature profile in the rod.

A recent environmental engineering graduate has proposed to estimate the temperature profile of the square pictured in Figure 27.1. She sets out to outline a calculational procedure to determine the temperature profile. One of the options available is to employ the Monte Carlo method. Provide an outline for such a procedure. (See also Illustrative Example 21.1 where this example was previously visited.)

Solution

One method of solution involves the use of the Monte Carlo approach, involving the use of random numbers. Consider the square (it could also be a rectangle)

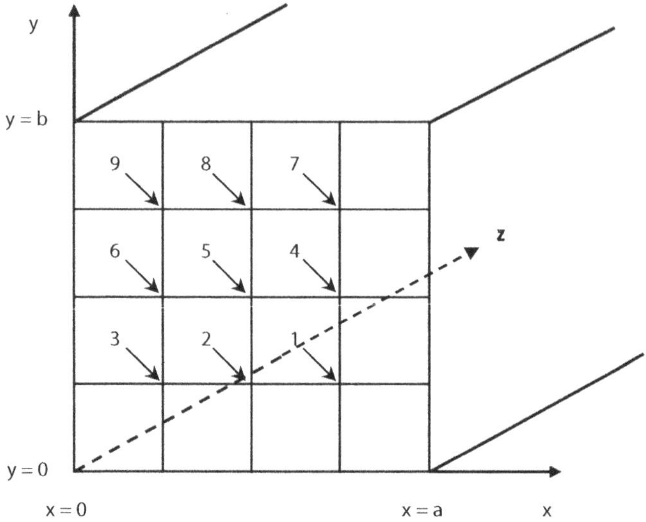

Figure 27.1 Monte Carlo grid (square surface).

pictured in Figure 27.1. If the describing equation for the variation of T within the grid structure is

$$\frac{\partial^2 T}{\partial x^2} + \frac{\partial^2 T}{\partial y^2} = 0$$

with specified boundary conditions (BC) for $T(x, y)$ of $T(0, y)$, $T(a, y)$, $T(x, 0)$, and $T(x, b)$, one may employ the following approach.

1. Proceed to calculate T at point 1, i.e., T_1.
2. Generate a random number between 00 and 99.
3. If the number is between 00 and 24, move to the left. For 25 to 49, 50 to 74, and 75 to 99, move upward, to the right, and downward, respectively.
4. If the move in step 3 results in a new position that is at an outer surface (boundary), terminate the first calculation for point 1 and record the T value of the boundary at the new position. However, if the move results in a new position that is not at a boundary and is still at one of the nine internal grid points, repeat step 2 and step 3. This process is continued until an outer surface or boundary is reached.
5. Repeat step 2 to step 4 numerous times e.g., 1000 times.
6. After completing step 5, sum all the T values obtained and divided that value by the number of times step 2 to step 4 have been

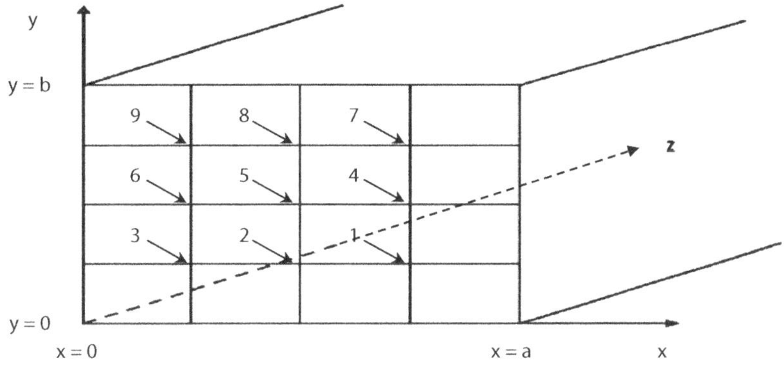

Figure 27.2 Rectangular grid.

repeated e.g., 1000. The resulting value provides a reasonable estimate of T_1.

7. Return to step 1 and repeat the calculations for the remaining 8 grid points.

Much of this material was presented earlier. As mentioned, this method of solution is not limited to square systems. Calculation details were provided in Illustrative Example 21.1.

Illustrative Example 27.7

Outline how to solve the problem in Illustrative Example 27.6 if the face surface is a rectangle as seen in Figure 27.2.

Solution

Consider the grid for the face surface in Figure 27.2. The outline of the calculation presented in the previous illustrative example applies. Some additional information is available in Part III, Chapter 21.

Illustrative Example 27.8

Consider the system provided in Figure 27.3. Outline a procedure to estimate the corresponding profile.

Solution

A "grid" of several annular circles must be set at points that produce equal annular areas. Unlike the previous two examples, this involves a one-dimensional

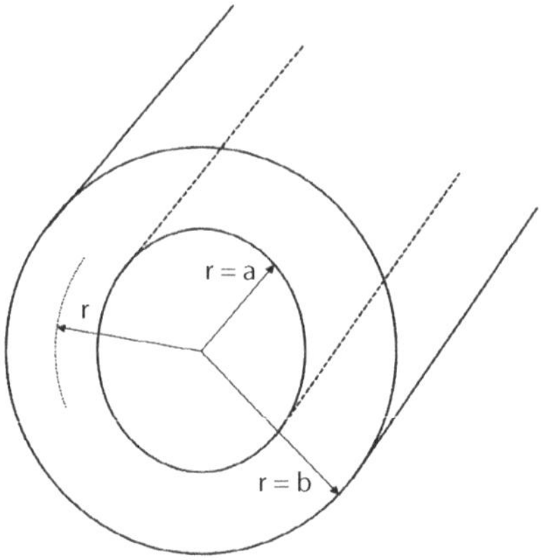

Figure 27.3 Hollow pipe grid.

calculation. The procedure essentially remains the same for determining the temperature at each grid. The analytical solution is simply given by [1,3]

$$T = T_a + (T_b - T_a)\left(\frac{b}{b-a}\right)\left(1-\frac{a}{r}\right);\ a = \text{inner radius},\ b = \text{outer radius}$$

References

1. L. Theodore, personal notes, East Williston, NY, 1970.
2. S. Shaefer and L. Theodore, *Probability and Statistics Applications for Environmental Science,* CRC Press/Taylor & Francis Group, Boca Raton, FL, 2007.
3. L. Theodore, *Heat Transfer Applications for the Practicing Engineer*, John Wiley & Sons, Hoboken, NJ, 2011.

28

Regression Analysis [1, 2]

It is not secret that statistical calculations are now performed with spreadsheets or packaged programs. This statement is particularly true with regression analysis. Microsoft Excel possesses an "add trend line" function that will fit scatter plot data to a function. The result of this has often been the reduction or elimination of one's fundamental understanding of this subject. Without a fundamental understanding, the user can easily become confused as to what type of function to use or how to interpret the data. This chapter attempts to correct this shortcoming.

Engineers and scientist often encounter applications that require the need to develop a mathematical relationship between data for two or more variables. For example, if Y (a dependent variable) is a function of or depends on X (an independent variable), i.e.,

$$Y = f(X) \tag{28.1}$$

one may be required to express this (X, Y) data in equation form. This process is referred to as *regression analysis*, and the regression method most often employed is the method of *least squares*.

Three sections compliment the presentation of this chapter. Section numbers and subject titles follow.

27.1: Scatter Diagrams
27.2: Method of Least Squares
27.3: The Correlation Coefficient

28.1 Scatter Diagrams

An important step in a regression analysis procedure – which is often omitted – is to prepare a plot of Y vs. X. The result, referred to as a scatter diagram, could take on any form. Three such plots are provided in Figure 28.1 (A to C). The first plot (A) suggests a linear relationship between X and Y, i.e.,

$$Y = a_0 + a_1 X \tag{28.2}$$

The second graph (B) appears to be represented by a second order (or parabolic) relationship, i.e.,

$$Y = a_0 + a_1 X + a_2 X^2 \tag{28.3}$$

The third plot suggests a linear model that applies over two different ranges, i.e., it could represent the data

$$Y = a_0 + a_1 X;\ X_0 < X < X_M \tag{28.4}$$

and

$$Y = a_0' + a_1' X;\ X_M < X < X_L \tag{28.5}$$

This multiequation model finds application in representing adsorption equilibria, multiparticle size distributions, quantum energy relationships, etc. In any event, a scatter diagram and individual judgment can suggest an appropriate model at an early stage in the analysis [1, 2].

Some of the models often employed by technical individuals are as follows:

$$\text{Linear: } Y = a_0 + a_1 X \tag{28.6}$$

$$\text{Parabolic: } Y = a_0 + a_1 X + a_2 X^2 \tag{28.7}$$

$$\text{Cubic: } Y = a_0 + a_1 X + a_2 X^2 + a_3 X^3 \tag{28.8}$$

$$\text{Quadratic: } Y = a_0 + a_1 X + a_2 X^2 + a_3 X^3 + a_4 X^4 \tag{28.9}$$

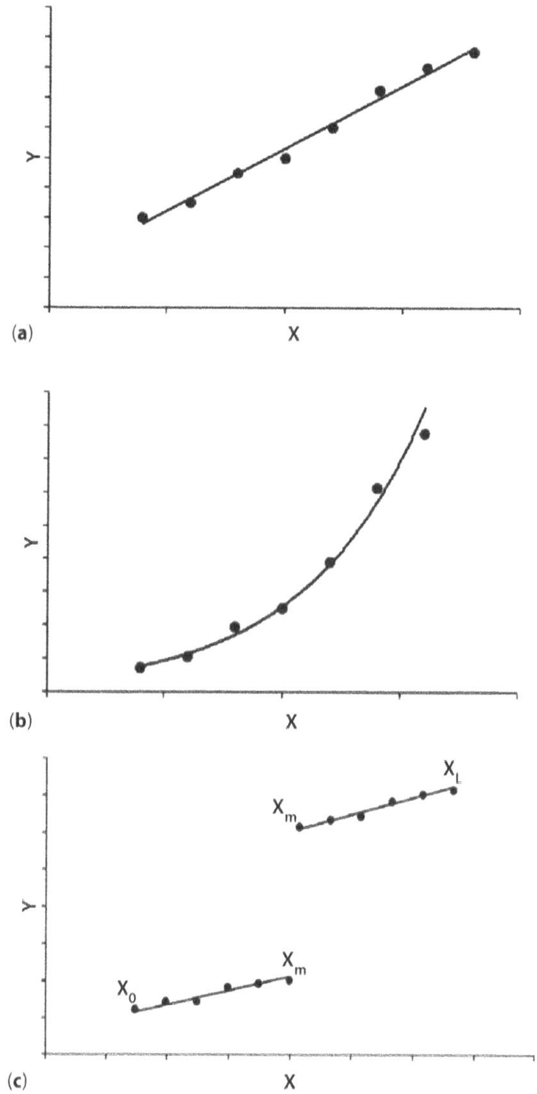

Figure 28.1 Scatter diagrams: (a) linear relationship, (b) parabolic relationship, and (c) dual-linear relationship.

28.2 Method of Least Squares

Procedures to evaluate the regression coefficients a_0, a_1, a_2, etc. referred to in the previous section are provided below. The reader should note that the analysis is based on the method of least squares. This technique provides numerical values for the regression coefficients a_i such that the sum of the square of the

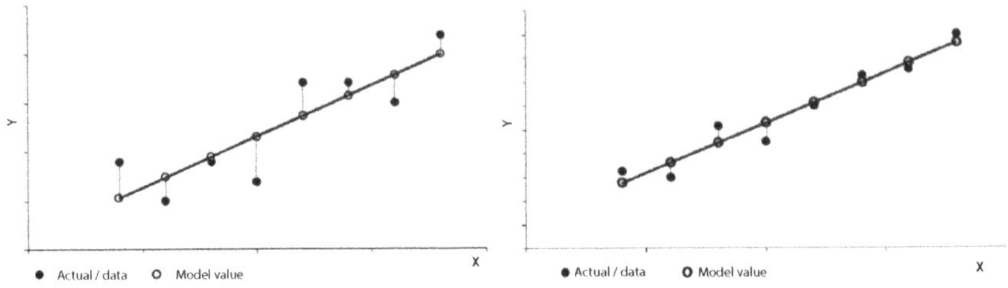

Figure 28.2 Error difference: actual and predicted values.

difference (error) between actual Y and the Y_e predicted by the equation or model is minimized.

Two sets of data are shown in Figure 28.2. In Figure 28.2, the solid dots (experimental value of Y) and open dots (equation or model value of Y, i.e., Y_e) represent the data and model values, respectively. On examining the two figures, one can immediately conclude that the error ($Y - Y_e$) squared and summed for the eight points is more for the first figure on the left. Note that a dashed line represents the error. The line that ultimately produces a minimum of the sum of the individual error squared, i.e., has the smallest possible value, is the regression model (based on the method of least squares). The proof is left as an exercise.

To evaluate a_0 and a_1 for a linear model (see Equation 28.6), one employs the following least squares algorithm for n data points of Y and X.

$$a_0 n + a_1 \Sigma X = \Sigma Y \tag{28.10}$$

$$a_0 \Sigma X + a_1 X^2 = \Sigma XY \tag{28.11}$$

All the quantities given, except a_0 and a_1, can be easily calculated for the data. Since there are two equations and two unknowns, the set of equations can be solved for a_0 and a_1. For this case,

$$a_1 = \frac{n\sum XY - \sum X \sum Y}{n\sum X^2 - (\sum X)^2} \tag{28.12}$$

Dividing numerator and denominator by n, and defining $\overline{X} = \Sigma X/n$ and $\overline{Y} = \Sigma Y/n$, leads to

$$a_1 = \frac{\sum XY - \dfrac{\sum X \sum Y}{n}}{\sum X^2 - \dfrac{(\sum X)^2}{n}} = \frac{\sum XY - n\overline{X}\overline{Y}}{\sum X^2 - n\overline{X}^2} \tag{28.13}$$

Using this value of a_1 produces the following equation for a_0:

$$a_0 = \bar{Y} - a_1\bar{X} \tag{28.14}$$

If the model (or line of regression) is forced to fit through the origin, then the calculated value of $Y_e = 0$ at $X = 0$. For this case, the line of regression takes the form

$$Y_e = a_1 X;\ a_0 = 0 \tag{28.15}$$

with

$$a_1 = \frac{\sum XY}{\sum X^2} \tag{28.16}$$

A cubic model takes the form (see Equation 28.8):

$$Y = a_0 + a_1 X + a_2 X^2 + a_3 X^3 \tag{28.8}$$

For n pairs for $X - Y$ values, the constants a_0, a_1, a_2, and a_3 can be obtained by the method of least squares so that $\Sigma(Y - Y_e)^2$ again has the smallest possible value, i.e., is minimized. The coefficients a_0, a_1, a_2, and a_3 are the solution of the following system of four linear equations (1, 2):

$$\begin{aligned}
a_0 n + a_1 \sum X + a_2 \sum X^2 + a_3 \sum X^3 &= \sum Y \\
a_0 \sum X + a_1 \sum X^2 + a_2 \sum X^3 + a_3 \sum X^4 &= \sum YX \\
a_0 \sum X^2 + a_1 \sum X^3 + a_2 \sum X^4 + a_3 \sum X^5 &= \sum YX^2 \\
a_0 \sum X^3 + a_1 \sum X^4 + a_2 \sum X^5 + a_3 \sum X^6 &= \sum YX^3
\end{aligned} \tag{28.17}$$

This set of equations can be solved for a_0, a_1, a_2, and a_3 since there are four equations and four unknowns. This development can be extended to other regression equations, e.g., exponential, hyperbola, higher order models, etc.

28.3 The Correlation Coefficient

The correlation coefficient provides information on how well the model, or line of regression, fits the data. It is denoted by r and is given by

$$r = \frac{\sum XY - \frac{\sum X \sum Y}{n}}{\sqrt{\left[\sum X^2 - \frac{(\sum X)^2}{n}\right]\left[\sum Y^2 - \frac{(\sum Y)^2}{n}\right]}} \tag{28.18}$$

or

$$r = \frac{n\sum XY - \sum X \sum Y}{\sqrt{[n\sum X^2 - (\sum X)^2][n\sum Y^2 - (\sum Y)^2]}} \tag{28.19}$$

or

$$r = \frac{\sum XY - n\bar{X}\bar{Y}}{\sqrt{(\sum X^2 - nX^{-2})(\sum Y^2 - nY^{-2})}} \tag{28.20}$$

This equation can also be shown to take the from

$$r = \pm\left[\frac{\Sigma(\bar{Y} - Y_e)^2}{\Sigma(Y - \bar{Y})^2}\right] \tag{28.21}$$

The correlation coefficient satisfies the following six properties:

1. If all points of a scatter diagram lie on a line, then $r = +1$ or -1. In addition, $r^2 = 1$. The square of the correlation coefficient is defined as the coefficient of determination.
2. If no linear relationship exists between the X's and Y's then $r = 0$. Furthermore, $r^2 = 0$. It can be concluded that r is always between -1 and $+1$, and r^2 is always between 0 and 1.
3. Values of r close to $+1$ and -1 are indicative of a strong, or positive linear relationship.
4. Values of r close to 0 are indicative of a weak linear relationship.
5. The correlation coefficient is positive or negative, depending on whether the linear relationship has a positive or negative slope. Thus, positive values or r indicate that Y increases as X increases; negative values indicate that Y decreases as X increases.
6. A value of $r = 0$ only indicates the lack of a linear correlation; X and Y might be strongly correlated by some nonlinear relation, as discussed earlier. Thus, r can only measure the strength of linear correlations; if the data are nonlinear, one should attempt to linearize before computing r.

Another measure of the model's fit to the data is the standard error of the estimate, or s_e. It is given as

$$s_e = \left[\frac{\Sigma(Y - Y_e)^2}{n}\right]^{-0.5} \tag{28.22}$$

Finally, the equation of the least-squares line of best fit for the given observations on X and Y can be written in terms of the correlation coefficient as follows:

$$Y - \overline{Y} = r\frac{s_y}{s_x}(X - \overline{X}) \tag{28.23}$$

where $\overline{X}$, $\overline{Y}$, s_x, s_y and r are the two sample means, the two sample standard deviations, and sample correlation coefficient obtained from the n pairs of observations on X and Y, respectively. It should be noted once again that the correlation coefficient only provides information on how well the model fits the data. It is emphasized that r provides no information on how good the model is, or whether this is the correct or best model to describe the functional relationship of the data.

Illustrative Example 28.1

Table 28.1 shows eight pairs of observations on X and Y where Y is the observed percentage yield of a biological reaction at various centigrade temperatures, X. Obtain the least-squares line of regression of Y on X.

Solution

The observed values of Y against the associated values of X are plotted in Figure 28.3. The scatter diagram appears to exhibit a linear pattern. Equation 28.1 is assumed to apply. The values of ΣX, ΣY, ΣXY, ΣX^2, ΣY^2, $(\Sigma X)^2$, $(\Sigma Y)^2$ and n are now calculated:

$$\sum X = 1900$$
$$\sum Y = 700$$
$$\sum XY = 169{,}547.5$$

Table 28.1 Temperature – biological yield data.

% Yield (Y)	**Temperature, °C (X)**
75.4	150
79.4	175
82.1	200
86.6	225
90.9	250
93.3	275
95.9	300
96.1	325

$$\sum X^2 = 477{,}500$$
$$(\sum X)^2 = 3{,}610{,}000$$
$$(\sum Y)^2 = 490{,}000$$
$$\sum Y^2 = 61{,}676.94$$
$$n = 8$$

The least-squares estimates, a_0 and a_1, are calculated using the equations provided above. Appling Equation (28.12),

$$\begin{aligned} a_1 &= \frac{n\sum XY - \sum X \sum Y}{n\sum X^2 - (\sum X)^2} \\ &= \frac{(8)(169547.5) - (1900)(700)}{(8)(477500) - (3610000)} \\ &= 0.126 \end{aligned} \tag{28.12}$$

The coefficient a_0 is calculated employing Equation (28.14).

$$a_0 = \bar{Y} - a_1\bar{X}$$
$$\begin{aligned} a_0 &= \frac{700}{8} - 0.126\left(\frac{1900}{8}\right) \\ &= 57.5 \end{aligned} \tag{28.13}$$

Thus, the describing equation is

$$Y = 57.5 + 0.126X;\ \text{consistent units}$$

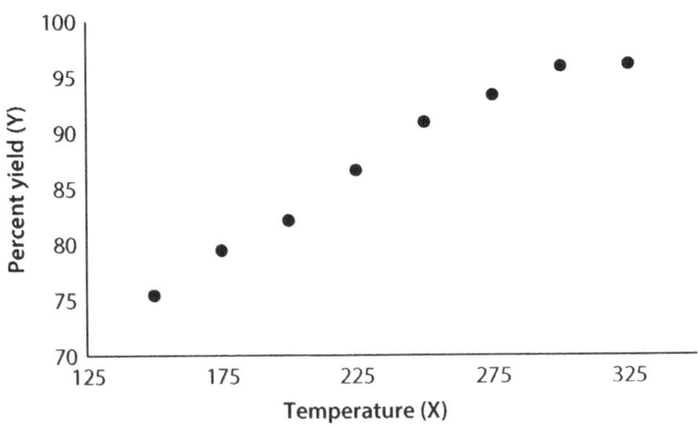

Figure 28.3 Scatter diagram of yield vs. temperature for a biological reaction; Illustrative Example 28.1.

The reader should note that if more temperature-yield data become available, thus increasing the number of points, then the calculated line may not be the best representation of all the data; the least-squares solution should then be recomputed using all the data. In addition, the assumed model, e.g., linear, may not be the "best" model.

Illustrative Example 28.2

Use the results from the previous illustrative example to estimate the "average" percentage yield of the biological reaction at 260 °C.

Solution

The result of

$$Y = 0.126\,X + 57.575$$

is employed to calculate the value of Y when $X = 260$. For $X = 260$,

$$\begin{aligned} Y &= 0.126(260) + 57.575 \\ &= 90.3\% \end{aligned}$$

Therefore, the estimated average percentage yield at 260 °C is 90.3%.

Illustrative Example 28.3

Pollutant A is undergoing a reaction in a specially controlled laboratory experiment. The data in Table 28.2 have been obtained for the reaction rate, $-r_A$, vs. concentration, C_A. Using the data, estimate the coefficient k_A and a in the equation below [3]:

$$-r_A = k_A C_A^a$$

Solution

As discussed in the introduction, given experimental data for Y measured at known values of X, an often-encountered problem is to identify the functional

Table 28.2 Reaction rate data.

Reaction rate, $-r_A$ (lbmol/ft^3 · sec)	**Concentration, C_A (lbmol/ft^3)**
48	8
27	6
12	4
3	2

relationship between the two. One invariably starts by constructing an X–Y plot. The goal is to find a simple relationship between X and Y, e.g., to seek a straight line. If the X–Y plot is not straight, one should try to "straighten it" by replotting, using other functions, other scales, or both.

Some functions readily linearize; examples include the exponential

$$Y = Ae^{mX} \tag{28.24}$$

and the power law

$$Y = X^{a} \tag{28.25}$$

These linearize by taking the natural logarithms:

$$\ln Y = \ln A + mX;\ \text{exponential} \tag{28.26}$$

and

$$\ln Y = a \ln X;\ \text{power law} \tag{28.27}$$

Thus, exponentials yield straight lines on semilog plots, whereas power laws yield straight lines on log-log plots.

For the problem at hand, linearize the equation by taking the natural logarithm (ln) of both sides of the equation.

$$-r_A = k_A C_A^{a}$$

$$\ln - r_A = \ln k_A + a \ln(C_A)$$

Change the variables to Y and X, so that

$$Y = A + BX$$

Regress the preceding four data points using the method of least squares, where $A = \ln(k_A)$ and $B = a$. Once again, the method of least squares requires that the sum of the errors squares between the data and the model is minimized.

$$\ln(3) = A + B\ln(2)$$
$$\ln(12) = A + B\ln(4)$$
$$\ln(27) = A + B\ln(6)$$
$$\ln(48) = A + B\ln(8)$$

Employ Equation (28.12) and Equation (28.14). The linear equation coefficients A and B are given by

$$A = -0.2878$$
$$B = 2.0$$

These may be obtained through longhand calculation. However, they are more often obtained with the aid of computer software.

Take the inverse natural logarithm of A to obtain k_A.

$$A = \ln k_A = 0.75$$
$$B = 2.0$$

The equation for the rate of reaction is therefore

$$-r_A = 0.75C_A^{2.0}$$

Illustrative Example 28.4

The resistance in ohms and failure time in minutes of a random sample of 20 transistors were recorded as shown in Table 28.3.

Calculate the linear correlation coefficient, r.

Solution

Calculate ΣX, ΣY, ΣXY, ΣX^2, ΣY^2, and n for an assumed linear model.

$$\sum X = 752$$
$$\sum Y = 692$$

Table 28.3 Transistor resistance – failure time data.

Resistance (X)	Failure time (Y)	Resistance (X)	Failure time (Y)
43	32	36	31
36	36	44	37
29	20	29	24
44	45	39	46
32	34	46	43
48	47	42	33
35	29	30	25
30	25	30	25
33	32	35	35
46	47	45	46

$$\sum XY = 26{,}952$$
$$\sum X^2 = 29{,}104$$
$$\sum Y^2 = 25{,}360$$
$$n = 20$$

Obtain r using Equation (28.19):

$$r = \frac{n\sum XY - \sum X \sum Y}{\sqrt{[n\sum X^2 - (\sum X)^2][n\sum Y^2 - (\sum Y)^2]}} \tag{28.19}$$

Substituting,

$$r = \frac{20(26952) - 752(692)}{\sqrt{(20(29104) - 752^2)(20(25360) - 692^2)}} = 0.86$$

Illustrative Example 28.5 [4]

A solution containing 50 moles of benzene and 50 moles of toluene is batch distilled at a constant pressure until such a time that only 50 moles of liquid is left. Determine the composition of the liquid residue at the conclusion of the distillation process. Benzene / toluene equilibrium data (x, y^*) is provided in Table 28.4. (Note that, as is common practice in batch distillation, all mole fractions refer to the lighter component, which in this case is benzene.)

Table 28.4 Benzene / toluene equilibrium data.

x	y^*	$y^* - x$	$1/(y^* - x)$
0.000	0.000	0.000	–
0.116	0.240	0.124	8.06
0.228	0.418	0.190	5.26
0.336	0.533	0.217	4.61
0.440	0.660	0.221	4.52
0.541	0.747	0.205	4.88
0.639	0.817	0.178	5.62
0.734	0.875	0.142	7.04
0.825	0.924	0.098	10.2
0.914	0.965	0.051	19.6
1.000	1.000	0.000	–

Solution

Write the Rayleigh equation for a simple batch distillation [4].

$$\ln\left(\frac{W_{final}}{F}\right) = -\int_{x_{W,final}}^{x_F} \frac{dx}{y^* - x}$$

The initial concentration of benzene (mole fraction) is:

$$x_F = \frac{50}{50+50} = 0.50$$

In addition,

$$\ln\left(\frac{W_{final}}{F}\right) = \ln\left(\frac{50}{50+50}\right) = \ln(0.5) = -0.693$$

A trial-and-error solution is now required to determine the benzene mole fraction in the liquid. All values are known in the above integral except for the integral itself, i.e., the value when $x_W = x_{W,\,final}$. The area under the curve is determined by guessing values of $x_{W,\,final}$ until the right-hand side of the integral is equal to –0.693. If a numerical package is not available, one may plot $1/(y^* - x)$ vs x as in Figure 28.4.

The equation appearing in Figure 28.4 is a polynomial regression fit to the data points, which can facilitate analytical, as opposed to numerical, integration. Upon integration of $f(x)$ at varying lower bounds, it can be shown that when $x_{W,\,final} = 0.35$,

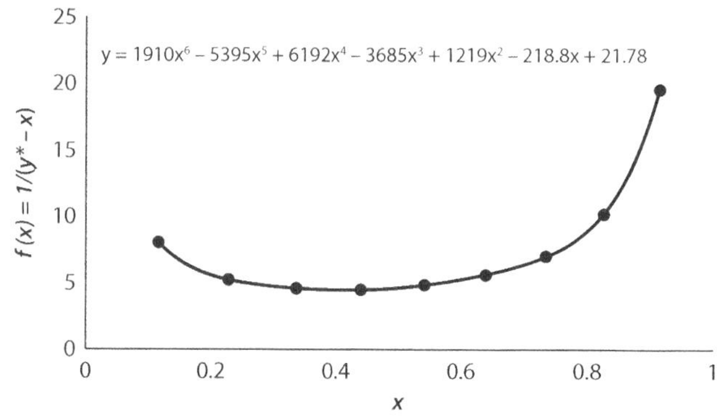

Figure 28.4 Benzene / toluene equilibrium Rayleigh diagram.

the area under the curve is approximately equal to 0.693, which satisfies the integral. (This calculation is left as an exercise for the reader.) Therefore,

$$x_{W,final} \approx 0.35(\text{benzene})$$

$$1 - x_{W,final} \approx 0.65(\text{toluene})$$

References

1. L. Theodore and F. Taylor, *Probability and Statistics*, A Theodore Tutorial, Theodore Tutorials, East Williston, NY, originally published by USEPA/APTI, RTP, NC,1996.
2. S. Shaefer and L. Theodore, *Probability and Statistics Applications for Environmental Science*, CRC Press/Taylor & Francis Group, Boca Raton, FL, 2007.
3. L. Theodore, *Chemical Reactor Analysis and Applications for the Practicing Engineer*, John Wiley & Sons, Hoboken, NJ, 2012.
4. L. Theodore and F. Ricci, *Mass Transfer Operations for the Practicing Engineer*, John Wiley & Sons, Hoboken, NJ, 2010.

Part V
OPTIMIZATION

As one might suspect, the term optimization has come to mean different things to different people. It has also come to mean different things for different applications, i.e., it could involve a simple two step calculation or one that requires the use of a detailed numerical method. To take this a step further, the authors were undecided to how to include optimization in this text, i.e., should it be a chapter – and in which Part, or does it deserve its own Part (in this case, Part V). After much deliberation and meditation, it was decided to present the material in the last Part (V) of the book, as opposed to preparing a chapter for Part III in Numerical Methods. The decision was primarily influenced by the desire that this be a book not only on the *introduction* to mathematical methods but also one that addresses a topic that is emerging as an important topic for environmental engineers and scientists.

There are seven chapters in Part V. The chapter numbers and accompanying titles are listed below.

Chapter 29: Introduction to Optimization
Chapter 30: Perturbation Techniques
Chapter 31: Search Methods
Chapter 32: Graphical Analysis

Chapter 33: Analytical Analysis
Chapter 34: Linear Programming
Chapter 35: Linear Programming Applications

The reader should note that some of the material in Chapter 32 - Graphical Analysis - appeared earlier in part I, Chapter 3.

29

Introduction to Optimization

The optimization problem has been described succinctly by Aris as "getting the best you can out of a given situation" [1]. Problems amenable to solution by mathematical optimization techniques generally have one or more independent variables whose values must be chosen to yield a viable solution and measure of "goodness" available to distinguish between the many viable solutions generated by different choices of these variables. Mathematical optimization techniques are also used for guiding the problem solver to the choice of variables that maximizes the aforementioned "goodness" measure (e.g., profit) or that minimizes some "badness" measure (e.g., cost).

One of the most important areas for the application of mathematical optimization techniques is in chemical engineering (the authors' supposed expertise) design. Applications include:

1. Generation of best functional representations (e.g., curve fitting)
2. Design of optimal control systems
3. Determining the optimal height (or length) of a mass (pollutant) transfer control (or recovery) unit
4. Determining the optimal diameter of a unit
5. Finding the best equipment material of construction

6. Generating operating schedules
7. Selecting operating conditions

In addition to the optimization definition presented above by Aris, one might offer the following generic definition for many environmental engineers: "Optimization is concerned with determining the 'best' solution to a given problem" [2]. Alternatively, a dictionary would offer something to the effect: "to make the most of… develop or realize to the utmost extent… often the most efficient or optimum use of." The process of optimization in environmental practice is required in the solution of many problems and often involves the maximization or minimization of a mathematical function. As one might suppose many of the applications involve economic consideration.

Three sections complement the presentation of this chapter. Section numbers and subject titles follow:

29.1: History of Optimization
29.2: The Computer Age
29.3: The Scope of Optimization

Illustrative Example 29.1

Qualitatively discuss optimization.

Solution

Optimization is viewed by many as a tool in decision-making. It often aids in the selection of values that allow the practicing environmental engineer to better solve a problem. This brief answer provides a qualitative look at optimization.

As noted above, in its most elementary and basic form, one may say that optimization is concerned with the determination of the "best" solution to a given problem. This process is required in the solution of many general problems in environmental engineering and applied science in the maximization (or minimization) of a given function(s), in the selection of a control variable to facilitate the realization of a desired condition, in the scheduling of a series of operations or events to control completion dates of a given project, in the development of optimal layouts of organizational units within a given design space, etc.

One of the most important areas for the application of mathematical optimization techniques is in engineering design. In engineering design, once a particular subject or process scheme has been selected for study, it is common practice to optimize the process from a capital cost and O&M (operation and maintenance) standpoint.

There are many optimization procedures available, most of them too detailed for meaningful application in a text of this nature. These sophisticated optimization techniques, some of which are routinely used in the design of conventional chemical and petrochemical plants, invariably involve computer calculations. Although the use of these techniques in the majority of environmental applications was not warranted in the past, more and more real-world problems are requiring the use of optimization techniques.

29.1 The History of Optimization

The subject of mathematics encompasses the study of relationships among quantities, magnitudes, and properties of logical operations by which unknown quantities, magnitudes, and properties may be deduced. In the past, mathematics was regarded as a science involving geometry, numbers, and/or algebra. Toward the middle of the 19th century, however, mathematics came to be regarded increasingly as the science of relations, or as the science that draws necessary conclusions. This latter view encompasses mathematical or symbolic logic, the science of using symbols to provide an exact theory of logical deduction and reference based on definitions, axioms, postulates, and rules for combining and transforming primitive elements into more complex relations and theorems. Enter optimization.

This brief survey of the history of optimization traces the evolution of mathematical ideas and concepts, beginning in prehistory. Indeed, mathematics is nearly as old as humanity itself; evidence of a sense of geometry and interest in geometrical patterns has been found in the designs of prehistoric pottery and textiles, and in cave paintings. Primitive counting systems were almost certainly based on using the fingers of both hands, as evidenced by the predominance of the number 10 as the base for many number systems employed today.

Interestingly, the mathematics of the late 19th and the 20th centuries is characterized by an interest in unifying elements across numerous fields of mathematical endeavor, especially in logic. For example, group theory has proven a particularly effective unifier. The amount of new math and the particular topics arising at that time were numerous and varied as the unified set theory, intuitive geometry, the development of the number systems, including methods of numeration, binary and other bases of notation, and modular arithmetic and measurement, with attention to accuracy, precision, and error study. It also included studies of algebraic systems, linear algebra, modern algebra, vectors, matrices, logic, truth tables, the nature of proofs, Venn and Euler diagrams, relations, functions, probability and statistics, linear programming, and computer programming.

As to the origin of optimization, it depends on who provides the response, because there are so many aspects of optimization of interest to the practitioner. For example, some claim it was Thomas Edison, when he developed a long-lasting, high-quality light bulb in the 1870s. His success was primarily the result of an extensive trial-and-error search for the optimum filament material. A few now refer to it as the Edisonian approach. One of the authors refers to it as the perturbation approach; on occasion, he has modestly termed it the *Theodore approach*.

Advances in computer calculation have not only reduced the cost and time required to perform each iteration but also provides a better understanding of how this method works. However, the success of this approach still relies heavily on the limitations of the user's intuition and experience, since one often cannot predict or even comprehend the effects of changing numerous variables in a complex system. Despite this barrier, the computer has expanded and improved the employment of the classical methods of optimization.

With the promise of reducing design time and cost while improving product quality, automated design optimization held tremendous potential. Starting with a sub-optimal design, a numerical optimization algorithm could be used to iteratively adjust a set of pre-selected design parameters in an attempt to achieve a set of design targets. This new class of optimization technology enables broader, more comprehensive and faster searches for innovative designs than was possible using previous generations of tools. Moreover, it requires no expertise in optimization theory, so it is easier to use for non-experts and experts alike. By leveraging an engineer's potential to discover new design concepts, this new class of optimization technology overcomes the limits of human intuition and extends the designer's professional capability to achieve break-through designs and accelerated innovation.

As noted earlier, today's computers now allow optimization (or mathematics, computer science, operations research, mathematical optimization of mathematical programming) to more easily select the best element (with regard to some criteria) from some set of available alternatives. In the simplest case, an optimization problem consists of maximizing (or minimizing) a real function by systematically choosing input values from within an allowed set of variables in order to compute the value of the function. Mathematically speaking, optimization allows one to find the best available value(s) of some objective function given a defined domain (or a set of constraints), including a variety of different types of objective functions and different types of domains. These various options are discussed in the last section of this chapter.

29.2 The Computer Age

Although digital computers are often viewed as fast calculators or special slide rulers, they are, in fact, rather general devices for manipulating *symbolic* information. As noted, the symbols being manipulated are numbers or digits (hence the name *digital computer*), and the operations being performed on the symbols are the standard arithmetical operations such as addition and subtraction. Famularo offered the following thoughts in the late 1960s: "The digital computer can be viewed as a high-speed calculator, which, with the availability of subroutines and a compiling language, is able to perform many mathematical operations such as add, multiply, generate analytic functions, logic decisions, etc." [3]. Although the details of coding are not in the scope of this text, there is an intermediate step between the equations and the coding program, and that is to arrange the computing procedure in block diagram or information flow form. The operation of each block is described inside the block, and the computer will perform the instructions around the loop until the condition in the decision block is satisfied, at which time the most recently computed values were then punched out on tape or card. For a large computer, this entire sequence, including a half dozen cycles around the loop, would be accomplished within a millisecond or so. Of course, as the number of equations to be solved increases, the time required to obtain a solution increases. For cases where the solution time becomes excessive, mathematics or other more sophisticated programming is used to accelerate the convergence and decrease the total computation time required.

High-speed computing equipment (today's computers) has had a tremendous impact on engineering design, scientific computation, and data processing. The ability of computers to handle large quantities of data and to perform the mathematical operations described above at tremendous speeds permits the analysis of many more applications and more engineering variables than could possibly be handled on the aforementioned slide ruler – the trademark of engineers of yesteryear. Scientific calculations previously estimated in lifetimes of computation time are currently generated in seconds and, on many occasions, microseconds, and in some rare instances, nanoseconds [4].

Increased growth is expected in the microcomputer field as inexpensive information-processing devices continue to be explored. The cost will continue to decrease, while the cost of the associated software will probably tend to increase; computerized processes will, therefore, be quite inexpensive unless customized programs must be written.

29.3 The Scope Of Optimization

One can conclude from the above that the theory and application of optimization is mathematical in nature, and it typically involves the maximization or minimization of a function (usually known), which represents the "performance" of some "system." This is carried out by the finding of values for those variables, which cause the function to yield an optimal value.

Perhaps the most important tool employed in optimization by environmentalists is linear programming. Linear programming consists of a linear, multivariable function, which is to be optimized (maximized or minimized), subject to a particular number of constraints. The constraints are normally expressed in linear form. Integer linear programming refers to optimization problems in which at least some of the variables must assume integer values. The reader should note that the terms linear programming and nonlinear programming are essentially similar from an application perspective. There were problems around the middle of the last century because of some difficulties that arose in attempting to solve nonlinear programming problems. However, the arrival of the modern-day computer and sophisticated software (e.g., EXCEL) have removed these problems. Unless a solution is presented graphically in this Part, it will be obtained directly from

EXCEL so that the reader need not be concerned with whether a system's pertinent describing equations and constraints are linear or nonlinear. Thus, details regarding the solution methodology for both linear programming and nonlinear programming problems plus the accompanying illustrative examples will not be presented later. They are simply not necessary and beyond the scope of this text.

One of the major responsibilities in optimization is to construct the correct objective function to be maximized. In a simple environmental problem, the objective function might be the profit associated with the operation of a landfill that treats two categories of waste: "inert" and "non-inert." One optimization problem could be to maximize the profit (from handling the waste) by employing an operating treatment schedule subject to the landfill's capacity and environmental regulatory requirements, or the problem might be rephrased in the following manner: what combination of the quantities of the two wastes will produce the maximum profit subject to the constraints connected with the landfill's processing capabilities and the different regulation requirements imposed on each waste.

The objective function may be quite simple and easy to calculate in some environmental applications, or it may be complicated and difficult to not only calculate but also specify and/or describe. The objective function may also be very illusive due to the presence of conflicting or dimensionally incompatible objectives; for example, one might be asked to optimize the profit for the aforementioned landfill that not only minimizes air and water contamination but also is aesthetically appealing. Thus, it may not be always possible to quantify an objective function and hence be able to use any of the mathematical optimization procedures available.

There are a large number of mathematical optimization methods available in practice. Some of the simple ones are listed below:

1. A function of one variable with *no* constraints
2. A function of two variables with *no* constraints
3. A function of more than two variables with *no* constraints
4. Simple perturbation schemes
5. A function of one variable *with* constraints
6. A function of two variables *with* constraints
7. A function of more than two variables *with* constraints

Several of the above methods will be reviewed in the chapters to follow.

Some optimization problems can be divided into parts, where each part is then optimized. In some instances, it is possible to attain the optimum for the original problem by simply realizing how to optimize these constituent parts. This process is very powerful, as it allows one to solve a series of smaller, easier problems rather than one large one. One of the best-known techniques to attack such problems is *dynamic programming*; this approach is characterized by a process which is performed in stages, such as manufacturing processes. Rather than solving the problem as a whole, dynamic programming thus optimizes one stage at a time to produce an optimal set of decisions for the whole process. Although dynamic programming has applicability in some systems/processes, it is not reviewed in this book. If the assumption of linearity cannot be made, as it was in the case of linear programming, there do exist some general procedures for nonlinear problems. The collection of techniques developed for these problems is called nonlinear programming. This too is addressed in this book.

Finally, most engineering optimization applications involve economics: maximizing profit, minimizing cost, or both. Environmental applications can involve minimizing toxic emissions, maximizing energy usage and conservation, minimizing health and safety risks, standard economic concerns, etc. The reader should not lose sight that most real-world industry applications involving optimization usually require simple solutions. Here is a comment from a retired engineer [4]: "Generally, the consumer goods industry where I worked does not rely on sophisticated models. Basically, optimization was conducted to determine overall consumer preference, or liking, to determine how to use minimum resources and lastly to determine how to minimize costs subject to constraints. In the determination of consumer preference, we typically used design of experiments to examine the broad range of attributes that led to maximum product liking or decision to purchase. Occasionally, we modeled processes and used regression models. In that case, we could search or compute the desired maximum. The last set of techniques used is what the business world calls Operations Research. Typically, linear or dynamic programming was used to evaluate an objective function subject to constraints. For example, we used linear programming in hot dog manufacturing. The goal was to come up with a mixture of meat cuts that minimized costs while meeting governmental requirements for protein, fat, and water content".

References

1. R. Aris, *Discrete Dynamic Programming*, Blaisdell, Boston, MA, 1964.
2. L. Theodore, personal notes, East Williston, NY, 2011.
3. J. Famularo, personal communication to L. Theodore (with permission), Englewood Cliffs, NJ, 1973.
4. R. Altomare, communication to L. Theodore, Bronx, NY, 2016.

30

Perturbation Techniques

A significant number of optimization problems face the environmental engineer. The optimal design of industrial processes as well as process equipment has long been of concern to the practicing engineer, and indeed, for some, might be regarded as a definition of the function and goal of applied engineering. The practical attainment of an optimum design is generally a result of factors that include mathematical analysis, empirical information, and both the subjective and objective experience of the engineer and/or scientist.

One simple approach for solving optimization problems that is recommended by the authors is a trial-and-error scheme that includes perturbation calculations. This involves a systematic change of variables, one by one, in an attempt to locate an optimum. To be practical, this can mean that one must limit the number of variables by assuming values to those (process) variables that are known beforehand to play an insignificant role. Reasonable guesses or simple or shortcut mathematical methods can further simplify the procedure. Much information can be gathered from this type of study since it usually identifies those variables that significantly impact the solution, such as on the overall performance of equipment or a process, and helps identify the major contributors effecting the optimization calculations.

Four sections complement the presentation of this chapter. Section numbers and subject titles follow:

30.1: One Independent Variable
30.2: Two Independent Variables
30.3: Three Independent Variables
30.4: The Heat Exchange Network Dilemma

30.1 One Independent Variable

Two illustrative examples highlight this section. These examples have a dependent variable that depends on one independent variable. One example involves a cyclone and air pollution recovery/control. The second example is concerned with minimizing the work requirement for a two-stage compressor.

Illustrative Example 30.1

A process emits 50,000 acfm of gas containing a dust (it may be considered ash and/or metal) at a loading of 2.0 gr/ft^3. A cyclone [1, 2] is employed for particle capture and the dust captured from the unit is worth \$0.03/lb of dust. Experimental data have shown that the collection efficiency, E, is related to the system pressure drop, ΔP, by the formula:

$$E = \frac{\Delta P}{\Delta P + 15.0}$$

where E = fractional collection efficiency
ΔP = pressure drop, lb$_f$/ft^2 (psf)

If the fan is 55% efficient (overall) and electric power costs \$0.18/kWh at what pressure drop and corresponding collection efficiency should the cyclone operate to maximize profits? Earlier studies suggest that the optimum pressure drop is in the 5–35 psf range.

Solution

The value of the recovered material (RV) may be expressed in term of the fractional collection efficiency E, the volumetric flowrate q, the inlet dust loading w, and the value of the dust (DV):

$$\text{RV} = (q)(w)(\text{DV})(E)$$

Substituting yields

$$\text{RV} = \left(\frac{50{,}000\ \text{ft}^3}{\text{min}}\right)\left(\frac{2.0\ \text{gr}}{\text{ft}^3}\right)\left(\frac{1\ \text{lb}}{7000\ \text{gr}}\right)\left(\frac{0.03\$}{\text{lb}}\right)E = 0.429E;\ \$/\text{min}$$

The recovered value can be expressed in terms of pressure drop, i.e., replace E by ΔP:

$$\text{RV} = \frac{(0.429)(\Delta P)}{\Delta P + 15.0}\ \$/\text{min}$$

The cost of power (CP) in terms of ΔP, q, the cost of electricity (CE) and the fan efficiency, E_f, is

$$CP = (q)(\Delta P)(CE)/(E_f)$$

Substitution yields

$$\text{CP} = \left(\frac{50{,}000\ \text{ft}^3}{\text{min}}\right)\left(\frac{\Delta P\ \text{lb}_f}{\text{ft}^2}\right)\left(\frac{0.18\$}{\text{kWh}}\right)\left(\frac{1\,\text{min}\cdot\text{kW}}{44{,}200\text{ft}\cdot\text{lb}_f}\right)\left(\frac{1}{0.55}\right)\left(\frac{1\text{h}}{60\ \text{min}}\right)$$
$$= 0.006\Delta P\$/\text{min}$$

The describing equation for the profit P is given by:

$$P = RV - CP$$

Substitution gives

$$P = \frac{0.429x}{x + 15.0} - 0.006x;\ x = \Delta P$$

Table 30.1 Calculations for Illustrative Example 30.1.

***x*, psf**	***P*, $/min**
5.0	0.0773
10.0	0.1116
15.0	0.1245
20.0	0.1251
25.0	0.1181
30.0	0.1060
35.0	0.0903

Since the pressure drop is believed to exist between 5 and 35 psf, employ increments of 5 psf over the range and calculate the profit P in \$/min. The results for P to four significant figures are recorded in Table 30.1. Using the above perturbation analysis, the results indicate that the profit achieves a maximum value of approximately 0.1251 \$/min at an operating pressure drop of approximately 20 psf. The reader may choose to resolve Illustrative Example 6.4.

Illustrative Example 30.2

Consider the two-stage reversible adiabatic compression of a gas from an initial pressure P_1 to a final pressure P_3. If the gas enters at T and is cooled to T between stages, the total work E is given by [3,4]

$$E = NRT\left(\frac{k}{k-1}\right)\left[\left(\frac{P_2}{P_1}\right)^{\frac{k-1}{k}} + \left(\frac{P_3}{P_2}\right)^{\frac{k-1}{k}} - 2\right]$$

where E = total work $\text{ft} \cdot \text{lb}_f$
R = molal gas constant
T = inlet absolute temperature
$k = 1.4$, the ratio of heat capacity at constant pressure to heat capacity at constant volume for the compressed gas

Determine the intermediate stage pressure if E is to be minimized and $P_1 = 1$ atm and $P_3 = 10$ atm.

Solution

The intermediate stage pressure of P_2 exists in the range $1 \leq P_2 \leq 10$. To apply the perturbation method, select increments of 2 atm between the range of 2 and 8, i.e. 2, 4, 6, 8. Calculated values of E at these intervals are presented in Table 30.2 assuming NRT = 1 and $k = 1.4$. The results in Table 30.2 indicate that for E to be minimized the intermediate pressure P_2 value is 4 atm with E at 2.748 $\text{ft} \cdot \text{lb}_f$. The results further suggest that the minimum energy expenditure occurs between 2 and 6 atm. Selecting the midpoint between 2

Table 30.2 Illustrative Example 30.2 calculation.

P_2, atm	E, $\text{ft} \cdot \text{lb}_f$
2	2.810
4	2.748
6	2.890
8	3.070

and 4 atm and 4 and 6 atm could provide an improved answer for the minimum E. The reader may choose to resolve Illustrative Example 11.2.

30.2 Two Independent Variables

This section presents two examples that illustrate the method solution when there are two independent variables. These examples introduce the reader to the concept of constraints.

Illustrative Example 30.3

Obtain the maximum and minimum for the following function describing the emission E from a stack.

$$E = 3x_1 + 5x_2$$

There are two constraints:

$$0 \leq x_1 \leq 3, \; 0 \leq x_2 \leq 4$$

Solution

Proceed as illustrated below. The function E can be calculated at the 20 points shown in parenthesis.

$$\begin{array}{cccc} (0,0) & (1,0) & (2,0) & (3,0) \\ (0,1) & (1,1) & (2,1) & (3,1) \\ (0,2) & (1,2) & (2,2) & (3,2) \\ (0,3) & (1,3) & (2,3) & (3,3) \\ (0,4) & (1,4) & (2,4) & (3,4) \end{array}$$

The value of the function for each point is shown immediately to the right of the point in parenthesis.

$$\begin{array}{cccc} (0,0)\;0 & (1,0)\;3 & (2,0)\;6 & (3,0)\;9 \\ (0,1)\;5 & (1,1)\;8 & (2,1)11 & (3,1)14 \\ (0,2)10 & (1,2)13 & (2,2)16 & (3,2)19 \\ (0,3)15 & (1,3)18 & (2,3)21 & (3,3)24 \\ (0,4)20 & (1,4)23 & (2,4)26 & (3,4)29 \end{array}$$

As can be seen from the above results, the maximum (emission) E is 29 and occurs at $x_1 = 3$ and $x_2 = 4$, and the minimum E is located at $x_1 = 0$ and $x_2 = 0$ with a minimum value of 0 for E.

Illustrative Example 30.4

Obtain the maximum and minimum for the following objective function with the same constraints employed in the previous example.

$$E = 3x_1 + 5x_2 + x_1^2 + 2x_2^2 - 5x_1x_2$$

Solution

The values for 20 points appear below,

(0, 0) 0	(1, 0) 4	(2, 0) 10	(3, 0) 18
(0, 1) 7	(1, 1) 6	(2, 1) 7	(3, 1) 10
(0, 2) 18	(1, 2) 12	(2, 2) 8	(3, 2) 6
(0, 3) 33	(1, 3) 22	(2, 3) 13	(3, 3) 6
(0, 4) 52	(1, 4) 36	(2, 4) 22	(3, 4) 10

For this example, the minimum is located at (0, 0) with a value of 0, and the maximum is located at the point (0, 4) with a value of 52.

30.3 Three Independent Variables

This section presents one illustrative example involving three independent variables.

Illustrative Example 30.5

Consider the following function:

$$P = 3x_1 - 5x_2 - 8x_3$$

Assume the same constraints as the previous two examples for x_1 and x_2; the constraint for x_3 is

$$1 \leq x_3 \leq 2$$

Obtain the maximum and minimum for the above function.

Solution

Apply the perturbation method again as shown below.

For $x_3 = 1$:

$$\begin{array}{llll}
(0,0)-8 & (1,0)-5 & (2,0)-2 & (3,0)\ 1 \\
(0,1)-13 & (1,1)-10 & (2,1)-7 & (3,1)-4 \\
(0,2)-18 & (1,2)-15 & (2,2)-12 & (3,2)-9 \\
(0,3)-23 & (1,3)-20 & (2,3)-17 & (3,3)-14 \\
(0,4)-28 & (1,4)-25 & (2,4)-22 & (3,4)-19
\end{array}$$

For $x_3 = 2$:

$$\begin{array}{llll}
(0,0)-16 & (1,0)-13 & (2,0)-10 & (3,0)-1 \\
(0,1)-21 & (1,1)-18 & (2,1)-15 & (3,1)-6 \\
(0,2)-26 & (1,2)-23 & (2,2)-20 & (3,2)-11 \\
(0,3)-31 & (1,3)-28 & (2,3)-25 & (3,3)-16 \\
(0,4)-36 & (1,4)-33 & (2,4)-30 & (3,4)-21
\end{array}$$

Based on the above two trials for x_3, the maximum is located at (3, 0, 1) and the minimum value is located at (0,4,2) with corresponding P values of 1 and −36, respectively. However, additional perturbation calculations should be performed for values of x_3 between 1 and 2.

30.4 The Heat Exchange Network Dilemma

This last section involves a heat exchange design project that is concerned with not only energy recovery but also energy conservation. The design is based on a network of heat exchangers at a refinery and the purpose of the project is to arrange a network consisting of an unknown (and unspecified in terms of area) number of exchangers in order to satisfy certain specified design conditions [5].

The remainder of this last section of the "Perturbation" chapter consists of two illustrative examples. The first example introduces the heat transfer system under study [6]. The last example requests the design of a heat exchanger network system to satisfy a specified requirement. This project demonstrates that despite the advances in analytical numerical methods, a perturbation method of solution may be the only viable option for the designer. For the second example, one out of a near infinite amount of possible solutions is presented. Theodore [7] provides two additional examples that ratchet up the complexity of other possible requirements by increasing the difficulty in obtaining the "best" solution. These examples present evidence of how difficult it may be to solve some real-word optimization problems.

Illustrative Example 30.6

A refinery has three streams to be heated (see Table 30.3) and three streams to be cooled (see Table 30.4). Cooling water (90 °F supply, 115 °F return) and steam (saturated at 250 psi) are available. Note that the saturated steam at 250 psi has a temperature of 401 °F. Calculate the heating and cooling duties and indicate what utility (or utilities) should be employed.

Solution

The sensible heating duties for all streams are first calculated. The total heating and cooling duties can now be computed and compared. The results are shown below.

$$\text{Heating: } 7{,}475{,}000 + 6{,}612{,}000 + 9{,}984{,}000 = 24{,}071{,}000 \text{ Btu/h}$$

$$\text{Cooling: } 12{,}600{,}000 + 4{,}160{,}000 + 3{,}150{,}000 = 19{,}910{,}000 \text{ Btu/h}$$

$$\text{Heating} - \text{Cooling: } 24{,}071{,}000 - 19{,}910{,}000 = 4{,}161{,}000 \text{ Btu/h}$$

As a minimum, 4,161,000 Btu/h will have to be supplied by steam or another hot medium.

Illustrative Example 30.7

Refer to the previous Illustrative Example. Devise a network of heat exchanges that will make full use of heating and cooling streams against each other, using utilities only if necessary [5].

Table 30.3 Streams to be heated.

Stream	Flowrate, lb/hr	C_p, Btu/lb °F	T_{in}, °F	T_{out}, °F
1	50,000	0.65	70	300
2	60,000	0.58	120	310
3	80,000	0.78	90	250

Table 30.4 Streams to be cooled.

Stream	Flowrate, lb/hr	C_p, Btu/lb °F	T_{in}, °F	T_{out}, °F
1	60,000	0.70	420	120
2	40,000	0.52	300	100
3	35,000	0.60	240	90

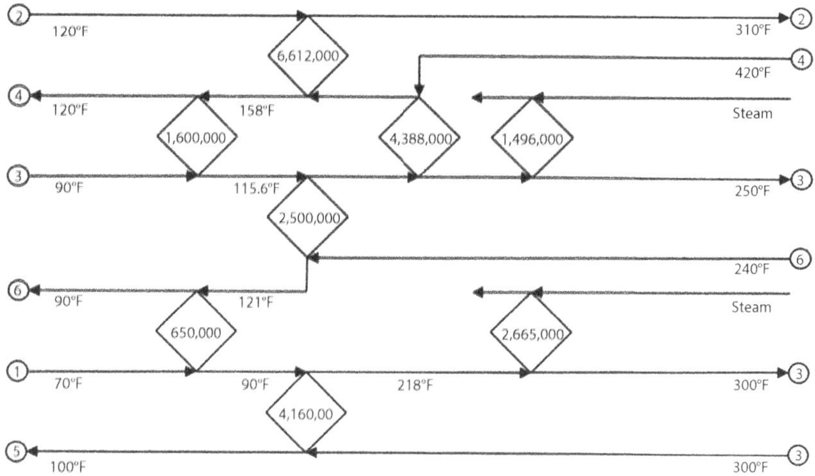

Figure 30.1 Heat exchanger network flow diagram.

Solution

Figure 30.1 represents a system of heat exchanges that will transfer heat from the hot streams to the cold ones in the amounts desired. It is important to note that this is but one of many possible schemes. The optimum system would require a trial-and-error procedure that would examine a host of different networks. Obviously, the economics would also come also into play.

It should also be noted that in many chemical and petrochemical plants there are cold streams that must be heated and hot streams that must be cooled. Rather than use steam to do all the heating and cooling water to do all the cooling, it is often advantageous, as demonstrated in this problem, to have some of the hot streams heat the cold ones. The problem of optimum heat exchanger networks to accomplish this has been extensively studied and is available in the literature. This example provides one simple illustration.

Finally, highly interconnected networks of exchangers can save a great deal of "*quality*" energy [3, 6, 8] in a refinery or chemical plant. This issue is also addressed by Theodore [7]. The more interconnected they are, however, the harder the refinery is to control, start-up, and shut down [5]. In addition, an economic analysis should be included in any study [5, 7]. Often auxiliary heat sources and cooling sources must be included in the plant design in order to ensure that the plant can operate smoothly [5].

References

1. L. Theodore, *Air Pollution Control Equipment Calculations*, John Wiley & Sons, Hoboken, NJ, 2008.

2. L. Theodore, *Chemical Engineering: The Essential Reference,* McGraw- Hill, New York City, NY, 2014.
3. L. Theodore, F. Ricci, and T. VanVliet, *Thermodynamics for the Practicing Engineer,* John Wiley & Sons, Hoboken, NJ, 2009.
4. P. Abulencia and L. Theodore, (adapted from) *Fluid Flow for the Practicing Chemical Engineer*, John Wiley & Sons, Hoboken, NJ, 2009.
5. D. Kauffman, *Process Synthesis and Design,* A Theodore Tutorial, Theodore Tutorials, East Williston, NY, originally published by the USEPA/APTI, RTP, NC, 1992.
6. A.M. Flynn and L. Theodore, "Kern's Process Heat Transfer," 2nd ed., Scrivener-Wiley, Salem, MA, (forthcoming) 2018.
7. L. Theodore, *Heat Transfer Applications for the Practicing Engineer,* John Wiley & Sons, Hoboken, NJ, 2011.
8. L. Theodore, R. Dupont, K. Ganesan, *Unit Operations in Environmental Engineering,* Scrivener-Wiley, Salem, MA, 2017.

31

Search Methods

The mathematicians' literature abounds with search methods, i.e., methods that allow one to search for a solution to an equation. There are both indirect and direct approaches, with indirect search methods generally defined as indirect methods. However, direct methods – those that produce an exact answer are generally preferred by environmental engineers and scientists. These methods have also been defined in the literature as *seek* methods, and usually apply to single-independent variable unimodal equations; they provide results in intervals (or limits) of uncertainty for the variable in a (minimum) number of trials during the search process. Two such procedures that are available include the interval halving (or bisection) and golden section search methods [1]. The unimodal function referred to above for single-independent variable equations contains a single optimization (maximum or minimum) point over a specified interval. Some applications can arise for multi-independent variable equations; perhaps the most popular approach to these equations is the steepest ascent/descent method [2]. It should also be noted that nonlinear algebraic equations are often solved using the Newton-Rhapson method [3]; however, this latter method requires that the derivative of the function is available.

When an initial interval has been established around the optimum, interval reduction schemes may be applied to obtain an improved estimate of the location of the optimum point. The amount of subinterval eliminated at each step depends on both the initial interval and the location of the initial trial points within the search interval. The location of subsequent trial points reduces the interval the same amount, and that reduction should be as large as possible with each step.

A general description of search methods for unimodal functions is now briefly described. As noted, these methods involve a "search" in which the optimum is found by eliminating sections of the original bounded interval in question. When the remaining subinterval is reduced to an acceptable one, the search is terminated. The main advantage of a search method is that it requires only the calculation of the function. In addition, the function need not be differentiable or be in a mathematical or analytical form. All that is required is that the function, $f(x)$, can be determined at a specified x.

Three sections complement the presentation of this chapter. Section numbers and subject titles follow:

31.1: Interval Halving
31.2: Golden Section
31.3: Steepest Ascent/Descent

The first two sections are concerned with unimodal functions. The last section provides a method for solving multi-variable independent equations.

31.1 Interval Halving

The method of interval halving depends on the same hypothesis as the method of false position [3]. An algorithm for determining the *minimum* for $f(x)$ over the initial interval (range) x_1 to x_2 follows [4]. (The procedure is essentially reversed for determining the *maximum* of a function [1, 4].)

1. Set an interval L from a low bound x_1 to a high bound x_2 where $x_2 > x_1$.
2. Calculate the midpoint of L, x_m between $x_1 - x_2$ using the formula

$$x_m = \frac{x_1 + x_2}{2}$$

3. Set $x_1^* = x_1 + \frac{L}{4}$ and $x_2^* = x_2 - \frac{L}{4}$.
4. Use the function being minimized to calculate $f(x_m)$, $f(x_1^*)$, and $f(x_2^*)$.

Continuing, follow the next three steps accordingly until the interval L has suitably converged.

I. Compare $f(x_1^*)$, $f(x_m)$, and $f(x_2^*)$. If $f(x_1^*) \geq f(x_m)$ and $f(x_2^*) \geq f(x_m)$, eliminate both intervals x_1 to x_1^* and x_2^* to x_2, if not, move to II.

The new range L, (x_1 to x_2) is the original x_1^* to x_2^* while x_m remains unchanged. The optimum is now known to be between the new x_1 and x_2. If the interval L has been reduced to a satisfactory level the calculation is stopped. If it is not, begin again at step 1 calculating new values based on the new interval L.

II. Compare $f(x_1^*)$ and $f(x_m)$. If $f(x_1^*) \le f(x_m)$, eliminate the interval from x_m to x_2 and set a new high interval x_2 to the old midpoint x_m. In addition, set the new midpoint x_m equal to x_1^*. The new range L is therefore x_1 to x_2 where x_2 is the old x_m. The optimum is now known to be between the new x_1 and x_2. If L has been reduced to a satisfactory level, the calculation is stopped. If the interval is not yet small enough, proceed to start again at step 1 calculating new values based on the new interval L. If, however $f(x_1^*) \ge f(x_m)$, proceed to III.

III. Compare $f(x_2^*)$ and $f(x_m)$. If $f(x_2^*) \le f(x_m)$, eliminate the interval x_1 to x_m and set a new low interval x_1 equal to the old midpoint x_m. In addition, set the new midpoint x_m equal to x_2^*. The new range L is therefore x_1 to x_2 where x_1 is now the old x_m. The optimum is now known to be between the new x_1 and x_2. If L has been reduced to a satisfactory level, the calculation is stopped. If the interval is not small enough, proceed to start again at step 1, calculating new values based on the new interval L.

Illustrative Example 31.1

Refer to Illustrative Example 30.1. Solve the problem using the interval halving method.

Solution

Once again, the describing equations (functions) for the profit P in \$/min is:

$$P = \frac{0.429x}{x + 15.0} - 0.006x$$

where x = pressure drop in psf. Employ the interval halving method over the range of 5 to 35 for $x_1 = 5$ and $x_2 = 35$ to obtain the *maximum* profit. Proceed as suggested above.

$$L = (x_2 - x_1) = 35 - 5 = 30$$

$$x_m = \frac{(x_2 + x_1)}{2} = \frac{35 + 5}{2} = 20$$

$$x_1^* = x_1 + \frac{L}{4} = 5 + \frac{30}{4} = 12.5$$

$$x_2^* = x_2 - \frac{L}{4} = 35 - \frac{30}{4} = 27.5$$

$$P(x^*, \text{lower}) = 0.1150$$

$$P(x^*, \text{upper}) = 0.1126$$

$$P(x_m) = P(20) = 0.1251$$

Since both $P\,(12.5) \leq P\,(20)$, *and* $P\,(27.5) \leq P\,(20)$ eliminate the intervals x_1^* to x_1, i.e. 5 to 12.5 and x_2^* to x_2, i.e., 27.5 to 35. The remaining interval is therefore 12.5 to 27.5.

Calculate an updated L, x_1^*, and x_2^*.

$$L = 27.5 - 12.5 = 15$$

$$x_1^* = 12.5 + \frac{15}{4} = 16.25$$

$$x_2^* = 27.5 - \frac{15}{4} = 23.75$$

Repeat the above calculations noting that x_m remains at 20.

$$P(16.25) = 0.1256$$

$$P(23.75) = 0.1204$$

Since $P\,(23.75) \leq P\,(20)$, eliminate the interval x_m to x_2, i.e., 20 to 27.5. The remaining interval is therefore 12.5 to 20. The calculation may be further updated (for L and x_m).

$$L = 20 - 12.5 = 7.5$$

$$x_m = \frac{20 + 12.5}{2} = 16.25$$

$$P(16.25) = 0.1256 \text{ \$/min}$$

Further calculations are optional. If one were to stop here, the maximum profit would approximately be 0.1256 $/min at a pressure drop of 16.25 psf.

Illustrative Example 31.2

Refer to Illustrative Example 30.2. Solve the problem employing the interval halving method.

Solution

Employing the interval halving approach between an initial range of $P_1 = 1$ atm and $P_3 = 10$ atm would proceed as follows.

$$L = P_3 - P_1 = 9$$

$$P_m = \frac{1+10}{2} = 5.5$$

$$P_1^*(lower) = 1 + \frac{9}{4} = 1 + 2.25 = 3.25$$

$$P_3^*(upper) = 10 - \frac{9}{4} = 10 - 2.25 = 7.75$$

$$E(P_m) = P(5.5) = 2.848$$

Evaluate E at P_1^* and P_2^*,

$$E(3.25) = 2.727$$

$$E(7.75) = 3.047$$

Since $E(3.25) \leq E(5.5)$, the correct P must therefore lie in the range $3.25 \leq P \leq 5.5$. Repeat the above calculations.

$$L = 5.5 - 1 = 4.5$$

$$P_m = \frac{1+5.5}{2} = 3.25$$

$$P_1^*(lower) = 1 + \frac{4.5}{4} = 1 + 1.125 = 2.125$$

$$P_3^*(upper) = 5.5 - \frac{4.5}{4} = 5.5 - 1.125 = 4.375$$

$$E(P_m) = E(3.25) = 2.727$$

$$E(P_1^*) = E(2.125) = 2.789$$

$$E(P_3^*) = E(4.375) = 2.768$$

Since P (2.125) and P (4.375) are $\geq P$ (3.25), eliminate the outer intervals, P_1 to P_1^* and P_2 to P_2^*. It appears the correct P is in the range $2.125 \leq P \leq 4.375$. The corresponding E at $P_m = 3.25$

$$E(3.25) = 2.727$$

The reader may choose to refine the calculation further for the optimum P and E.

31.2 Golden Section

An improvement on the interval halving method can be had by employing the golden section method. Consider the two trial points located between the initial interval x_1 and x_2 as pictured in Figure 31.1. Assume the initial interval, x_1 to x_2, has arbitrarily been set equal to 1.0 units and the *minimum* of $f(x)$ in the range (x_1, x_2) is desired. If the right side of the interval, i.e., $1 - a$ (where a is a fractional quantity) is discarded based on the calculations for $f(x_1)$, $f(x_2)$, $f(x_3)$, and $f(x_4)$, then the new interval that would remain is a with x_3 located a from x_1; this is pictured in Figure 31.2. The question that arises is: what value of a should be selected? The golden selection method is based on $a = 0.618$. The basis of this choice is provided in the literature [5].

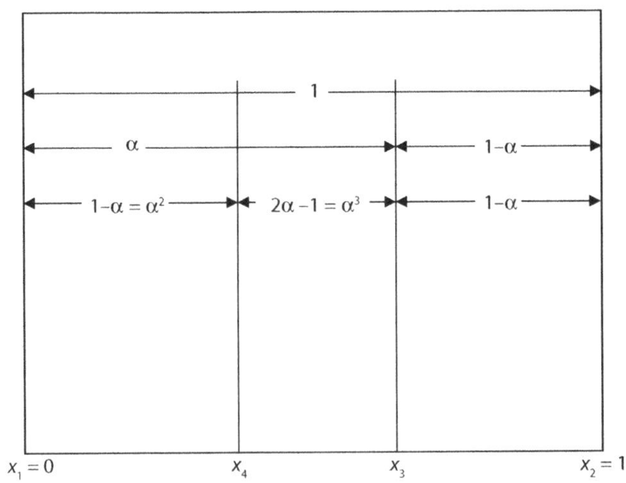

Figure 31.1 Golden section initial search.

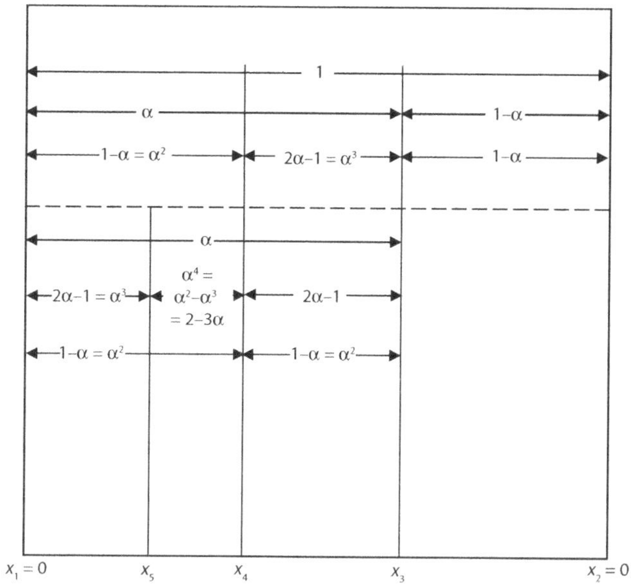

Figure 31.2 Golden Section; remaining interval.

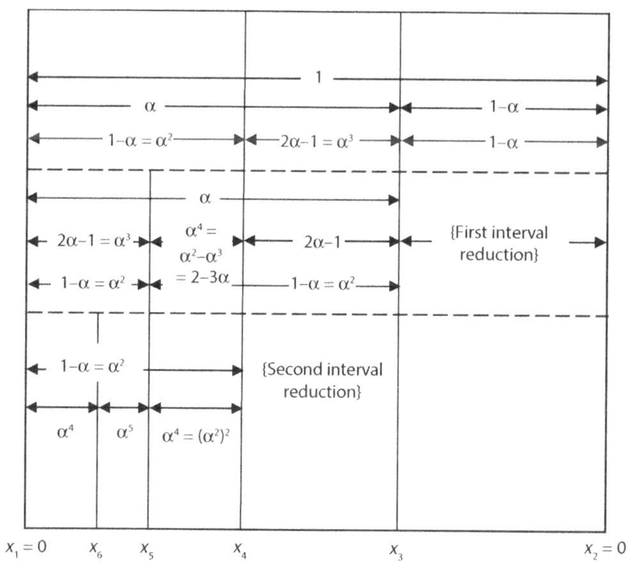

Figure 31.3 Golden Selection; Subsequent Interval.

The result of the first calculation has reduced the first (original) interval from 1 to a. In effect a is now the new interval. This may be viewed as a fraction of the original interval, often given as the product of the first interval and a. Refer once again to Figure 31.2. The next trial point would then be located at x_5 which is $2a - 1$ from x, the same distance from x_4 to x_5. [Note: It can be shown that in the golden section method $a^2 = 1 - a$. The solution of this quadratic equation – $a^2 + a - 1 = 0$ – is simply

$$a = \left[\frac{(-1) \pm \sqrt{(1)(1) - (4)(1)(-1)}}{2}\right]$$

the positive solution of which is 0.618.] Following the second trial with x_5 results in a further reduction of the interval by $2a - 1$ which can be shown to equal $a(1 - a)$. The interval has thus been reduced by the product of the interval (a) and ($a - 1$). Further, the new or remining interval is once again the product of the previous interval (a) and (a) i.e., $(a)(a) = a^2$. Each successive trial would result in an interval reduction given by multiplying the interval by a. For n trials, the final interval would be $(a)^n$. If $n = 5$, the final interval would be $(0.618)^5 = 0.0901$; the interval is therefore reduced by more than 90% after 5 trials.

Illustrative Example 31.3

Refer to Illustrative Example 31.1. Solve this maximization problem employing the golden section search method.

Solution

The function to be maximized is

$$P = \frac{0.429x}{x + 15.0} - 0.006x$$

Once again, assume an initial range of (5,35 with an interval of 30). Based on the material provided above obtain P at the following values of x.

$$P(5)$$

$$P(35)$$

Noting that $(0.618)(30) = 18.54$ and $(0.382)(30) = 11.46$,

$$P(11.46 + 5) = P(16.46)$$

$$P(18.54 + 5) = P(23.54)$$

The calculated results are:

$$P(5) = 0.03025$$

$$P(35) = 0.0903$$

$$P(16.46) = 0.2302$$

$$P(23.54) = 0.1208$$

Since the function is to be maximized, eliminate the interval (23.54, 35). The interval has therefore been reduced from (5,35) to (5, 23.45). Calculate P at the location of the next trial, i.e., at $a - a^2$. However, and as shown in Figure 31.3,

$$a - a^2 = a(1-a) = a(a^2) = a^3 = 0.236$$

the new location then becomes 5 + (0.236)(30) = 5 + 7.08 = 12.08. Continuing, the interval (5, 12.08) may be eliminated, narrowing the interval where the maximum is located to (12.08, 23.54). The calculation may be continued as demonstrated above... or simply assume an average over the new interval x and calculate the corresponding P

$$x_m = \frac{12.08 + 23.54}{2} = 17.81$$

$$P(12.08) = 0.1189$$

$$P(17.81) = 0.1260$$

$$P(23.54) = 0.1208$$

This represents an estimate of the solution of the example.

31.4 Steepest Ascent/Descent

The method of steepest ascent/descent involves calculations where a step size is selected to achieve the maximum amount of increase/decrease of the function at each succeeding step. Thus, the step size gradient varies as the calculation continues, settling on zero as the maximum/minimum point is approached. Perhaps Happel [2] provided a satisfactory explanation and illustrative example involving steepest ascent/descent calculations from an engineering perspective. His description follows.

> "There are many situations, especially where experimentation is involved, in which neither an analytical nor graphical representation is available for the function to be optimized for the entire range of the variables involved. The question then arises as to what course the experimenter should follow, be it

in the research laboratory or full-scale commercial plant. He may employ any of the several trial-and-error procedures previously discussed. However, if results are known only for a small area, some efficient method of extrapolation will help to minimize the amount of trial and error necessary. Such a scheme is the method of steepest ascents. The concept behind the idea of predicting a path of steepest ascent is simply that a plane does a good job of approximating a curved surface over a limited area. Thus, for a situation involving the maximizing of a function of *two* independent variables, a contour map with the variables as horizontal co-ordinates and the (profit) function as the vertical coordinate will illustrate the plan. The best-fitting plane is obtained in a small area under investigation. Then, from the tilt of the fitted plane, the direction of further experimentation is established as the direction of steepest ascent. Experiments are performed along this path until a decline in response is noted. Additional observations/ calculations are taken around this point to confirm whether a maximum has indeed been reached or whether a new path of steepest ascent should be predicted. The method, of course, does not eliminate the possibility that more than one optimum may exist in the area of interest.

In practice, the method of steepest ascents is generally applied to calculation involving more than two variables. It is difficult, but the mechanics is entirely similar to the two-dimensional approach. More specifically, the partial derivatives of the response $f(x)$ will indicate its rate of change with respect to a series of variables, $x_1, x_2, x_3, \ldots$. From these derivatives the gradient [6] $\nabla[f(x)]$ is computed

$$\nabla[f(x)] = i_1\frac{\partial f(x)}{\partial x_1} + i_2\frac{\partial f(x)}{\partial x_2} + i_3\frac{\partial f(x)}{\partial x_3} + \ldots \tag{31.1}$$

where $i_1, i_2, i_3, \ldots$ are unit vectors in the directions of coordinate axes representing the variables, each variable being associated with one dimension in a multidimensional space. The gradient $\nabla[f(x)]$ is a vector which has the property of being directed along the path on which $f(x)$ increases most rapidly (i.e. the path of "steepest ascent"). The direction is along a path proportional to the rate of increase for a unit change of each variable."

The following example, similar to the two stage compressor problem of Illustrative Example 30.2, illustrates the way the procedure works. By applying it in a case where the result is known, it is possible to readily note its limitations. Technically, it is correct only at the base chosen for each ascent. As exploration moves from this point, the shape of the response surface must change.

Illustrative Example 31.4

Consider the case of a *three*-stage compressor. It is desired to demonstrate the method of steepest ascents for $P_1 = 1$ and $P_4 = 10$ atm. Since the procedure implies knowledge of the work expended corresponding to only a limited area, rather than the complete range, the arbitrary assumption will be made that the intermediate pressures are taken at the start as $P_2 = 4$ and $P_3 = 7$ atm.

Solution

In this example, the optimum will be reached by the steepest *descent* to *minimum* total work, which corresponds to *maximum* profit for the operation employing such a compressor. The describing equation for the work E is [2]

$$E = NRT\left(\frac{k}{k-1}\right)\left[\left(\frac{P_2}{P_1}\right)^{\frac{(k-1)}{k}} + \left(\frac{P_3}{P_2}\right)^{\frac{(k-1)}{k}} + \left(\frac{P_4}{P_3}\right)^{\frac{(k-1)}{k}} - 3\right]$$

Once again, NRT is set equal to 1, $k = 1.4$, $P_1 = 1$, and $P_4 = 10$. It will be shown in Chapter 33 that the minimum work is

$$E = 2.575 \text{ ft} \cdot \text{lb}_f$$

and the corresponding optimum values of P_2 and P_3 are

$$P_2 = 2.154 \text{ atm}$$

$$P_3 = 4.642 \text{ atm}$$

The direction of steepest ascent at $P_2 = 4$, $P_3 = 7$ is obtained by differentiating the describing equation for E (differentiation and substitution steps have been omitted for simplicity),

$$\frac{\partial E}{\partial P_2} = \left[(4)^{\frac{(-1)}{1.4}} - (7)^{\frac{(1.4-1)}{1.4}}(4)^{\frac{(1-2.8)}{1.4}}\right]$$
$$= 0.079$$

Similarly,

$$\frac{\partial E}{\partial P_3} = 0.009$$

To obtain the path of steepest descent, increments of pressure must be taken proportional to these rates of change; thus

$$\frac{\Delta P_2}{\Delta P_3} = \frac{0.079}{0.009} = 8.8$$

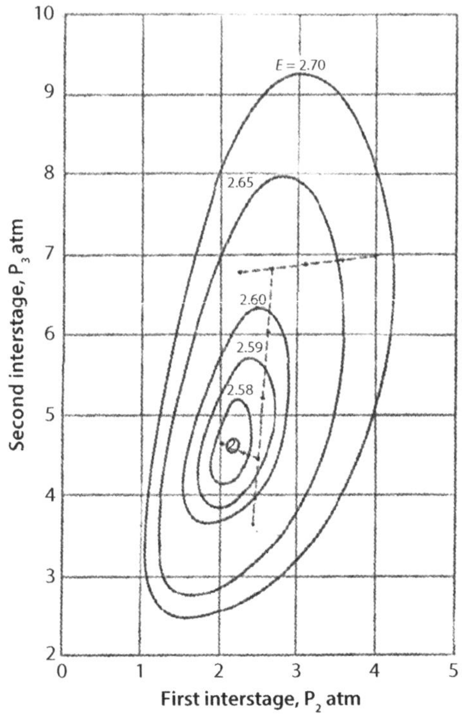

Figure 31.4 Contours of constant work Note: Path of steepest descent shown dotted.

If increments of P_3 are taken equal to 0.05, then the increments of $\Delta P_2 = (0.05 \cdot 8.8) = 0.44$. In this way the *first descent* is established starting at $P_2 = 4$ and $P_3 = 7$:

Using the same procedure following the first decent, a second derivative is made starting at $P_2 = 2.68$ and $P_3 = 6.85$. This results in establishing the condition $P_2 = 2.50$ and $P_3 = 4.45$ with $E = 2.587$ as the minimum point. A third descent from this point establishes $P_2 = 2.14$ and $P_3 = 4.60$ with $E = 2.576$. These conditions are shown on Figure 31.4. Three descents are sufficient to establish the optimum condition.

It should be noted that considerable accuracy in computation is required to establish the optimum. This frequently occurs in applying the method so that it is finally necessary to use an analytical expression to establish the optimum accurately. Numerical results are provided in Table 31.1. Another limitation in the procedure, using the method of steepest ascents and the general differentiation procedure previously discussed is that they do not take into account constraints imposed on the system. The gradient may lead into an area in which operations are not possible."

Table 31.1 Results of Illustrative Example 31.4.

P_2, atm	P_3, atm	E, ft · lb_f
4.00	7.00	2.681
3.56	6.95	2.653
3.12	6.90	2.629
2.68	6.85	2.615 (min)
2.24	6.80	2.622

References

1. R. Ketter and S. Prawel, *Modern Methods of Engineering Computation*, McGraw-Hill, New York City, NY, 1969.
2. J. Happel, *Chemical Process Economics*, John Wiley & Sons, Hoboken, NJ, 1958.
3. L. Theodore, *Chemical Engineering: The Essential Reference*, McGraw-Hill, New York City, NY, 2014.
4. L. Theodore, personal notes, East Williston, NY, 2010.
5. G. Reklaitis, A. Ravindran, and K. Ragsdell, *Engineering Optimization*, John Wiley & Sons, Hoboken, NJ, 1983.
6. L. Theodore, *Transport Phenomena for Engineers*, Theodore Tutorials, East Williston, NY, originally published by the International Textbook Company, Scranton, PA, 1971.

32

Graphical Approaches

The general subject of graphical analysis serves as an excellent starting point before proceeding to analytical analysis, the subject title for the next chapter. The overlap between these two topics will become apparent in the both the developmental material and illustrative examples to follow. Most of the early presentations in this chapter will involve the repetition and extension of some of the mathematical methods reviewed earlier in Parts I, II, and III; hopefully, this duplication will assist the reader during his/her review of this material as it relates to optimization.

Engineering data are often best understood when presented in the form of graphs or mathematical equations. The earlier preferred method of obtaining the graphical relations between variables was to plot the data as straight lines and use the *slope-intercept* method to obtain coefficients and exponents; therefore, it behooves the environmental engineer and scientist to be aware of the methods of obtaining straight lines on the various types of graph paper, and of determining the equations of such lines [1]. One always strives to plot data as straight lines because of the simplicity of the curve, ease of interpolation, and ease of extrapolation. Graphical methods proved invaluable in the past in the analysis of the relatively complex processes of that era. Much of the basic physical and chemical data are still best represented graphically.

One or more of the many types of graphical representations may be employed for the following purposes:

1. as an aid in visualizing a process for the representation of quantitative data
2. for the representation of quantitative data
3. for the representation of a theoretical equation
4. for the representation of an empirical equation
5. for the comparison of experimental data with a theoretical expression
6. for the comparison of experimental data with an empirical expression
7. for solving some optimization problems

The relation between two quantities, y-the dependent variable and x-the independent variable, is commonly obtained as a tabulation of values of y for a number of different values of x. The relation between y and x may not be easy to visualize by studying tabulated results and is often best seen by plotting y vs. x. If the conditions are such that y is known to be a function of x only, the functional relation will be indicated by the fact that the points may be represented graphically by a smooth curve. Deviations of the points from a smooth curve can indicate the reliability of the data. If y is a function of two variables, x_1 and x_2, a series of results of y in terms of x_1 may be obtained for a (definite) constant x_2. When plotted, the data will be represented by a family of curves, each curve representing the relation between x_1 and y for a constant value of x_2. If another variable x_3 is involved, one may have separate graphs for constant values of x_3, each showing a family of curves of y vs. x_1. One may expand this method of representing data to more than three independent variables, although this rarely occurs in practice. In addition, graphs may be employed to perform common mathematical manipulations such as integration and differentiation, operations that usually arises in analytical analyses.

Data may be of almost any type encountered in environmental engineering and science practice. They may be physical property data, such as the variation in density and viscosity with temperature along a tube in a heat exchanger. One of the things that should be kept in mind is the loss of accuracy arising due to the use of graphs whenever one works with plotted data. In general, the number of significant figures may be only as great as the size of the divisions on the graph.

Five sections complement the presentation of this chapter. Section numbers and subject titles follow:

32.1: Rectangular Coordinates
32.2: Logarithmic-Logarithmic (Log-Log) Coordinates
32.3: Semi- Logarithmic (Semi-Log) Coordinates
32.4: Methods of Plotting Data
32.5: Optimization Illustrative Examples

32.1 Rectangular Coordinates

Rectangular (sometimes referred to as Cartesian) coordinates graph paper is most generally used to represent equations of the form

$$y = mx + b \tag{32.1}$$

where

y = variable represented on the ordinate
x = variable represented on the abscissa
m = slope of the line
b = y-intercept at $x = 0$

The most common form of graph paper for engineering use is the $8\frac{1}{2} \times 11$-inch sheet, having 20 lines per inch, with every fifth line accented, and every tenth line heavily accented. This type of coordinate graph may be obtained on drawing or tracing paper, with lines in black, orange, or green, and with or without accented and heavy lines; it is also available in other sizes. Also note that the bulk of the development in the last section of this chapter will key on rectangular coordinates.

All graphs are created by drawing a line, or lines, about one or more axes. For the purposes of this chapter the presentation will be primarily concerned with graphs built around two axes, the x axis and the y axis. The graph data may be presented as coordinates of the x and y axis in the form (x, y), or they may be presented as the abscissa (distance from the y axis or the x coordinate) and the ordinate (distance from the x axis or the y coordinate). Referring to Figure 32.1, point L5 has an abscissa (x coordinate) of +60 and an ordinate of +40 (y coordinate). Using coordinate notation, the point can be described as (60, 40) where the points are listed as (x, y). Point L2 can be described as (−6.7,0).

All coordinates are produced as the result of solving an equation by assigning different values to x or y, and solving for the value of the other. By convention, equations are usually stated in the following form: y equals some value(s) of x and other constants, e.g., $y = mx + b$.

Equations are identified by the type of line (or curve) they produce when plotted on a graph. A linear equation, as the name implies, will result in a plot of a straight line on a graph. With the linear equation, each incremental change of x (or y) will result in an incremental change of a fixed ratio in y (or x). The ratio of change between y and x is known as the *slope* of the line. A linear equation is also referred to as a ***first-degree equation***. The remainder of this section provides methods briefly introduced in Chapter 2.

In Figure 32.1, the line plotted between L1 and L5 is linear. A *non-linear equation*, when plotted, will result in a line having one or more curves, and the ratio of change between x and y is *not* constant. The line between points N0 and N5 in Figure 32.2 represents a plot of a non-linear equation. Non-linear equations come in a variety of forms and may be quickly identified since they have at least one term

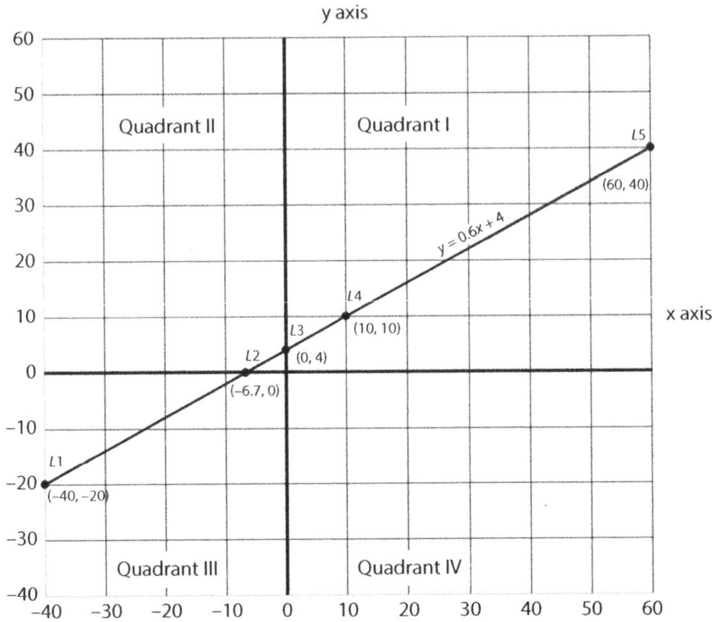

Figure 32.1 Linear coordinates.

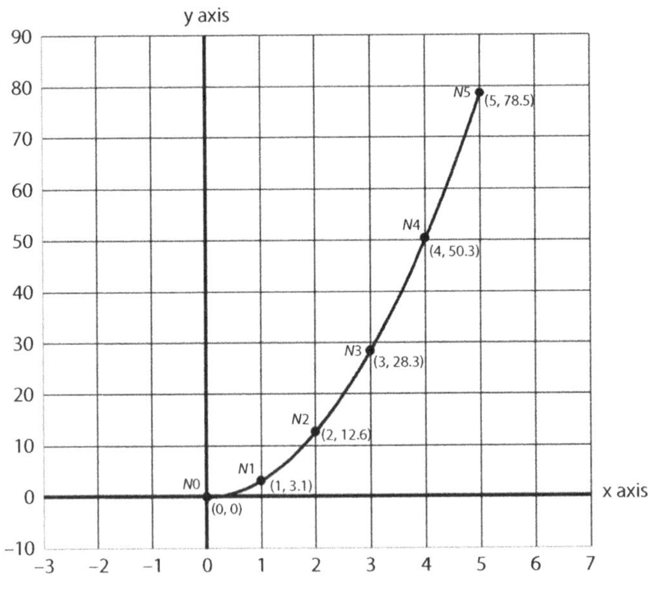

Figure 32.2 Non-linear equation.

containing an exponent. The equation $y = \pi x^2$ (see Figure 32.2) is representative of a non-linear equation.

Intercepts

Reference to the x and y intercepts are often made when working with graphs. An *intercept* is the point at which a line crosses the x or y axis. The *x-intercept* is the point where a line crosses the x axis ($y = 0$), and the *y-intercept* is the point where a line crosses the y axis ($x = 0$). In Figure 32.1, the x-intercept occurs at (−6.7,0) and the y-intercept occurs at (0,4).

In a linear equation there can be only one x and one y intercept. With non-linear equations, there may be no x or y intercept, *or* there may be multiple intercepts. Note that in Figure 32.2, an intercept occurs when $x = 0$ or $y = 0$.

32.2 Logarithmic-Logarithmic (Log-Log) Coordinates

A logarithmic scale is easily constructed by plotting (on rectangular paper) numbers from one to ten on the ordinate versus the logarithm of the number on the abscissa. If the points representing the numbers on the ordinate are projected to the resultant curve and then upward, the scale formed by the vertical lines will be a logarithmic scale. The construction is illustrated in Figure 32.3. The utility of

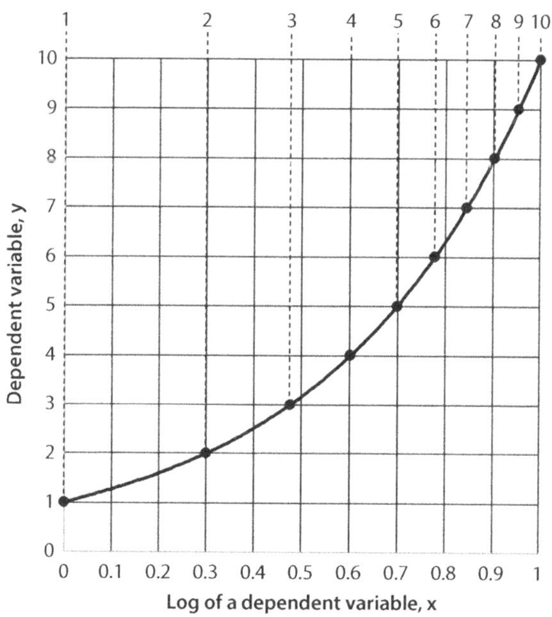

Figure 32.3 The construction of a logarithmic scale.

the logarithmic scale lies in the fact that one can use actual numbers on the scale instead of the logarithms. When two logarithmic scales are placed perpendicular to each other and lines are drawn vertically and horizontally to represent major divisions, full logarithmic graph paper results, i.e., a log-log graph.

The term *log-log* or *full logarithmic* is used to distinguish between another kind of graph or relationship yet to be discussed, i.e., semi-logarithmic (see next section). The full logarithmic may be obtained in many types depending on the scale length, sheet size, and number of cycles desired. The paper is specified by the number of cycles of the ordinate and abscissa; for example, *logarithmic* 2×3 *cycles* means two cycles on the ordinate and three on the abscissa. Cycle lengths are generally the same on both axes of a particular graph. In addition, the distance between numbers differing by a factor of 10 is constant on a logarithmic scale.

Equations of the general form

$$y = bx^m \tag{32.2}$$

where

y = a variable
x = a variable
b = a constant
m = a constant

will plot as straight lines on (full) logarithmic paper. A form of the equation analogous to the slope-intercept equation for a straight line is obtained by writing and then subsequently plotting the equation in logarithmic form.

$$\log y = \log b + m \log x \tag{32.3}$$

If log y versus log x were plotted on rectangular coordinates, a straight line of slope m and y-intercept log b would be obtained. An expression for the slope is obtained by differentiating Equation (32.3).

$$\frac{d(\log y)}{d(\log x)} = m \tag{32.4}$$

When $x = 1$, $m \log x = 0$, Equation (32.3) reduces to

$$\log y = \log b \tag{32.5}$$

Hence, log b is the y-intercept.

To illustrate the above, a linear plot of the equation

$$y = 2x^{1.3} \tag{32.6}$$

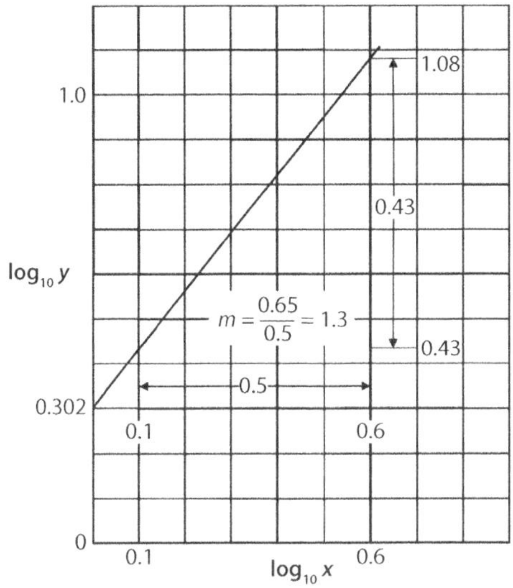

Figure 32.4 A linear plot of $y = 2x^{1.3}$ on rectilinear graph paper.

on rectangular coordinate paper is given in Figure 32.4. It should be noted that although the slope was obtained from a ratio of logarithmic differences, the same result could have been obtained using measured differences if the scale of ordinate and abscissa were the same (in this case they were).

Inspection of Figure 32.5 reveals that equal measured distances on the logarithmic scale are equivalent to equal logarithmic differences on the abscissa of that plot. It therefore follows that since the slope of Equation (32.3) is a ratio of coordinate differences, the slope on full logarithmic paper could be obtained using measured distances. The plot of Equation (32.6) on 2×2 cycle logarithmic paper is given in Figure 32.5. The slope using logarithmic differences is

$$m = \frac{1.08 - 0.43}{0.60 - 0.10} = 1.3 \tag{32.7}$$

and the slope using measured differences is also

$$m = \frac{a}{b} = 1.3 \tag{32.8}$$

32.3 Semilogarithmic (Semi-Log) Coordinates

This section is a review on Semi-Log coordinates visited earlier in Chapter 3, section 3.3 and includes their application with regard to optimization. Graph

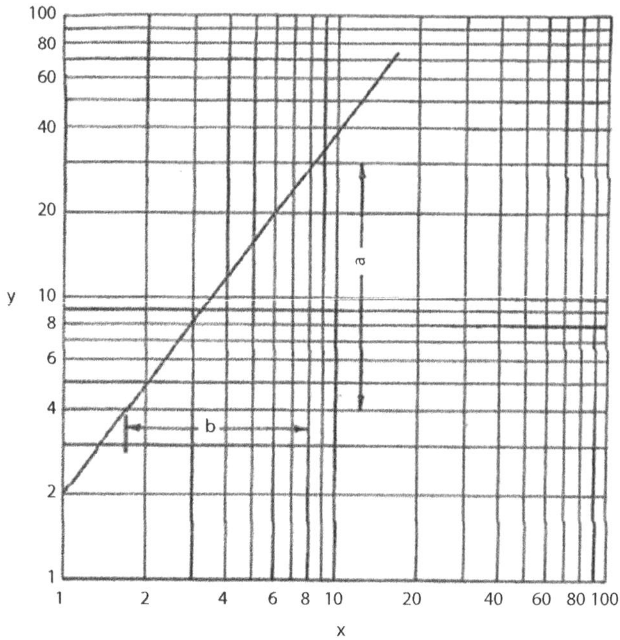

Figure 32.5 The Equation $y = 2x^{1.3}$ Plotted on Logarithmic Graph Paper.

paper made with one logarithmic scale and one arithmetic scale is termed *semilogarithmic.* The logarithmic scale is normally the ordinate and the arithmetic scale is the abscissa on semilogarithmic paper. The paper is available in many styles depending on the sheet size, type of paper, color of lines, number of cycles, length of cycles, and the divisions on the uniform scale. The designation *semilogarithmic, 7 × 5 to the ½ inch* refers to semilogarithmic paper whose logarithmic scale contains seven cycles whose uniform scale contains five divisions to the half inch. The designation *10 division per inch (70 divisions) by two 5-inch cycles* refers to paper whose arithmetic scale is seven inches long and contains ten division per inch, and whose logarithmic scale contains two-cycles for each five inches. Thus, a semilogarithmic graph uses a standard scale for one axis and a logarithmic scale for the other axis, i.e., the graph uses scales of different mathematical proportions. The reason for this use of scales is that a logarithmic (exponential) plot would quickly exceed the physical boundaries of the graph if one were to plot values of some base having exponents of 2, 3, 4, 5, 6, 7 and 8 against an ordinary numeric scale.

Semilogarithmic paper can be used to represent as straight lines equation of the form

$$y = ne^{mx} \tag{32.9}$$

or

$$y = n10^{mx} \tag{32.10}$$

where
y = dependent variable
x = independent variable
n = a constant
m = a constant
e = the natural logarithm base

Here again a form of the equation analogous to the slope-intercept equation can be obtained by placing the above equation in logarithmic form. For example, Equation (32.9) may be written as

$$\log y = b + mx \tag{32.11}$$

where $b = \log n$

An expression for slope is obtained from the differential form of Equation (32.11):

$$\frac{d(\log y)}{dx} = m \tag{32.12}$$

so that the y-intercept is $\log n$ or b.

One can illustrate the use of semilogarithmic paper by plotting a segment of the curve representing the equation

$$\log y = 2.2 + \frac{x}{46.3} \tag{32.13}$$

The line may be constructed (see also Figure 32.6) by choosing values of x, solving for y, and plotting the results. Note that values of y *and not* $\log y$ are plotted on the ordinate. The slope is a logarithmic difference divided by an arithmetic difference and in this case may be calculated from the two indicated points (see Figure 32.6) as follows:

$$m = \frac{\log(1587) - \log(158.7)}{43.6 - 0} = \frac{1}{46.3} \tag{32.14}$$

The slope of a line on semilogarithmic coordinates may be calculated from any two points, but the simplest method is the one used above in which the slope is the logarithmic difference on one cycle (equal to log 10 or 1.0) divided by the arithmetic difference of one cycle. Note that if the cycle had been taken between $y = 300$ and $y = 3000$, the slope would be

$$m = \frac{\log(3000) - \log(300)}{59 - 12.7} = \frac{1}{46.3} \tag{32.15}$$

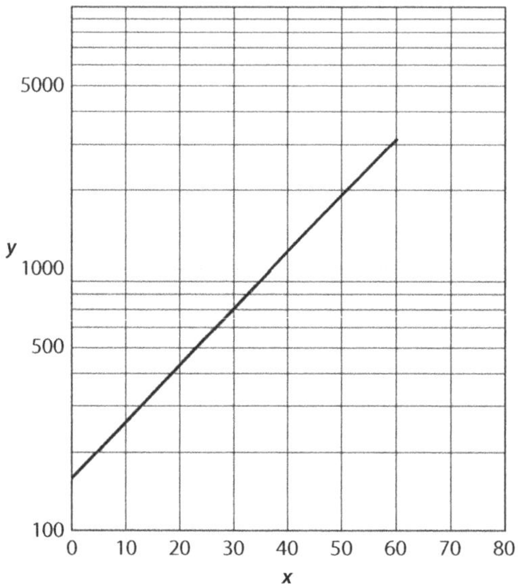

Figure 32.6 The equation $\log y = 2.2 + x/46.3$ plotted on semi-logarithmic paper.

It should also be noted that the y-intercept is 2.2, as required by Equation (32.13).

In addition to the more common types of paper previously discussed, the practicing environmental engineer and scientist may also have need for probability versus arithmetic, probability versus logarithmic, logarithmic versus reciprocals, and other special type of graphs. Non-standard scales, if desired, may be constructed in a similar manner to the logarithmic scale shown earlier. For example, logarithmic-probability (log-normal) graphics find applications in air pollution studies [1] to describe particle size distribution.

32.4 Methods of Plotting Data

The simplest procedure to employ in plotting equations of various forms is provided below in Table 32.1. Various additional forms are available in the literature. Details on statistical methods for calculating the coefficients in the 7 equations are provided in the literature [2].

32.5 Optimization Illustrative Examples

This last section contains five illustrative examples. Some are extensions of problems discussed earlier with the material keying on solutions employing graphical approaches.

Table 32.1 Plotting procedure.

$y = a + bx$	Plot y vs. x.
$y = ax^n$	Plot log y vs. log x or y vs. x on logarithmic coordinates.
$y = c + ax^n$	First obtain c as the intercept on a plot of y vs. x; then plot log $(y - c)$ vs. x on logarithmic coordinates.
$y = ab^{cx}$	Plot log y vs. x or y vs. x on semilogarithmic coordinates.
$y = ab^x$	Plot log y vs. x or y vs. x on semilogarithmic coordinates.
$y = a + \frac{b}{x}$	Plot y vs. $\frac{1}{x}$.
$y = \frac{x}{a + bx}$	Plot $\frac{x}{y}$ vs. x or $\frac{1}{y}$ vs. $\frac{1}{x}$.

Illustrative Example 32.1

Some optimization problems require the solution to a problem where the dependent variable is a function of *two* independent variables, e.g.,

$$y = f(x) = f(x_1, x_2)$$

Outline a simple graphical procedure that one can employ to obtain the minimum (or maximum) of this type of function.

Solution

The optimum values of x_1 and x_2 can be found graphically on a two-dimensional plot by employing the following method. See Figure 32.7. In this figure, y is plotted against one of the independent variables (x_1) with the second variable (x_2) held at a constant value. A series of such plots can be made with each dashed curve representing a different constant value of the second variable. As shown in Figure 32.7, each of the curves (*a, b, c, d,* and *e*) gives one value of the variable x_1 at the point where $f(x)$ is a minimum. The curve *AB* represents the locus of all these minimum points, and the optimum value of x_1 and x_2 occurs at the minimum point on the *AB* curve.

The reader should also note once again that similar graphical procedures can be used when there are more than two independent variables, e.g., x_1, x_2, and x_3. The first step would be to make a plot similar to Figure 32.7 at one constant value of x_3. Each plot would give an optimum value of x_1, x_2, and $f(x)$ for a particular x_3. Finally, the overall optimum value of x_1, x_2, x_3, and $f(x)$ could be obtained by plotting x_3 versus the individual optimum values of $f(x)$. This procedure is rarely employed by the practicing environmental engineer.

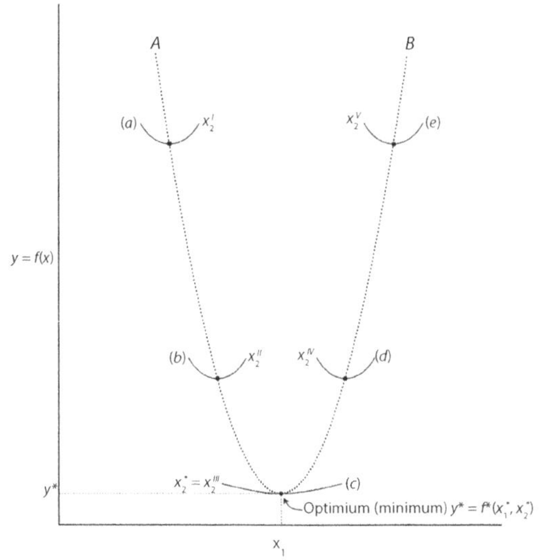

Figure 32.7 Graphical determination of optimum conditions with two independent variables.

Illustrative Example 32.2

The following equation describes the effect of the variables $T(K)$ and P (psia) on the pollutant emission E(lb/hr) for a particular operation

$$E = 2.33\,T + \frac{11{,}900}{TP} + 1.86P + 10$$

Determine the operating values of T and P which will produce the minimum emission.

Solution

The following constant values of P (psia) are chosen arbitrarily

$$P_1 = 32 \qquad P_2 = 26 \qquad P_3 = 20 \qquad P_4 = 15 \qquad P_5 = 12$$

at each constant value of P, i.e., P_1, P_2, P_3, P_4, and P_5, a plot of E vs. T is prepared. These plots are presented as solid lines in Figure 32.8 as curves *a, b, c, d,* and *e*. A summary of these results is presented in Table 32.2 with T_1, T_2, T_3, T_4, and T_5 representing the minimum values of T for the solid curves *a, b, c, d,* and *e*, respectively. The dashed curve AB represents the relationship for the five T_i^*'s as a function of E. The point T^* represents the minimum E which occurs at approximately P_3^*; the corresponding value of E is 121.6 lb/hr.

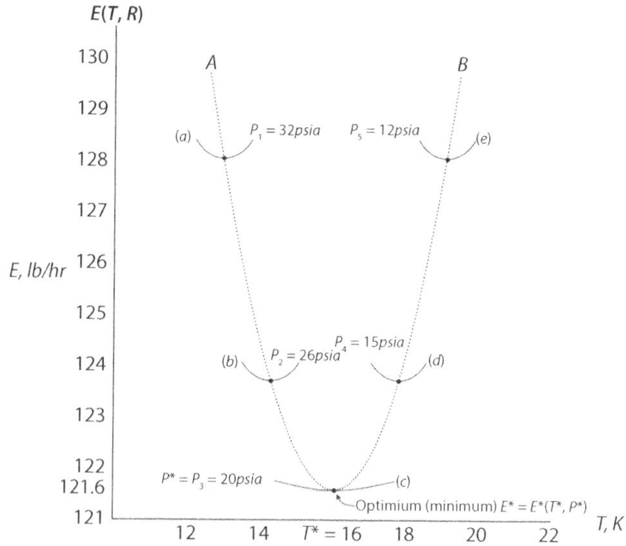

Figure 32.8 Graphical determination of emission with two independent variables (T, P).

Table 32.2 Optimal T and P results

P, psia	Optimum T, K	Minimum E, lb/hr
$P_1 = 32$	$T_1^* = 12.7$	128.3
$P_2 = 26$	$T_2^* = 14.1$	123.6
$P_3 = 20$	$T_3^* = 16.0$	121.6
$P_4 - 15$	$T_4^* - 18.5$	123.9
$P_5 = 12$	$T_5^* = 20.7$	128.5

The curve AB in Figure 32.8 passing through the various optimum points shows that the overall minimum emission occurs at approximately

$$T = 16\text{ K}$$

with corresponding

$$P = 20\text{ pisa}$$

and

$$E = 121.6\text{ lb/hr}$$

Illustrative Example 32.3

Refer to Illustrative Example 30.1. Solve the problem using a graphical approach.

Solution

To solve the program graphically, plot P versus x on arithmetic graph paper. The calculated values are shown in Figure 32.9. The maximum profit appears approximately at 0.125 \$/min of a pressure drop of 20 psf.

Illustrative Example 32.4

Refer to Illustrative Example 30.2. Solve the problem employing a graphical approach.

Solution

Figure 32.10 provides a graphical plot of the four E vs P calculations from Illustrative Example 30.2 The results indicate that for minimum work expenditure

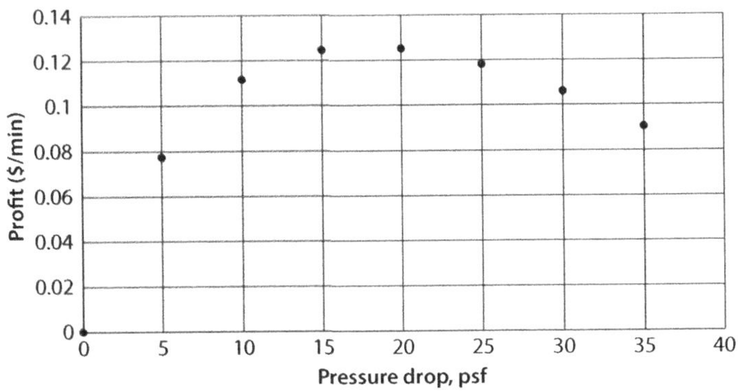

Figure 32.9 P vs. x for Illustrative Example 32.3.

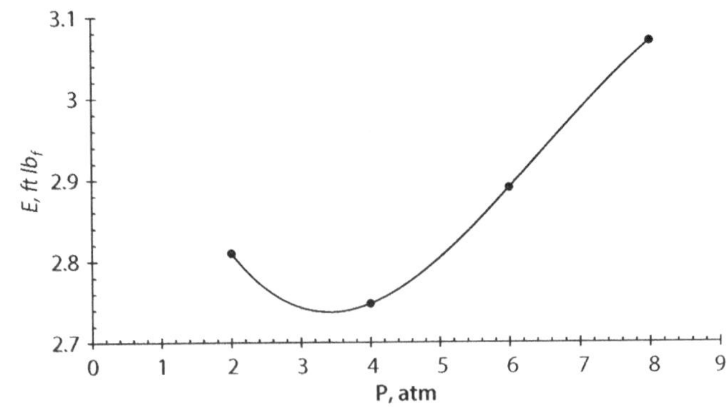

Figure 32.10 E vs x for Illustrative Example 32.4.

the intermediate stage pressure appears to be in the neighborhood of 4 atm with a corresponding work load of 2.75 $ft \cdot lb_f$.

References

1. L. Theodore, *Air Pollution Control Equipment Calculations*, John Wiley & Sons, Hoboken, NJ, 2008.
2. S. Shaefer and L. Theodore, *Probability and Statistics Applications for Environmental Science*, CRC Press/Taylor & Francis Group, Boca Raton, FL, 2007.

33

Analytical Approaches

The subject of analytical analysis serves as an excellent starting point before preceding to the last subject of optimization in this Part – linear programming. In terms of introduction, once again consider the following. Since the derivative dy/dx or $f'(x)$, represents the rate of change of y with respect to the change in x, it is evident that the derivative will be zero if the function passes through a maximum or minimum. If the occurrence of such a maximum or minimum is to be determined and located, the derivative must be equated to zero, thus providing the condition for which the maximum or minimum exists. This analytical procedure is of considerable value in environmental engineering calculations since the location of a maximum or minimum is frequently of practical importance. As one might suppose, many of the applications are related to economics, e.g., maximizing profit and/or minimizing cost (see also the first section in this chapter).

Five sections complement the presentation of this chapter. Section numbers and titles follow:

33.1: Breakeven Considerations
33.2: One Independent Variable
33.3: General Analytical Formulation of the Optimum
33.4: Two Independent Variables
33.5: Three Independent Variables

33.1 Breakeven Considerations

In solving many environmental optimization problems, one often selects optimum conditions from the standpoint of economics. Consider, a simple case in which there is only one independent variable affecting profit P, i.e., income I minus cost C. Refer to Figure 33.1. The term y is the dependent variable that will later be referred to as the objective function and x is the independent variable. The former term could represent the aforementioned profit and x could be the operating temperature. As can be seen from Figure 33.1, the "system" operates at a loss when operating below x^* and operates at a profit when above x^*. The term x^* is located at the *breakeven point*, i.e. where the profit is zero.

The above breakeven problem can be solved by three methods.

1. Solve the problem analytically as follows. Assume I and C can be represented in equation form, i.e.,

$$I = I(x) \tag{33.1}$$

$$C = C(x) \tag{33.2}$$

Since

$$P = I - C \tag{33.3}$$

the solution for x when $P = 0$ provides the breakeven point (of the operation).

2. Select various values of x and calculate both the income and cost, and profit. The breakeven point arises when the profit is zero.
3. Graph the results. A typical solution is presented in Figure 33.1. See also the previous chapter.

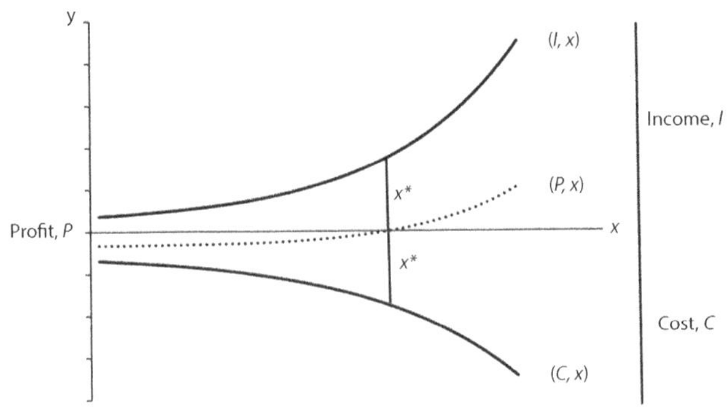

Figure 33.1 Breakeven figure.

Economic studies often reveal that certain cost factors increase while others decrease relative to an independent variable. For example, as the thickness of the insulation (x) on a steam pipe increases, the cost of the insulation increases but correspondingly the heat loss (and steam-generating cost) decreases [1–3]. Here the thickness of the insulation serves as the independent variable and if the installed cost of the insulation and the steam-generating cost (the independent variable) could both be determined as functions of the independent variable, a condition would exist for which the total cost is a minimum. This solution is pictured in Figure 33.2 and the optimum income is located at the point, x^*. Here, the total cost is plotted against the insulation thickness.

33.2 One Independent Variable

The case of a dependent variable and a single (one) independent variable was reviewed in the thee previous chapters. Here

$$y = f(x) \tag{33.4}$$

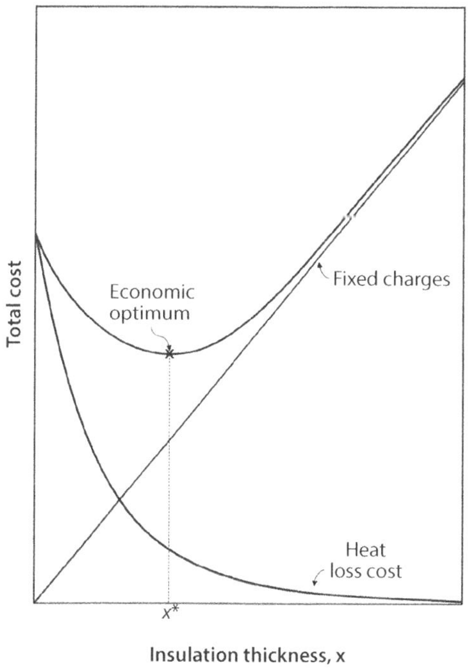

Figure 33.2 Total cost vs insulation thickness.

As noted above, the optimization of y can be generated by setting

$$\frac{dy}{dx} = 0 \tag{33.5}$$

The optimum y is obtained by substituting the value of x from Equation (33.5) into Equation (33.4) and, y is a maximum if

$$\frac{dy}{dx} < 0 \tag{33.6}$$

When x from Equation (33.5) is substituted into Equation (33.4) the calculated y is a minimum if

$$\frac{d^2y}{dx^2} > 0 \tag{33.7}$$

To summarize, a maximum occurs when $dy/dx = 0$ and $d^2y/dx^2 < 0$; a minimum occurs when $dy/dx = 0$ and $d^2y/dx^2 > 0$; this development will be revisited later in this chapter in section 33.3.

Illustrative Example 33.1

Consider the following function

$$y = 162 - 2.33x + \frac{11{,}900}{x}$$

Determine the maximum (or minimum) of the above function by analytical means.

Solution

Apply Equation (33.5).

$$\frac{dy}{dx} = -2.33 - \frac{11{,}900}{x^2} = 0$$

Solution for x,

$$x^2 = \frac{11{,}900}{2.33} = 5{,}107$$

$$x = 71.5$$

The second derivative dy^2/dx^2 is negative since

$$\frac{dy^2}{dx^2} = -\frac{23800}{x^3}$$

The calculated value of y at $x = 71.5$ is therefore a maximum with a value of

$$y = 162 - (2.33)(71.5) + \frac{11,900}{71.5}$$
$$= 162 - 166.6 + 166.4 = 161.8$$

Illustrative Example 33.2

Refer to Illustrative Example 30.1. Solve the problem analytically.

Solution

Once again, the describing equation is

$$P = \frac{0.429x}{x + 15.0} - 0.006x$$

To proceed analytically, calculate dP/dx and set it equal to zero.

$$\frac{dP}{dx} = \frac{0.429}{x + 15.0} - \frac{0.429x}{(x + 15.0)^2} - 0.006 = 0$$

Solve for x by any suitable method.

$$0.429 - \frac{0.429x}{x + 15.0} - 0.006(x + 15.0) = 0$$

$$x = 17.5 \text{ psf}$$

The corresponding maximum profit is then

$$P = \frac{(0.429)(17.5)}{17.5 + 15.0} - (0.006)(17.5)$$
$$= 0.231 - 0.105 = 0.126; \ \$/\text{min}$$

The above analytical result compares formally with the results provided earlier in Illustrative Examples 30.1, 31.1, 31.3 and 32.3.

Illustrative Example 33.3

Refer to Illustrative Example 30.2. Solve the problem employing an analytical approach.

Solution

As noted in Illustrative Example 31.2 the total work is given by

$$E = NRT\left(\frac{k}{k-1}\right)\left[\left(\frac{P_2}{P_1}\right)^{\frac{k-1}{k}} + \left(\frac{P_3}{P_2}\right)^{\frac{k-1}{k}} - 2\right]$$

If this quantity is to be a optimized, then the derivative must be zero(s), i.e.,

$$\frac{dE}{dP_2} = \left(\frac{k}{k-1}\right)\left[\left(\frac{k-1}{k}\right)P_1^{\left(\frac{1-k}{k}\right)}P_2^{-\left(\frac{1}{k}\right)} - \left(\frac{k-1}{k}\right)P_3^{\left(\frac{1-k}{k}\right)}P_2^{\left(\frac{1-2k}{k}\right)}\right] = 0$$

NRT has once again been arbitrarily set equal to unity. Solving for P_2 yields

$$P_2^{(k-1)/k} = P_1^{(k-1)/k}P_3^{(k-1)/k}$$

Rearranging yields

$$P_2 = (P_1P_3)^{1/2} = \sqrt{10} = 3.162 \text{ atm}$$

The corresponding minimum value of *E* is then

$$E(P_2 = 3.162) = 2.726 \text{ ft}\cdot\text{lb}_\text{f}$$

This result compares favorably with the answers provided earlier in Illustrative Examples, 31.2, and 32.2.

Interestingly, Happel [3] addresses the problem of maximizing the work requirement for a three-stage compressor operating between 1 and 10 atm. This problem was solved using an analytical approach along with the method of steepest ascent [3] (See also the previous development in Illustrative Example 3.4). Applying the analytical method for the two intermediate pressures P_2 and P_3 yields the equations (see Illustrative Example 31.4 for details)

$$P_2 = (P_1^2P_4)^{1/3}$$

$$P_3 = (P_4^2P_1)$$

With $P_1 = 1$ and $P_2 = 10$ atm,

$$P_2 = 10^{1/3} = 2.154 \text{ atm}$$

$$P_3 = 10_{2/3} = 4.642 \text{ atm}$$

and

$$E = 2.575 \text{ ft} \cdot \text{lb}_f$$

Happel [3] also noted that the location of a maximum or minimum becomes obvious if the function is plotted, and this graphical procedure may be followed in economic-balance calculations. The analytical method has the advantage of directness and simplicity in many cases. The formal method involves the calculation of the function at a number of points and requires more time to obtain the result. When the curve is obtained, however, its shape indicates the importance of operating at the exact optimum. If the curve is flat, a considerable variation in operating conditions will not appreciably affect the costs or profits. If the maximum or minimum is not flat, it may be quite important to operate at the optimum point. This problem will be visited later for the case of optimizing a four-stage compressor [3]. In some situations, the dependent variable may be a function of several independent variables. The determination of a maximum or minimum value under such circumstances is reviewed later in this chapter.

33.3 General Analytical Formulation of the Optimum

When more than one independent variable is involved in determining a function, e.g., profitability, a more elaborate treatment is necessary than was employed above. Thus, for a system of $x_1, x_2, x_3, \ldots, x_n$ independent variables, the function y will depend on these variables.

$$y(\text{profit}) = f(x_1, x_2, x_3, \ldots, x_n) \tag{33.8}$$

Instead of profit, y may naturally be expressed in other convenient and equivalent forms. The problem is to specify $x_1, x_2, x_3, \ldots, x_n$ so that y will be a maximum (or minimum). In general, this will involve some type of trial-and-error calculation.

It is possible in some cases to set up an explicit function which will relate the objective function y to the independent variables. Differentiation will then result in a series of partial derivations

$$\frac{\partial y}{\partial x_1}, \frac{\partial y}{\partial x_2}, \text{ etc.}$$

Setting these partial derivatives equal to zero will result in the same number of simultaneous equations as the variables involved. These equations, subject to limitations, can be solved for the corresponding optimum values of these variables. The necessary conditions for a maximum at a point where the first derivatives become zero include restrictions on the second derivatives, continuity of the function involved throughout the range of independent variables, and absence of optimum conditions at limiting values of one or more of the variables. These conditions can on occasion restrict the applicability of this approach [3].

Thus, for the case of a two-variable function, $y = f(x_1, x_2)$ a point $x_1 = a$, $x_2 = b$ will have a maximum value $y = f(x_1, x_2)$ if the following conditions hold:

$$\frac{\partial y}{\partial x_1} = \frac{\partial y}{\partial x_2} = 0 \tag{33.9}$$

with

$$\frac{\partial^2 y}{\partial x_1^2} < 0; \quad \frac{\partial^2 y}{\partial x_2^2} < 0 \tag{33.10}$$

and

$$\frac{\partial^2 y}{\partial x_1^2}\frac{\partial^2 y}{\partial x_2^2} > \left(\frac{\partial^2 y}{\partial x_1 \partial x_2}\right)^2 \tag{33.11}$$

Similarly, if $\partial^2 y/\partial x_1^2$ and $\partial^2 y/\partial x_2^2$ are positive and Equation (33.11) holds, there will be a minimum. But, if

$$\frac{\partial^2 y}{\partial x_1^2}\frac{\partial^2 y}{\partial x_2^2} < \left(\frac{\partial^2 y}{\partial x_1 \partial x_2}\right)^2 \tag{33.12}$$

then the point concerned will be a *saddle point*. The case where

$$\frac{\partial^2 y}{\partial x_1^2}\frac{\partial^2 y}{\partial x_2^2} = \left(\frac{\partial^2 y}{\partial x_1 \partial x_2}\right)^2 \tag{33.13}$$

is open and the value may be a maximum, a minimum, or neither. Happel also provided the following three general comments on the above as they relate to both economics and engineering practices [3].

1. For a process with a fixed output, involving a number of pieces of equipment (for example, an absorber, a heat exchanger, a preheater, a stripper, a cooler, ect. (4)), there are a number of design variables, some which affect the cost of several of the operations. For the moment assume that there are two major independent variables -x_1

and x_2. For a given value of x_2 the total costs are tabulated for various values of x_1, and the minimum total cost, C, is noted; this procedure is repeated for other values of x_2 and that combination of values of x_1 and x_2 is selected which gives the minimum of the minima. If the values of x_1 and x_2 can be varied by differential increments, the same result can be obtained by first setting the partial derivatives $\partial C/\partial x_1$ and $\partial C/\partial x_2$ equal to zero, and solving simultaneously for the optimum values of x_1 and x_2 (these procedures where discussed earlier). Similar procedures could be used where there are more than two design variables. Careful study of the particular process may show that certain design variables affect costs of several but not all parts of the process. In such cases, the problem is thereby simplified.

2. In the important problem of estimating the optimum production of a plant, similar methods of attack may be followed, although in this case maximum annual profit (or minimum annual loss) are the criteria, since here the output can be varied, whereas in the cases consided in [1], the output was fixed. Hence, the minimum sum of variable costs could be used. In case of operating at a loss, the plant should not be shut down until the out-of-pocket income would fall below the out-of-pocket expenses, excluding fixed charges. If continuity of supply is stipulated in a contract, the plant may be operated at a loss on an out-of-pocket basis.
3. In the case of a proposed expenditure that is not essential to the success of an environmental process, the decision is often based on whether or not the pay-off time is attractive. The pay-off time is defined as that period in which the actual savings due to the proposed change would equal the original investment. The novice sometimes credits more towards savings than actually would be realized. For example, the total cost of 1,000 lb of steam might be 35 cts. If the bare steam mains were insulated, the required steam load on the power plant might be reduced by 1,000 lb/h. At first inspection, the insulation is credited with a saving of 35 c/h. Further analysis of the steam costs might show that the 35 c/1,000 lb consisted of 10 cts for fuel and 25 cts for fixed charges. Since the latter would not be changed by insulating the steam lines, the true saving is only the cost of fuel based on the value of the fuel saved.

33.4 Two Independent Variables

Illustrative Example 33.4

An objective function is specified as:

$$f(x_1,x_2) = x_1^2 - 2x_1 + x_2^2 + x_1x_2 + 1$$

Determine if the function has a maximum or a minimum. Also, calculate that value.

Solution

For the above equation

$$\frac{\partial f}{\partial x_1} = 2x_1 - 2 + x_2 = 0$$

$$\frac{\partial f}{\partial x_2} = 2x_2 + x_1 = 0$$

The following solution results, when the above two equations are set equal to zero

$$x_1 = -\frac{4}{3}$$

$$x_2 = \frac{2}{3}$$

Since

$$\frac{\partial^2 f}{\partial x_1^2} = 2$$

$$\frac{\partial^2 f}{\partial x_2^2} = 2$$

the solution provided for x_1 and x_2 represents a minimum with

$$y = f(x_1, x_2) = 8$$

Illustrative Example 33.5

A profit equation is given by:

$$P = 162 - 2.33x_1 - \frac{11{,}900}{x_1^2 x_2} = 0$$

Calculate values of x_1 and x_2 that would either maximize or minimize P.

Solution

Begin by generating the first derivative and setting them equal to zero.

$$\frac{\partial P}{\partial x_1} = -2.33 + \frac{11,900}{x_1^2 x_2} = 0$$

$$\frac{\partial P}{\partial x_2} = -1.86 + \frac{11,900}{x_1 x_2} = 0$$

Solving simultaneously yields

$$x_1 = 16$$

$$x_2 = 20$$

The corresponding $P(16,20)$ is a maximum since both second partial derivatives can be shown to be negative. Thus,

$$P = P(16,20) = 50.3$$

The reader may choose to revisit Illustrative Example 32.2.

Illustrative Example 33.6

Obtain values for T and P which will yield the minimum emissions in the equation below by employing an analytical method of analysis.

$$E = 2.33T + \frac{11,900}{TP} + 1.86P + 10$$

Solution

Begin once again by generating the first derivatives and setting them equal to 0.

$$\frac{\partial E}{\partial T} = 2.33 - \frac{11,900}{T^2 P}$$

$$\frac{\partial E}{\partial P} = 1.86 - \frac{11,900}{TP^2}$$

The simultaneous solution to the above two equations yields

$$T = 16$$

$$T = 20$$

Furthermore,

$$E = E\ (16,20) = 121.6$$

Since

$$\frac{\partial^2 E}{\partial T^2} = +\frac{(2)(11,900)}{(16)^2(20)}$$

$$\frac{\partial^2 E}{\partial P^2} = +\frac{(2)(11,900)}{(16)(20)^2}$$

the optimum conditions represent a minimum emission.

Illustrative Example 33.7

Refer to Illustrative Example 31.4. Resolve the "search" problem employing a purely analytical approach.

Solution

The describing equation is once again.

$$E = NRT\left(\frac{k}{k-1}\right)\left[\left(\frac{P_2}{P_1}\right)^{\frac{k-1}{k}} + \left(\frac{P_3}{P_2}\right)^{\left(\frac{k-1}{k}\right)} + \left(\frac{P_4}{P_3}\right)^{\left(\frac{k-1}{k}\right)} - 3\right]$$

If this quantity is to be a minimum, $\partial E/\partial P_2$ and $\partial E/\partial P_3$ must be zero.

$$\frac{\partial E}{\partial P_2} = P_1^{\left(\frac{1-k}{k}\right)} P_2^{-\left(\frac{1}{k}\right)} - P_3^{\left(\frac{1-k}{k}\right)} P_2^{\left(\frac{1-2k}{k}\right)}$$

$$\frac{\partial E}{\partial P_3} = P_2^{\left(\frac{1-k}{k}\right)} P_3^{-\left(\frac{1}{k}\right)} - P_4^{\left(\frac{1-k}{k}\right)} P_3^{\left(\frac{1-2k}{k}\right)}$$

Setting $\partial E/\partial P_2$ and $\partial E/\partial P_3 = 0$, and solving the resulting equations simultaneously ultimately leads to

$$P_2^{\,2} = P_1P_3 \text{ and } P_3^{\,2} = P_2P_4$$

so that

$$P_2 = (P_1^{\,2}P_4)^{1/3}$$
$$P_3 = (P_4^{\,2}P_1)^{1/3}$$

Setting $P_1 = 1$ and $P_4 = 10$ yields the solution

$$P_2 = 2.154 \text{ atm}$$

$$P_3 = 4.642 \text{ atm}$$

and

$$E = E\,(2.154,\, 4.642) = 2.572\ \text{ft} \cdot \text{lb}_\text{f}$$

One can further show that

$$\frac{\partial^2 E}{\partial P_2} = +\frac{(2)(11{,}900)}{(16)^2(20)}$$

$$\frac{\partial^2 E}{\partial P_2} = +\frac{(2)(11{,}900)}{(16)(20)^2}$$

$$\frac{\partial^2 E}{\partial P_2 \partial P_3} = +\frac{(2)(11{,}900)}{(16)^2(20)}$$

and

$$\left(\frac{\partial^2 E}{\partial P_2}\right)\left(\frac{\partial^2 E}{\partial P_3}\right) > \frac{\partial^2 E}{\partial P_2 \partial P_3}$$

Therefore, the above calculated E is a *minimum.*

33.5 Three Independent Variables

This section presents one illustrative example involving three independent variables.

Illustrative Example 33.8

Consider now a traditional mathematician's exercise to determine the maximum or minimum of a function dependent on three variables.

$$P = 8x + 6y + 8z + xy + xz + yz - 2x^2 - 3y^2 - 4z^2$$

Solution

$$\left.\frac{\partial P}{\partial x}\right|_{y,z} = 8 + y + z - 4x = 0$$

$$\left.\frac{\partial P}{\partial y}\right|_{x,z} = 6 + x + z - 6y = 0$$

$$\left.\frac{\partial P}{\partial z}\right|_{x,y} = 8 + y + x - 8z = 0$$

The solution to the above three linear simultaneous equations is

$$x = 2.8256;\ y = 1.7326;\ z = 1.5698$$

for which

$$P = 22.78$$

To determine whether the above P is a maximum or minimum, calculate the three second order partial derivatives.

$$\frac{\partial^2 P}{\partial x^2} = -4$$

$$\frac{\partial^2 P}{\partial y^2} = -6$$

$$\frac{\partial P}{\partial z^2} = -8$$

Since the three derivatives are negative, the value of P above (22.78) is a maximum, i.e., any other calculated P for a different x, y, and z would produce a value of P less than 22.78. The "proof" is left as an exercise for the reader. However, if one were to select values of x, y, and z straddling the maximum set of values, P should be below 22.78. For example, the value of P for $x = 2$, $y = 2$, and $z = 2$ is 20. For $x = y = z = 10$, the value of P is -380.

References

1. L. Theodore, *Heat Transfer Applications for the Practicing Engineer,* John Wiley & Sons, Hoboken, NJ, 2011
2. A.M. Flynn, T. Akashige L. Theodore, *Kern's Process Heat Transfer*, 2nd ed., Scrivener-Wiley, Salem, MA, (forthcoming) 2018.
3. J. Happel, *Chemical Process Economics*, John Wiley & Sons, Hoboken, NJ, 1958.

34

Introduction to Linear Programming

Linear programming is one of several techniques for optimization which has increased in usefulness because of the availability of digital computers. The title results from the assumptions in the method that linear relationships describe the problem/system/relationship under consideration. For example, each production element per day increase in a process could produce a corresponding linear profit increase per day. However, there are many instances where a linear model does not accurately describe the totality of the problem/system/relationship. It may then be necessary to resort to other procedures such as nonlinear programming; some brief details will be presented for handling the nonlinear section in the next paragraph.

The reader should note that the terms linear programming and nonlinear programming are essentially similar from an applications perspective. There were problems around the middle of the last century because of some difficulties that arose in attempting to solve nonlinear programming problems. However, the arrival of the modern day computer (see Chapter 29) and sophisticated software (e.g., EXCEL) have removed these problems. Unless a solution is presented graphically in this and the next chapter, it will be obtained directly from EXCEL so that the reader need not be concerned with whether a system's pertinent describing equations and constraints are linear or nonlinear. Thus, details regarding the solution methodology for both linear programming and nonlinear programming problems and illustrative examples will not be presented. They are simply not necessary and beyond the scope of this introductory optimization Part.

Other than linear programming, *dynamic programming* is probably the more often used optimization method by the practicing environmental engineer and scientist. As described earlier, it is a process involving decisions made at multiple stages. Just as an optimization model is defined by its variables, constraints, and objective function, a multistage decision process is defined by its stages and decisions. At any stage, there is enough information about decisions made at preceding stages to make informed decisions at the present stage. Thus, some optimization problems can be reformulated as multistage decision process problems. One then proceeds through subsequent stages until the final stage of the process is reached. Some additional details are provided in the next section.

As one might suppose, there are numerous other optimization methods available to the practicing environmental engineer and scientist. Several of these are detailed in Perry's Handbook [1].

1. Mixed Integer
2. Mixed Integer Nonlinear Programming
3. Mixed Integer Linear Programming
4. Quadrative Programming
5. Differential/Nondifferentiable Systems
6. Convex/Nonconvex

The above topics are *not* reviewed in this text. The reader is referred to the optimization literature [2–6] for details.

Regarding contents, Section 34.1 provides definitions as they apply to optimization and, in particular, to linear programming – the title of this chapter. Section 34.2 examines linear programming from an elementary principles perspective while Section 34.3 addresses the topic from a mathematician's viewpoint. Section 34.4 is concerned with an engineer's approach to linear programming. Linear programming examples that are not applications-related highlight the last section (34.5).

34.1 Definitions

Seven key definitions – as they apply to optimization and linear programming are provided below (in alphabetical order).

1. **Algebra:** A branch of mathematics in which letters are used to represent basic arithmetic relations. As in arithmetic, the basic operations of algebra are addition, subtraction, multiplication, and division. Arithmetic, however, cannot generalize such

mathematical relations as the Pythagorean theorem (which states that the sum of the squares of the sides of any right triangle is also a square). Arithmetic can produce specific instances of this relations (for example 3, 4, and 5, where $3^2 + 4^2 = 5^2$). Algebra can make a purely general statement that fulfills the conditions of the theorem, i.e., $a^2 + b^2 = c^2$. Any number multiplied by itself is termed squared and is indicated by a superscript number 2. For example, 3×3 is notated as 3^2 and is termed three squared; similarly, $a \times a$ is equivalent to a^2. Classical algebra, which is concerned with solving equations, uses symbols instead of specific numbers and employs arithmetic operations to establish procedures for manipulating symbols while modern algebra has evolved from classical algebra by placing more attention to the structures within mathematics.

2. **Constraint:** "A constraining or being constrained... confinement or restrictions... forced... something that constrains." The term constraints is defined as "to bond together, draw together... to force into or hold in, close bounds, confine... restrain."
3. **Dynamic Programming:** The dimesionality of a particular operation is reduced from $(m)(n)$ variables to m repeated n times. As noted above, if it is not practical or possible to solve a large complex system of equations, the system is reduced in complexity in a manner that requires the solution of a smaller subset of the original problem, noting that for each n (for example, each stage in a mass transfer operation) [7] only m variables are involved in the optimization process. One can then work backwards and combine the solutions generated at each n.
4. **Linear Algebra:** A branch of mathematics that is concerned with systems of linear equations, linear transformations, vectors, vector spaces, and related topics. The major interest is with linear equations.
5. **Linear Programming:** A mathematical and operations research technique used in engineering, science, administrative engineering, and economic planning to maximize the linear function(s) of a large number of variable(s) subject to certain constraint(s). It is basically used to find a set of values, chosen from a prescribed set of numbers, that will maximize or minimize, i.e. optimize, a given equation.
6. **Operations Research:** A confusing term that most often has been used to describe linear programming.
7. **Optimization:** As noted earlier in this Part, optimization has come to mean different things to different people. Its definition has also varied in the literature. In its most elementary and basic form, *optimization* is concerned with the determination of the optimum solution to a given problem. This process is required in

the solution of many problems in environmental engineering that involve the maximization or minimization of a given function(s). Some specific examples include:

a. environmental engineering research
b. applied environmental science
c. in the selection of *control variables* to accomplish a desired operation
d. in the scheduling of a series of operations or events to control dates of a given project
e. the development of *optimal* layouts of organizational units within a given design space, etc.

34.2 Basic Concepts of Optimization

The term linear programming defines a particular class of optimization problems in which the constraints of the system can be expressed as linear equations or inequalities and the function is linear with respect to the independent design variable. These techniques are widely used to solve a number of engineering, environmental, chemical, structural, economic, and societal problems. The primary reasons for its wide use today are the availability of a host of commercial software to solve complex problems and the ease with which any data variation that could involve sensitivity analysis can be handled.

As noted in the earlier chapters in this Part, the optimization problem in which the performance measure is a function of one variable is the most elementary type of optimization problem since it is a type of problem that the environmental engineer commonly encounters in practice. It also allows single-variable optimization to be employed as a sub-problem within iterative procedures for solving multivariable optimization problems.

Constraints, constraints, constraints. Constraints continue to be imposed on individuals, organizations, groups, governments, religions, relatives, friends, etc. It seems to be getting worse each year. And, one of the hallmarks of linear programming is the inclusion of constraints in the problem description, i.e., the objective function. Linear programming involves linear constraints. For example, the function

$$f(x) = x^4 + 3x^2 - 2x + 5;\ \text{all values of } x \tag{34.1}$$

is an unconstrained function. However, the function

$$f(x) = x^6 - 4x + 6;\ -3 \le x \le 6 \tag{34.2}$$

is a constrained function. When constraints are present, the function being optimized is usually referred to as the *objective function* to distinguish it from any other functions that may be used to define the constraints or the problem. Although the emphasis to follow in this and the next chapter is on linear constraints, many real-world applications also involve non-linear constraints *and* nonlinear objective functions.

From a basic perspective, the problem is to optimize the objective function

$$y = f(x_1, x_2) \tag{34.3}$$

The constraints can include

$$g(x_1, x_2) < 0 \tag{34.4}$$

$$x_1 \geq 0 \tag{33.5}$$

$$x_2 \geq 0 \tag{34.6}$$

This is a simple yet typical problem in optimization – the maximization or minimization of a real-value function of a number of real variables (occasionally just a single variable) subject to a number of constraints (occasionally the number is zero). However, the mathematical model normally contains several (n, for example) variables known as the *independent design variables* denoted as x_1, x_2, x_3, …, x_n. This set of variables has been described by some as a vector, X. One may view each x as one dimension in the set (space) of variables, and any particular set of variables, and any particular set of values for these variables in this system as a solution.

One of the more important recent areas for the application of mathematical optimization techniques is in environmental engineering and science. Many environmental problems require an iterative and/or trial-and-error solution and the problem of concern is often stated in an ambiguous and open-ended way. The first task is to decide what the problem is and to identify the requirements or specifications for any proposed and/or feasible solution. One then generates a concept, perhaps in the form of a rough configuration, for the object, system, or process to be optimized. This phase of the problem is usually the most difficult and can require the greatest ingenuity and engineering judgement.

The next step attempts to *model* the object, system, or process. This model may assume a wide variety of forms, but the usually preferred ones are mathematical in nature. One may then use many mathematical, scientific, and engineering tools. These can involve not only the most empirical of data correlations but also the most rigorous mathematical procedures to the least formal cut-and-try methods for assisting in the analysis of this aspect of the solution.

It may happen that certain functional *equality* constraints also apply.

$$\begin{matrix} f_1(x_1, x_2, x_3, \ldots, x_n) = 0 \\ \vdots \qquad\qquad \vdots \\ f_k(x_1, x_2, x_3, \ldots, x_n) = 0 \end{matrix} \tag{34.7}$$

There may also be certain inequality constraints such as

$$\begin{aligned} &g_1(x_1, x_2, x_3, \ldots, x_n) < 0 \\ &\vdots \qquad\qquad\qquad\qquad \vdots \\ &g_m(x_1, x_2, x_3, \ldots, x_n) < 0 \end{aligned} \tag{34.8}$$

that are specified on the space so that only points in a portion of the total space may be considered as acceptable values for the solution of the problem. This region is defined as the *feasible space* and points in the region are defined as *feasible solutions*.

The above equality constraints – Equation (34.7) – come from functional relationships that must be satisfied among the aforementioned n variables. If there is to be a choice in the values for which at least some of the variables may assume, then k must be smaller than n. The inequality constraints in Equation (34.8) usually arise from some specified limitations (maximum permissible stress, minimum emissions, minimum allowable temperatures, etc.). Note that there is no upper limit on the value of m.

Finally, the following comments are provided on the objective function. It may be quite simple and easy to determine, or complicated and difficult to calculate. The objective function may also be very illusive due to the presence of conflicting or dimensionally incompatible sub-objectives; for example, one might be asked to design an absorber [7] at minimum cost and make it aesthetically appealing. In some cases, it may not be possible to find a *quantitative* objective function. In this case, one can assume that a suitable qualitative objective function can be formulated.

34.3 Applied Mathematics Concepts on Linear Programming

As noted earlier, linear programming deals with the determination of an optimum solution of a problem expressed in linear relationships where there are a large number of possible solutions. In general, the methods of solution employed are of a trial-and-error nature, where the procedure systemically follows a mathematical treatment which minimizes one's labor and insures that a correct result has been obtained. The following graphical approach to a linear programming problem illustrates some of the mathematics employed in the solution.

Consider now an example that involves minimizing an objective variable E (e.g., expenditure or emission)

$$E = a_1x_1 + a_2x_2 \tag{34.9}$$

subject to two constraints

$$B \geq b_1x_1 + b_2x_2 \tag{34.10}$$

$$C \geq c_1x_1 + c_2x_2 \tag{34.11}$$

In addition,

$$x_1 \geq 0 \tag{34.12}$$

$$x_2 \geq 0 \tag{34.13}$$

In terms of a solution, first examine the above system of equations analytically. The problem can be viewed in Figure 34.1 and 34.2. First note that only the first quadrant need be considered since Equations (34.12) and (34.13) must hold. Equations (34.10) and (34.11) are rewritten in the following form

$$x_2 = \frac{B}{b_2} - \left(\frac{b_1}{b_2}\right)x_1 \tag{34.14}$$

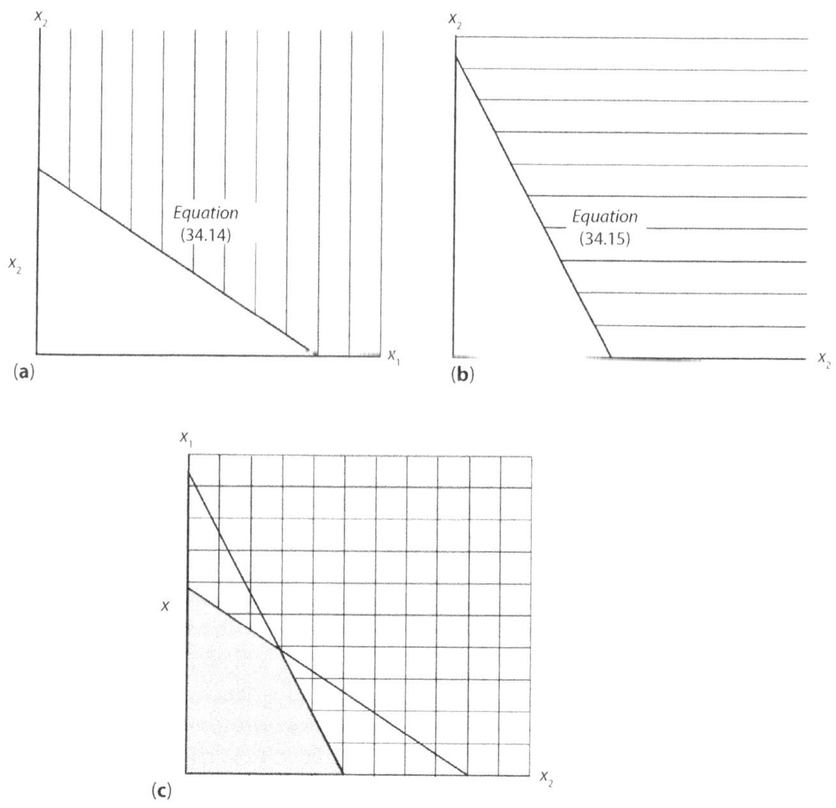

Figure 34.1 Two variable minimization problems.

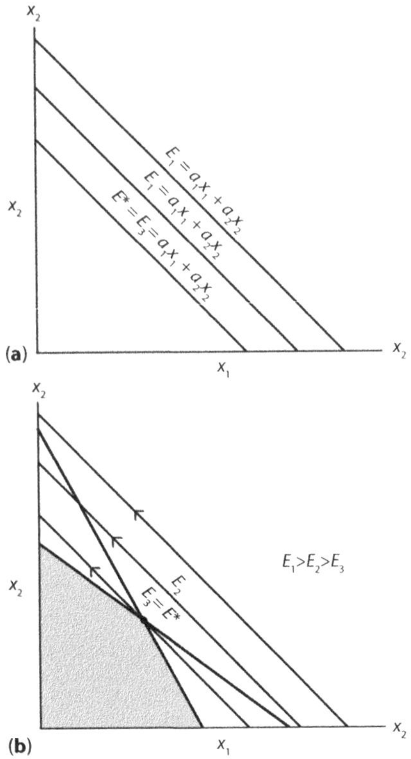

Figure 34.2 Solution to a two variable minimization problem.

and

$$x_2 = \frac{C}{c_2} - \left(\frac{c_1}{c_2}\right)x_1 \tag{34.15}$$

It is apparent that the only values of x_1 and x_2 which will satisfy Equations (34.14) and (34.15) lie *above* the solid lines (the cross-hashed area) in Figure 34.1 (c). This area is termed the constraint set. Now place in Figure 34.1 a series of straight lines (or contours) corresponding to Equation (34.9) with a different value of E used for each line (see Figure 34.2 (a)). Of all these lines, the smallest E *for* which x_1 and x_2 remain in the specified region (shaded area) will be the minimum E (see Figure 34.2b). This will yield that proportion of x_1 and x_2 which minimize the objective function. Stated in other words, one desires the lowest-value contour having some point in common with the constraint set. For the particular value of slope chosen for Equation (34.9) the minimum point is seen to correspond to E where $E_3 = E^*$.

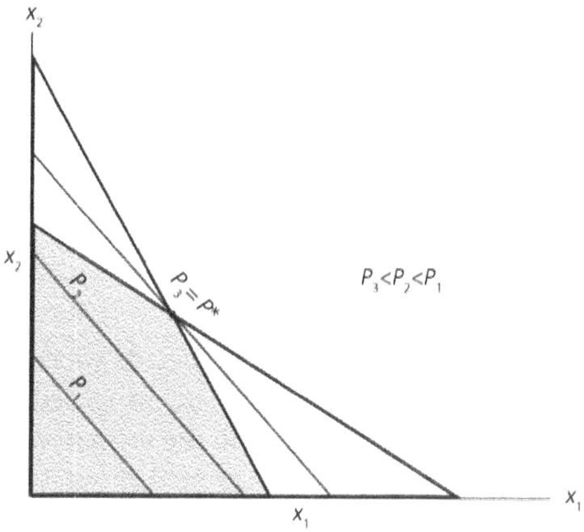

Figure 34.3 Possible solution to a two variable maximization problem.

The point which represents the minimum is therefore determined by constraints a_1 and a_2 of the problem. Further, if the constants c_1 and c_2 in Equation (34.11) are changed, it is possible that the minimum point would occur elsewhere. The situation is essentially reversed for a maximization problem, i.e., for a profit P. This is demonstrated in Figure 34.3.

It should be pointed out that if the objective function – Equation (34.7) – was nonlinear the contour lines would be curved and if the constraint equations were nonlinear the constraint lines would not be straight. It should be intuitively obvious that locating the minimum point graphically is more complicated in this case.

The above methodology can be extended to more than two variables and two constraints. The following would apply for three variables and four constraints.

$$y = f(x_1, x_2, x_3) = a_1x_1 + a_2x_2 + a_3x_3 \tag{34.16}$$

and

$$B = b_1x_1 + b_2x_2 + b_3x_3 \tag{34.17}$$

$$C = c_1x_1 + c_2x_2 + c_3x_3 \tag{34.18}$$

$$D = d_1x_1 + d_2x_2 + d_3x_3 \tag{34.19}$$

$$E = e_1x_1 + e_2x_2 + e_3x_3 \tag{34.20}$$

and

$$x_1 \geq 0 \tag{34.21}$$

$$x_2 \geq 0 \tag{34.22}$$

$$x_3 \geq 0 \tag{34.23}$$

Naturally, the above can be rewritten for n objective functions and m constraints.

34.4 Applied Engineering Concepts in Linear Programming

An engineer's first task in problem solving is to decide what the problem is and to identify the requirements or specifications for any proposed and/or feasible solution. One then generates a concept, perhaps in the form of a rough configuration, for the object, or process to be optimized. As noted above, this phase of the problem is often the most difficult and requires the greatest ingenuity and engineering judgement.

Perhaps Happel [8] best described linear programming from a practicing engineer's perspective (the reader should note the overlap with some of the material in section 34.5). His edited write up follows.

> "The general problem of finding a maximum usually assumes that some partial derivative of the function $y = f(x_1, x_2, x_3)$ vanishes within the range of interest of the variables. In practical problems it often happens, however, that the optimum solutions will be obtained at the boundary or limiting values of the variables. This is, of course, always true where the function to be optimized is linear. In such cases, explicit analytical solution is not possible and the optimum must be obtained by some other method or iteration.
>
> In general, even the simplest type of formulations may require a great deal of computation in practical applications which is facilitated by the availability of high-speed computers. Linear programming, discussed in greater detail below, is beginning to be used to a substantial extent by practicing engineers.
>
> In order to apply linear programming to a problem, the relationships between variables must be expressed as a set of linear equations or inequalities. The variables must be otherwise independent and must exceed the number of equations. Mathematically expressed,
>
> $$\sum_i a_{ij} X_i = b_i (i = 1, 2, \ldots, m;\ J = 1, \ldots, n;\ n > m) \tag{34.24}$$
>
> where the sum of the coefficient a_{ij} times the value of each variable X_i equals a requirement b_i in the ith equation. The number of variables must exceed the number of equations, so that an infinite number of solutions is possible. The "best" one is selected. Often the relationships in equation (34.24) are expressed as inequalities. "Dummy" additional variables are introduced in order to convert these inequalities into appropriate equations. Another requirement is that no quantity can be negative. Thus, mathematically

$$x_i \geq 0 \text{ and } b_i \geq 0 \tag{34.25}$$

This requirement will insure that only useful answers are obtained.

The solution of a problem by linear programming also requires a profit function or objective function to be maximized (or minimized). This requirement is stated

$$P = \sum_i c_i X_i = \text{maximum (or minimum)} \tag{34.26}$$

where c_i is the profit (or cost) per unit of X_i used. The solution is carried out by an iterative process, which at each stage of calculation assigns either zero or positive values to all X_i variables. As the calculation proceeds, the values selected, while meeting the constraints imposed, will tend to increase the summation to be maximized at each stage. If a solution exists, a maximum profit case will finally be realized."

Figure 34.4 illustrates a simple situation involving the operation of an absorber[7, 9, 10] with temperature and pressure limits for which each depends on the magnitude of the other. The former might come from limitations due to equilibrium consideration, since "the higher the temperature in the absorber the lower the absorptive capacity of the absorbing liquid while the higher the pressure in the absorber increases the absorption" [11]. The region of feasible solutions for the absorber problem is the region enclosed by the constraints. The optimization problem is to find the one set of variables or point in the feasible solution space

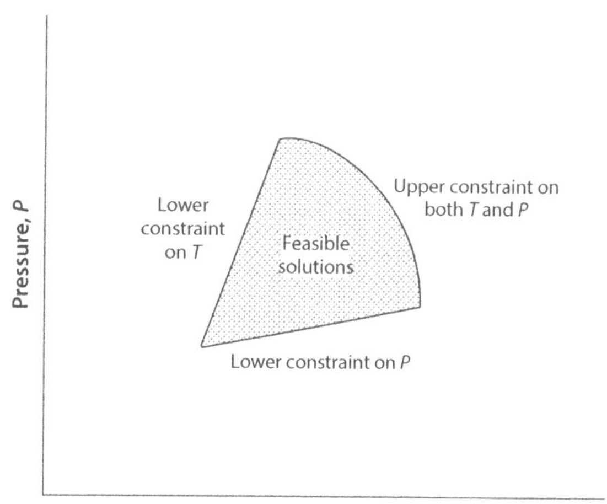

Figure 34.4 Feasible solution for operation of an absorber.

that corresponds to the so-called best solution - in this case, minimizing the gaseous pollutant emission.

In order to locate the best feasible solution of the potentially infinite number of feasible solutions using a mathematical optimization technique it is essential that a meaningful objective function *f(x)* of the independent design variables be specified for an operation concerned with - for example - the control and/or recovery of a particular component of gaseous emission. For simplicity, assume that for the aforementioned application, one must choose two independent variables – temperature T and pressure P. The space is, therefore, two dimensional, with temperature and pressure as the coordinates. In practice, there usually are certain inequality constraints placed upon the range on both T and P. As shown in Figure 34.4 there could be lower limits for the temperature and pressure, and an upper limit for each that depends on the magnitude of the other.

Once again, the optimization problem is to find the one set of variables or point, X, in the feasible solution space that corresponds to the best solution. In order to locate the best feasible of the potentially infinite number of feasible solutions using a mathematical optimization technique, it is essential that a meaningful objective function $f(x)$ of the design variables be specified to provide for a comparison of alternative solutions.

34.5 Applied Engineering Concepts in Linear Programming

This last section attempts to clarify the ideas presented above via three illustrative examples. An attempt will also be to relate the variables to real-world quantitative factors. That will occur in the next and last chapter. In fact, the first illustrative example will be revisited as it applies to pollutant emission concerns from a coal-fired boiler.

Illustrative Example 34.1

You have been requested to obtain the maximum value of P subject to the following conditions.

Objective function:

$$P = 0.3x_1 + 5x_2 \tag{1}$$

Constraints:

$$x_2 \le -0.5x_1 + 3 \tag{2}$$

$$x_2 \le -3x_1 + 6 \tag{3}$$

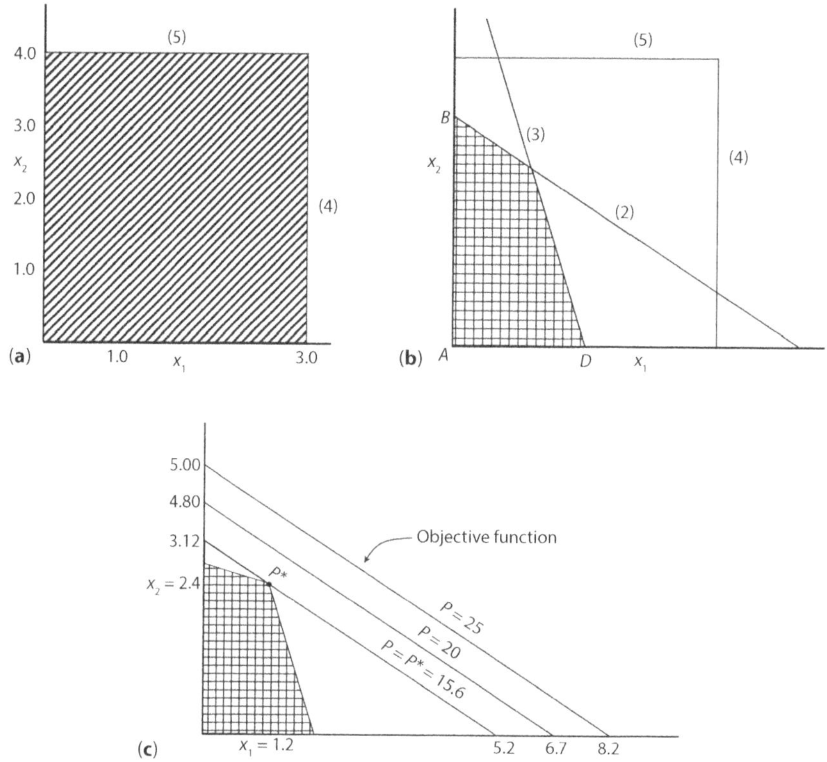

Figure 34.5 Graphical solution for Illustrative Example 34.1.

$$x_1 \leq 3 \tag{4}$$

$$x_2 \leq 4 \tag{5}$$

Once again, the solution is provided in graphical form. Refer to Figure (34.5). Conditions (4) and (5) require that the solutions, i.e., the values x_1 and x_2 that will maximize P lie within the rectangle provided in Figure (34.5). Conditions (2) and (3) require that the solution be within the area ABCD in Figure 34.5 (b). When the objective function equation is superimposed on Figure 34.5 (b), Figure 34.5 (c) results. The maximum P is located at $x_1 = 1.2$ and $x_2 = 2.4$ for which $P = 15.6$. EXCEL provides the same solution.

Illustrative Example 34.2

A company manufactures two products, A and B, with an accompanying unit profit of P_A and P_B, respectively. A minimum number of N products can be produced on a daily basis. The daily production rates of A and B is x_A and x_B, respectively. In addition, the fractional defective rate of producing A and B are a and b respectively. Each defective product reduces the profit associated with product A and B by c and d, respectively. Determine the describing equations (the model for the production rate of A and B) that will maximize daily profits P.

Solution

As noted, x_A and x_B are the daily production rates of A and B, respectively. The function P is then

$$P = x_A P_A + x_B P_B - x_A ac - x_B bd$$

The constraints are

$$x_A + x_B \geq N$$

$$x_A, x_B \geq 0$$

This problem will be evaluated again in the next chapter.

Illustrative Example 34.3

A batch chemical reactor [12] can produce either product A, B, or C. The daily production levels of each are represented by x_A, x_B, and x_C with an accompanying unit profit of P_A, P_B, and P_C, respectively. The two unit expressions of concern for each product are manpower (M) and energy costs (E), neither of which can exceed Y and Z, respectively. Develop the linear programming model to optimize the profit P for this system.

Solution

Based on the problem statement one concludes:

$$P = P_A x_A + P_B x_B + P_C x_C$$

with

$$M_A x_A + M_B x_B + M_C x_C \leq Y$$

$$E_A x_A + E_B x_B + E_C x_C \leq Z$$

$$x_A, x_B, x_C \geq 0$$

Numerical problems of this nature will also be revisited in the next/last chapter.

References

1. D. Green and R. Perry (editors), *Perry's Chemical Engineers' Handbook*, 8th ed., McGraw-Hill, New York City, NY, 2007.
2. R. Ketter and S. Prawel, *Modern Methods of Engineering Computation*, (adapted from), McGraw-Hill, New York City, NY, 1969.
3. L. Lapidus, *Digital Computations for Chemical Engineers*, (adapted from) McGraw-Hill, New York City, NY, 1962.
4. G. Reklaitis, A. Ravindran, and K. Ragsdell, *Engineering Optimization: Methods and Applications*, John Wiley & Sons, Hoboken, NJ, 1983.
5. E. Chong and S. Zak, *An Introduction to Optimization*, 4th ed., John Wiley & Sons, Hoboken, NJ, 2013.
6. H. Mickley, T. Sherwood, and C. Reed, *Applied Mathematics in Chemical Engineering*, McGraw-Hill, New York City, NY, 1957.
7. L.Theodore, F. Ricci, and T. VanVliet, *Thermodynamics for the Practicing Engineer*, John Wiley & Sons, NJ, 2009.
8. J. Happel, *Chemical Process Economics*, John Wiley & Sons, Hoboken, NJ, 1958.
9. L. Theodore, *Air Pollution Control Equipment Calculations*, John Wiley & Sons, Hoboken, NJ, 2008.
10. L. Theodore, *Chemical Engineering: The Essential Reference*, McGraw-Hill, New York City, NY 2014.
11. J. Reynolds, J. Jeris, and L. Theodore, *Handbook of Chemical and Environmental Engineering Calculations*, John Wiley & Sons, Hoboken, NJ, 2004.
12. L. Theodore, *Chemical Reactor Analysis and Applications for the Practicing Engineer*, John Wiley & Sons, Hoboken, NJ.

35

Linear Programming Applications

This is the last chapter of not only this Part (V) but also of the book. Part V is devoted to optimization, but the emphasis is on linear programming – a subject that the authors view as a subset of optimization. Since the earlier material in the previous five chapters has for the most part – relied on optimization methods, it is appropriate that this last chapter compliment the presentation with applications-oriented illustrative examples.

It should be noted once again that formal optimization techniques have as their goal the development of procedures for the attainment of an optimum in a system which can be characterized mathematically. The mathematical characterization may be:

1. partial or complete,
2. approximate or exact, and/or
3. empirical or theoretical.

The resulting optimum may be a final implementable solution or a guide to a practical solution and a criterion by which practical solutions are to be judged. In either case, the applications of these optimization techniques should serve as an important part of the effort in the solution of some environmental problems [1].

Nine illustrative examples follow. Three are concerned with environmental matters, five deal with the chemical process industries, and one involves a Monte Carlo exercise. The order of presentation and the accompanying titles are:

Illustrative Example 35.1: Vitamin Caloric Requirement
Illustrative Example 35.2: Operation of a Natural Gas Boiler
Illustrative Example 35.3: Maximizing Coal Usage Profits
Illustrative Example 35.4: Maximizing Crude Oil Profits
Illustrative Example 35.5: Tube-and-Bundle/Finned Heat Exchangers
Illustrative Example 35.6: Minimizing Gaseous Pollution Emissions from an Oil-Fired Burner
Illustrative Example 35.7: Pharmaceutical Cancer Drug
Illustrative Example 35.8: Optimum Operation of a Butane Plant
Illustrative Example 35.9: Monte Carlo Application

Illustrative Example 35.1: Vitamin Caloric Requirement

Vitamin *(1)* costs \$0.20 per mg and contains 50 cal/mg and 20 mg protein/mg. Vitamin *(2)* costs \$0.15 per mg and contains 40 cal/mg and 10 mg protein/mg. What amounts of vitamins x_1 and x_2 would meet the minimum regulatory requirement while minimizing production cost, C. The USDA provided the following objective function for C:

$$C = 20x_1 + 15x_2$$

along with the constraints

$$50x_1 + 40x_2 \geq 300$$

$$20x_1 + 10x_2 \geq 120$$

and

$$x_1 \leq 6$$

$$x_2 \leq 6$$

Solution

EXCEL provides the following solution:

$$x_1 = 6$$

$$x_2 = 0$$

and

$$C = 120$$

The reader should note before leaving this problem that this illustrative example is similar to that presented in Section 34.3 of the last chapter, only this time with numbers specified for the variables and with the problem related to a real-world situation.

Illustrative Example 35.2: Operation of a Natural Gas Boiler

The cost associated with the operation of a gas-fired boiler R has been found to be linearly related to both the operating temperature T and operating pressure P. That relationship (in consistent *absolute* units) is given by

$$R = 0.18T + 0.05P; \text{ consistent units, \$/hr}$$

However, for ignition to occur, the following constraints must be satisfied

$$a_1T + b_1P \geq I$$

along with the energy component of the gaseous fuel

$$a_2T + b_2P \geq E$$

If

$$a_1 = 0.2, \; b_1 = 0.05$$

$$a_2 = 0.4, \; b_2 = 0.02$$

and

$$I = 600$$

$$E = 450$$

$$T_1 \geq 0$$

$$P_1 \geq 0$$

calculate the operating temperature and pressure that minimizes the cost of operation and satisfy the constraints associated with ignition and energy content.

Solution

This is a typical linear programming problem. For this example,

$$R = 0.18T + 0.05P$$

The constraints are

$$0.2T + 0.05P \geq 450$$

and

$$0.4T + 0.02P \geq 600$$

with

$$T \geq 0$$

$$P \geq 0$$

EXCEL provides the following solution

$$T = 2250$$

$$P = 0$$

and

$$R = 40.5 \text{ \$/hr} = 972\text{\$/day}$$

Illustrative Example 35.3: Maximizing Coal Usage Profits

A coal manufacturer mines two grades of coal – *(1)* and *(2)*. Coal *1* emits 25 mg NO_2 per ton combusted, 6 mg SO_4 per ton combusted, and 68 mg particulate per ton combusted. Each ton of Coal *1* sold yields a \$30 profit. Coal *2* emits 75 mg NO_2, 60 mg SO_4, and 34 mg particulate per ton combusted. Each ton of Coal *2* sold yields a profit pf \$40. Due to stringent state air pollution regulatory requirements no more than 450 mg NO_2, 480 mg SO_4, and 374 mg particulate may be potentially emitted from the coal sold by the manufacturer. How many tons of Coal (x_1) and (x_2) should be mined daily to maximize profits and satisfy regulatory rules? What is the corresponding maximum daily profit?

Solution

The objective function for the coal operation is

$$P = 30x_1 + 40x_2$$

subject to the constraints

$$a_1x_1 + a_2x_2 \leq E_1$$

$$b_1x_1 + b_2x_2 \leq E_2$$

$$c_1x_1 + c_2x_2 \leq E_2$$

where (in consistent units)

$$a_1 = 25, \ a_2 = 75$$

$$b_1 = 60, \ b_2 = 60$$

$$c_1 = 68, \ c_2 = 34$$

$$E_1 = 450, \ E_2 = 480, \ E_3 = 374$$

and

$$x_1 \geq 0$$

$$x_2 \geq 0$$

Solution

EXCEL provides the following solution.

$$x_1 = 3$$

$$x_2 = 5$$

and

$$P = \$290$$

The low value of P suggests that the coal manufacturer is operating at or near breakeven economics.

Illustrative Example 35.4: Maximizing Crude Oil Profits [2,3]

A refinery has two catcrackers that can produce various grades of hydrocarbon products. Due to consumer fuel demand, the production of gasoline, home heating oil, and diesel must be limited. This information is provided in Table 35.1. The profit on processing U.S. crude oil is \$2.00/gal and on Venezuelan crude is \$1.60/gal. Find the approximate daily processing rate of the two crudes in order to maximize profits.

Table 35.1 Catcracker information

Product grades	U.S. Crude (% distribution)	Venezuelan crude (% distribution)	Max. production rate (gal/day)
Gasoline	8	11	1,500
Home Heating Oil	29	54	5,500
Diesel Oil	63	35	11,000

Solution

Set

$$x_1 = \text{gallons of U.S. crude}$$

$$x_2 = \text{gallons of Venezuelan crude}$$

The objective function for the daily profit, P, that is to be maximized is:

$$P = 2.0x_1 + 1.6x_2$$

The constraints are:

$$0.08x_1 + 0.11x_2 \leq 1500$$

$$0.29x_1 + 0.54x_2 \leq 5500$$

$$0.63x_1 + 0.35x_2 \leq 11000$$

with

$$x_1 \geq 0$$

$$x_2 \geq 0$$

The solution to the above from EXCEL is: 16,820 gal of U.S. crude per day should be processed and 1152 gal per day should be derived from Venezuelan crude. This results in a total daily profit of $35,484 [1, 2]. Thus,

$$x_1 = 16{,}820 \text{ gal U.S. crude per day}$$

$$x_2 = 1152 \text{ gal Venezuelan crude per day}$$

$$P = \$35{,}484$$

Illustrative Example 35.5: Tube-and-Bundle/Finned Heat Exchangers

Ricci Consultants were requested to develop a mathematical model involving heat exchangers at the Bayway Refinery in Linden, NJ. The purpose of the assignment was to develop a heat exchanger network (see Illustrative Examples 30.6 and 30.7 in Chapter 30) that maximizes both the profit associated with the recovery of both energy *and* the "quality" energy [3] by employing a number of only tube-and-bundle (x_1) and finned (x_2) exchangers. The proposed model for the objective function takes the form:

$$RC = e^{(x_1+0.25x_2)} + x_1x_2 + x_1^2;\ \$$$

The constraints specified were

$$x_1^2 + x_2 \leq 200$$

$$x_1 + 4x_2 \leq 50$$

$$x_1 \geq 0$$

$$x_2 \geq 0$$

Determine RC, x_1, and x_2,

Solution

Note that his is a "nonlinear" programming problem. EXCEL provides the following solution:

$$x_1 = 13$$

$$x_2 = 9$$

$$RC = \$4.2 \times 10^6$$

Thirteen tube-and-bundle and 9 finned heat exchangers are recommended.

Illustrative Example 35.6: Minimizing Gaseous Pollutant Emissions from an Oil- Fired Utility Boiler

A state-wide utility has proposed to convert some of its coal-fired boilers to oil-fired. DuPont Associates was hired to generate a model (objective function) that would minimize the cost of the replacement relative to the quantity of oil to be combusted. Two oils – *(1)* and *(2)* – are available to supply the necessary energy for all the new boilers. The project also recognized that an attempt be made to salvage as many of the old boilers currently burning coal [3]. The cost associated with each oil varies with its content, regulatory requirements, heating value, etc. The engineers at DuPont Associates proposed the following cost model, DA.

$$DA = 3x_1 + 4x_2 - x_3; 10^{-4}\$$$

(Salvaging the existing boilers resulted in a credit.) The constraints, primarily due to emission rules for NOX, SOX, and ROX were

$$x_2 + x_3 \leq 100 \text{ (nitrous oxide, NOX)}$$

$$x_1 + x_3 \leq 50 \text{ (sulfur dioxide, SOX)}$$

$$x_1 + x_2 x_3 + x_3^2 \leq 1000 \text{ (particulate, ROX)}$$

with

$$x_1 = x_2 = x_3 \leq 50$$

Solution

EXCEL provides the following solution:

$$x_1 = 4$$

$$x_2 = 10$$

$$x_3 = 26$$

with

$$DA = \$26 \times 10^4 = \$260{,}000$$

Thus, the solution requires the purchase of 4 new boilers that burn Oil *(1)*, 10 new boilers that will burn Oil *(2)*, and the reassignment of 26 of the old coal-fired boilers (*3*).

Illustrative Example 35.7: Pharmaceutical Cancer Drug

A new consortium of drug laboratories has combined their resources to produce four variations of the same drug. Initial estimates of the predicted hourly profit from the production of drugs *A*, *B*, *C*, and *D* were \$100, \$150, \$200, and \$250, respectively. Table 35.2 below shows how much time is required in each of the three research labs to make one unit of each of the four products.

Table 35.2 Drug laboratory production time expenditure (hours); Illustrative Example 35.7.

Lab	Drug A	Drug B	Drug C	Drug D
1	10	4	8	9
2	6	5	3	9
3	3	9	7	12

Laboratory *1* has a maximum of 300 man-hours available per day. Lab *2* is limited to no more than 400 man-hours per day. Similarly, Lab 3 is limited to 425 man-hours per day.

As a recent environmental engineering graduate, you have been hired to determine the number of *A*, *B*, *C*, and *D* units that should be produced to maximize *daily* profit, *P*. The following information is provided.

$$P = 100x_1 + 150x_2 + 200x_3 + 250x_4; \text{ \$/hr}$$

where x_1, x_2, x_3 and x_4 represent the number of drugs *A*, *B*, *C*, and *D*, respectively, produced per hour. The constraints are,

$$10x_1 + 4x_2 + 8x_3 + 9x_4 \leq 300$$

$$6x_1 + 5x_2 + 3x_3 + 9x_4 \leq 400$$

$$3x_1 + 9x_2 + 7x_3 + 12x_4 \leq 425$$

and

$$x_1 \geq 0$$

$$x_2 \geq 0$$

$$x_3 \geq 0$$

$$x_4 = 0$$

Solution

EXCEL provides the following solution

$$x_1 = 0$$

$$x_2 = 29$$

$$x_3 = 23$$

$$x_4 = 0$$

with a profit of

$$P = \$8950/\text{hr} = \$214{,}900/\text{day (around-the-clock operation)}$$

Illustrative Example 35.8: Optimum Operation of Butadiene Plant [1]

Note: Illustrative Example 35.8 was drawn (with permission) from the classic work of Happel titled "Chemical Process Economics" [1].

It is desired to be determined the optimum operation of a plant for the production of butadiene by catalytic dehydrogenation of butylene. The steps involved are shown schematically in Figure 35.1.

It is assumed that such items as labor, steam, and fuel consumption will remain relatively constant so that the profitability of one system of operation as compared with another will be a function of the feed required (at \$8.40/bbl), the fuel gas produced (equivalent to \$0.90 of butane degraded), the recycle from the purification unit (at \$6.32/bbl), and the butadiene product (at \$32.80/bbl). The profit function, Z, to be maximized is therefore

$$Z = 32.8D + 6.32C + 0.90B - 8.40A$$

The operation of the plant is assumed to involve only two variables subject to independent selection; namely, the per cent conversion of butylenes per pass through the catalytic dehydrogenation system, and the per cent of normal butane (which is assumed to be inert) allowed to build in the total feed to the dehydrogenation unit (i.e., fresh feed plus recycle). Additional restrictions due to process limitations are that the fresh feed A cannot exceed 9000 bbl/day, that the buildup F of n-butane in the recycle stream must not exceed 17% and that the conversion

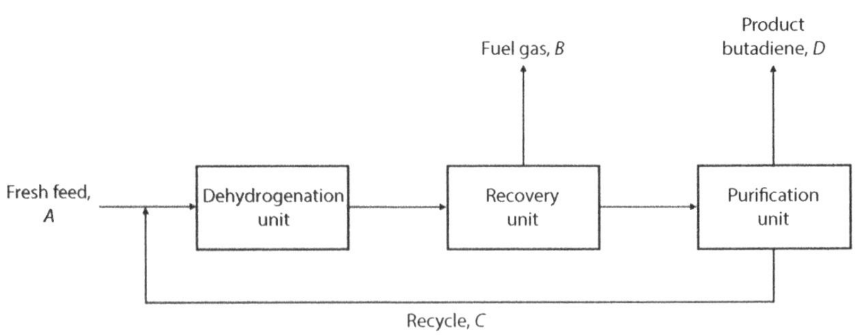

Figure 35.1 Butadiene plant; Illustrative Example 35.8.

E of n-butylenes per pass will be at least 25%. Calculations based on plants tests gave the results in Table 35.3.

If it is assumed that over the range of interest, variation in the dependent variables A, B, C, and D will be linear with respect to changes in the independent variables E and F, the following four equations results:

$$A = 7880 - (7880 - 6740)\left(\frac{30-E}{5}\right) + (12900 - 7880)\left(\frac{15-F}{4}\right)$$

$$B = 1300 - (1300 - 900)\left(\frac{30-E}{5}\right) + (1380 - 1300)\left(\frac{15-F}{4}\right)$$

$$C = 3500 - (3500 - 3120)\left(\frac{30-E}{5}\right) + (8460 - 3500)\left(\frac{15-F}{4}\right)$$

$$D = 2960 - (2960 - 2630)\left(\frac{30-E}{5}\right) + (3034 - 2960)\left(\frac{15-F}{4}\right)$$

with the additional restrictions

$$A \le 9000$$

$$F \le 17$$

$$E \ge 25$$

Since the problem involves only two independent variables, it can readily be solved without elaborate mathematical tools. The values of A, B, C, and D above are inserted into the relationship defining the profit function to be maximized, obtaining

$$\text{Profit} = W = -1024 + 800E + 2080F$$

Table 35.3 Process data for butadiene plant; Illustrative Example 35.8.

A Feed, bbl/day	*B* Fuel, bbl/day	*C* Recycle, bbl/day	*D* Butadiene, bbl/day	*E* Percent Conversion Of C_4	*F* Percent n-C_4 in Total Feed
7,88	1300	3500	2960	30	15
6,740	900	3120	2630	25	15
12,900	1380	8460	3034	20	11

Thus, it is clear that E and F should be selected as large as possible without violating the restriction $A \leq 9000$. The equation for A shows that, in fact increasing F decreases A, so it is immediately clear that $F = 17$ is the optimum. The value of E corresponding to $A = 9000$ is then determined from the same equation: i.e., $F = 46.0$.

Another way of visualizing the problem is to estimate the variable A from the restrictions by making use of the value of A in terms of E and F to obtain

$$288E - 1255F \leq -10{,}865$$

The above equation, together with $F \leq 17$ and $E \geq 25$, constitutes a statement of the constraints in terms of the independent variables E and F. These relationships are plotted on Figure 35.2 and indicate the area, which is shaded, in which the desired solution must be located. The locus of the profit function on this figure consists of straight lines parallel to each other, shown as a dashed line on Figure 35.2. The maximum occurs where the profits function intersects one of the extreme points (vertices of the triangle in Figure 35.2): i.e., $E = 46.0$.

If more than two independent variables were involved, it would not be possible to employ a geometrical construction, and more elaborate mathematical techniques would be required which in effect accomplish the same thing analytically.

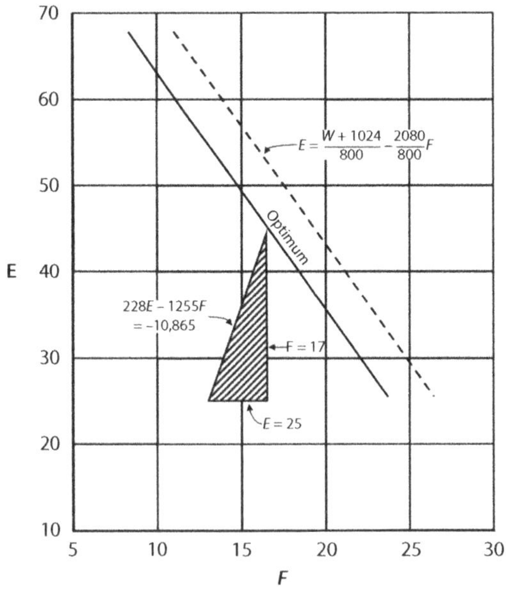

Figure 35.2 Illustration of solution by geometry of Illustrative Example 35.8.

To solve this problem by linear programming the constraints must be changed to equalities by the use of *slack* [1] variables S_1, S_2, and S_3 to obtain the following form:

$$228E - 1255F + S_1 = -10{,}865$$

$$-E + S_2 = -25$$

$$F + S_3 = 17$$

A base-case matrix table is set up, showing these equations and the profit function in Table 35.4. It will be noted that this matrix table consists of the coefficients in the preceding equations under the appropriate column for each variable. The columns S_1, S_2, S_3, and Z are "identity columns," since the only element in each is a single +1. The row corresponding to each of these unity coefficients is known by the same designation as the column, and the quantity Q is the actual value assigned to this variable. The variables S_1, S_2, S_3, and Z are said to be "in the basis," and the variables E and F are not in the basis and have zero values. The problem is to establish variables of E and F that will maximize W.

Variable F has the largest negative value in the W row and therefore has the greatest profit associated with it. Table 35.5 shows how it is brought into the basis to replace S_1.

Now F has been put into the basis, and column F is an identity column. The equations represented by the matrix still satisfy the necessary restrictions. The profit can be further increased by putting E into the basis. Table 35.6 shows this transformation, replacing S_3. Table 35.6 represents the solution to the problem. The values of the variables required are shown in the Q column, in agreement with results by other methods. The profit function shows that \$71,073/day is the optimum.

Illustrative Example 35.9: Monte Carlo Approach

Maximize the following objective function utilizing a Monte Carlo approach [4, 5]. This model was recently developed by Kourtakis Engineering via a regression analysis to describe the profit MC generated from operating 6 different production plants producing a new product x. Outline how to calculate the profit product x from each of the pilot processes.

Table 35.4 Profit function table; Illustrative Example 35.8.

	Z	E	F	S_1	S_2	S_3	Q
S_1	–	228	–1255	1	–	–	–10,865
S_2	–	–1	–	–	1	–	–25
S_3	–	–	1	–	–	1	17
Z	1	–800	–2080	–	–	–	–1,024

Table 35.5 Establish maximum profit.

	Z	E	F	S_1	S_2	S_3	Q
Divide equations S_1 by −1255, and replace in this position as equation F	F	−0.182	1	$-\frac{1}{1255}$			8.65
Leave S_2 unchanged	S_2	−1			1		−25
Subtract the new equation for F from S_3 equation to obtain new S_3 equation.	S_3	+0.182		$-\frac{1}{1255}$		1	8.35
Multiply the new F equation by 2080, and add to Z equation to obtain new Z equation	Z	−1178		−1.652			16,866

Table 35.6 Problem solution; Illustrative Example 35.8.

	Z	E	F	S_1	S_2	S_3	Q
Divide equations S_2 by 0.182, replace as equation E	E	1		$\frac{5.5}{1255}$		5.50	45.9
Add 0.182 multiplied by equation E to equation F to obtain new Equation F	F		1			1	17.0
Add equation E to equation S_2 to obtain new equation S_2	S_2			$\frac{5.5}{1255}$	1	5.50	20.9
Multiply the new equation E by 1178 and add to equation Z to get new equation Z	Z	1		3.51		6483	71,073

$$MC = x_1^3 + x_2^3 - x_3^3 + x_1^2 x_2^2 x_3^2 - x_4^2 x_5^2 x_6^2 + (\ln(x_1))(\ln(x_2 x_3 x_4)) + e^{-x_6 \tanh\left(\frac{x_3 - x_6}{x_1}\right)};\ \$$$

The constraints are:

$$4 \le x_1 \le 99$$

$$4 \le x_2 \le 99$$

$$4 \le x_3 \le 99$$

$$4 \le x_4 \le 99$$

$$4 \le x_5 \le 99$$

$$4 \le x_6 \le 99$$

Note that there are $(96)^6$ *MC* integer values to be checked. Outline a method of solution. (The reader to referred to Ch. 24, Part IV – Monte Carlo Methods).

Solution

Apply the Monte Carlo method in the solution to the problem. Here is one of the author's approaches [6]. It involves the reading/sampling of 10,000 (it could be more or less) possible solutions and selecting the maximum value derived from the objective function. First set MC = 0.

1. Generate a random 2-digit number from 4 – 99 and assign it to x_1.
2. Generate a random 2-digit number from 4 – 99 and assign it to x_2.
3. Generate a random 2-digit number from 4 – 99 and assign it to x_3.
4. Generate a random 2-digit number from 4 – 99 and assign it to x_4.
5. Generate a random 2-digit number from 4 – 99 and assign it to x_5.
6. Generate a random 2-digit number from 4 – 99 and assign it to x_6.
7. Substitute the above x_1, x_2, x_3, x_4, x_5, and x_6 values into the objective function *MC* and calculate *MC*.
8. Store (retain) this *MC* value.
9. Compare this new value of *MC* with the previous value of *MC*. Store or retain the larger of the two.
10. Repeat steps (1–9) 9,999 times.
11. The last man (or is it person) standing, i.e., the last retained *MC* is the maximum value of *MC*.

The answer in (11) represents the maximum value of the sample of the 10,000 solutions totaled. The "*actual*" or "*true*" maximum of the entire population of $(96)^6$ *MC's* is probably slightly higher than the value in (11).

The reader should consider what other approaches could be employed to improve on the accuracy of the answer provided by the Monte Carlo method.

References

1. J. Happel, *Chemical Process Economics*, John Wiley & Sons, Hoboken, NJ, 1958.
2. L. Theodore, personal notes, East Williston, NY, 2011.
3. L. Theodore, *Heat Transfer Applications for the Practicing Engineer*, John Wiley & Sons, Hoboken, NJ, 2011.
4. L. Theodore, personal notes, East Williston, NY, 2016.
5. S. Shaefer and L. Theodore, *Probability and Statistics Applications for Environmental Science*, CRC Press/Taylor & Francis Group, Boca Raton, FL, 2007.
6. Lou Theodore, personal notes, East Williston, NY, 2017.

Index

Printed and bound by CPI Group (UK) Ltd, Croydon, CR0 4YY

17/09/2026

14964241-0008